ADVANCES IN MODELING
AGRICULTURAL SYSTEMS

Springer Optimization and Its Applications

VOLUME 25

Aims and Scope
Optimization has been expanding in all directions at an astonishing rate during the past few decades. New algorithmic and theoretical techniques have been developed, the diffusion into other disciplines has proceeded at a rapid pace, and our knowledge of all aspects of the field has grown even more profound. At the same time, one of the most striking trends in optimization is the constantly increasing emphasis on the interdisciplinary nature of the field. Optimization has been a basic tool in all areas of applied mathematics, engineering, medicine, economics, and other sciences.

The *Springer Optimization and Its Applications* publishes undergraduate and graduate textbooks, monographs, and state-of-the-art expository works that focus on algorithms for solving optimization problems and also study applications involving such problems. Some of the topics covered include nonlinear optimization (convex and nonconvex), network flow problems, stochastic optimization, optimal control, discrete optimization, multiobjective programming, description of software packages, approximation techniques, and heuristic approaches.

For other books in this series:
http://www.springer.com/series/7393

ADVANCES IN MODELING AGRICULTURAL SYSTEMS

Edited by

PETRAQ J. PAPAJORGJI
University of Florida, Gainesville, Florida

PANOS M. PARDALOS
University of Florida, Gainesville, Florida

 Springer

Editors
Petraq J. Papajorgji
University of Florida
Gainesvile, FL
USA
petraq@ifas.ufl.edu

Panos M. Pardalos
University of Florida
Gainesvile, FL
USA
pardalos@ufl.edu

ISSN: 1931-6828
ISBN: 978-1-4419-4525-9   e-ISBN: 978-0-387-75181-8
DOI 10.1007/978-0-387-75181-8

AMS Subject Classifications: 49XX, 46N60

Printed on acid-free paper

springer.com

To our children:

Dea Petraq Papajorgji
and
Miltiades Panos Pardalos

Preface

Agriculture has experienced a dramatic change during the past decades. The change has been structural and technological. Structural changes can be seen in the size of current farms; not long ago, agricultural production was organized around small farms, whereas nowadays the agricultural landscape is dominated by large farms. Large farms have better means of applying new technologies, and therefore technological advances have been a driving force in changing the farming structure.

New technologies continue to emerge, and their mastery and use in requires that farmers gather more information and make more complex technological choices. In particular, the advent of the Internet has opened vast opportunities for communication and business opportunities within the agricultural community. But at the same time, it has created another class of complex issues that need to be addressed sooner rather than later. Farmers and agricultural researchers are faced with an overwhelming amount of information they need to analyze and synthesize to successfully manage all the facets of agricultural production.

This daunting challenge requires new and complex approaches to farm management. A new type of agricultural management system requires active cooperation among multidisciplinary and multi-institutional teams and refining of existing and creation of new analytical theories with potential use in agriculture. Therefore, new management agricultural systems must combine the newest achievements in many scientific domains such as agronomy, economics, mathematics, and computer science, to name a few.

This volume came to light as the result of combined efforts by many researchers in different areas with the goal of providing the readers with a wide spectrum of advanced applications in agriculture. Readers will find new software modeling approaches such as Unified Modeling Language (UML), Object Constraint Language (OCL), model driven architecture (MDA), and ontologies. Readers will also find a large arsenal of advanced mathematical tools used to study the multiple aspects of agricultural production such as calculation of leaf area index (LAI) from Earth observation (EO) data, accurate estimation of chlorophyll level by remote sensing methods, data mining techniques applied in machine vision, analysis of remotely sensed forest images, fruit and wine classification, and finally packaging of agricultural products. The main message authors would like to transmit through the chapters of this book is that modeling is a very serious activity, and many types of models must be developed to cope with the complexity of agricultural systems.

Two years ago, we published the book *Software Engineering Techniques Applied to Agricultural Systems: An Object-Oriented Approach and UML*, whose main goal was to provide researchers and students in agricultural and environmental areas with new developments in the field of software engineering. At the time of working on the current volume, the first book has been used as the basic text for the course "Biological Simulation" at the Agricultural and Biological Engineering Department, University of Florida, thanks to the kind decision of Dr. Gregory Kiker with whom we co-teach this course.

While teaching this course, students provided many useful comments and suggestions. They asked questions regarding issues we never thought about before. After 2 years of teaching, the issue that came to us for discussion over and over again was that students wished to see more applications developed using this modeling paradigm. They wished to see more examples of modeling complex agricultural systems. We hope this additional volume will satisfy some of their wishes.

We would like to take this opportunity to thank all contributors and referees for their valuable contributions. Without them, this volume would not have been possible. We would like to thank our students for their valuable feedback and comments; we taught them the science, and they taught us how to improve our work. Last but certainly not least, we would like to thank Springer for giving us another opportunity to work with them.

Gainesville, Florida   Petraq J. Papajorgji
Gainesville, Florida   Panos M. Pardalos

Contents

ix

Contributors

Vincent Abt
Cemagref, Clermont Ferrand, France, LAMSADE, Université
Paris-Dauphine, Paris, France, Vincent.abt@cemagref.fr

Valérie Auffray
SAS ITK, Montpellier, France

François Barnabé
Cemagref, Clermont Ferrand, France, Francois.barnabe@cemagref.fr

Amalia Barone
Department of Soil, Plant, Environmental, and Animal Production Sciences,
University of Naples Federico II, Naples, Italy, amalia.barone@unina.it

Howard Beck
Department of Agricultural and Biological Engineering, University of Florida,
Gainesville, Florida, USA, hwb@ufl.edu

Jevin Bhorania
Department of Industrial and Systems Engineering, University of Florida,
Gainesville, Florida, jevin@ufl.edu

Ismail Bogrekci
Department of Agricultural and Biological Engineering, University of Florida,
Gainesville, Florida, USA, bogrekci@ufl.edu

Nikita Boyko
Department of Industrial and System Engineering, University of Florida,
Gainesville, Florida, USA, nikita@ufl.edu

Thomas Brun
Cemagref, Clermont Ferrand, France, Thomas.brun@cemagref.fr

Domenico Carputo
Department of Soil, Plant, Environmental, and Animal Production Sciences,
University of Naples Federico II, Naples, Italy, domenico.carputo@unina.it

Silvana Cavella
Department of Food Science, University of Naples Frederico II, Naples, Italy,
silvana.cavella@unina.it

Radnaabazar Chinchuluun
Department of Agriculture and Biological Engineering, University of Florida,
Gainesville, Florida, USA, radnaach@ufl.edu

Maria Luisa Chiusano
Department of Soil, Plant, Environmental, and Animal Production Sciences,
University of Naples Federico II, Naples, Italy, marialuisa.chiusano@unina.it

Claudio Cifarelli
Department of Industrial and System Engineering, University of Florida,
Gainesville, Florida, USA, claudioc@ufl.edu

Ryan Clark
Sosy Inc., San Francisco, CA, USA, rclark@integranova.net

Laurence Cohen-Jonathan
SAS ITK, Montpellier, France, Laurence.cohen-jonathan@itkweb.com

M. Ricardo Cunha
College of Economics and Management, Catholic University of Portugal
and Lancaster University, Porto, Portugal, mrcunha@porto.ucp.pt

Nunzio D'Agostino
Department of Soil, Plant, Environmental, and Animal Production Sciences,
University of Naples Federico II, Naples, Italy

Guido D'Urso
Department of Agriculture, Engineering, and Agronomy, University of Naples
Federico II, Naples, Italy, durso@unina.it

Birgit Demuth
Department of Computer Science, Dresden University of Technology,
Dresden, Germany, Birgit.Demuth@inf.tu-dresden.de

Luigi Dini
Geodesy Centre, Earth Observation Unit, Italian Space Agency, Matera, Italy

Magali Duboisset
Cemagref, Clermont Ferrand, France, Magali.duboisset@cemagref.fr

Pierre Escande
SAS ITK, Montpellier, France

Dalila B. M. M. Fontes
Faculdade de Economia da Universidade de Porto, Rua Dr. Roberto Frias
Porto, Portugal, Fontes@fep.up.pt

Melanie Fritz
International Research Center on Food Chain and Network Research,
Department of Food and Resource Economics, University of Bonn, Bonn,
Germany, m.fritz@uni-bonn.de

Luigi Frusciante
Department of Soil, Plant, Environmental, and Animal Production Sciences,
University of Naples Federico II, Naples, Italy, luigi.frusciante@unina.it

Maria de Sousa Gallagher
Department of Process and Chemical Engineering, University College, Cork,
Ireland, m.desousagallagher@ucc.ie

Osvaldo Gargiulo
Agricultural Engineering and Biological Department, University of Florida,
Gainesville, Florida, USA, ogargiul@ufl.edu

Susana Gomez
Institute of Applied Mathematics, National University of Mexico, Mexico City,
Mexico, susanag@servidor.unam.mx

Sabine Grunwald
Soil and Water Science Department, University of Florida, Gainesville,
Florida, USA, sabgru@ufl.edu

Vianney Houlès
SAS ITK, Montpellier, France

Eric Jallas
SAS ITK, Montpellier, France, Eric.jallas@wtkweb.com

James W. Jones
Department of Agricultural and Biological Engineering, University of Florida,
Gainesville, Florida, USA, ifur@ufl.edu

Yunchul Jung
Department of Agricultural and Biological Engineering, University of Florida,
Gainesville, Florida, USA

Gregory A. Kiker
Department of Agriculture and Biological Engineering, University of Florida,
Gainesville, Florida, USA, gkiker@ufl.edu

Ho-young Kwon
Soil and Water Science Department, University of Florida, Gainesville,
Florida, USA

Won Suk Lee
Department of Agricultural and Biological Engineering, University of Florida,
Gainesville, Florida, USA, Wslee@ufl.edu

Thérèse Libourel
National Center for Scientific Research, Montpellier Laboratory of Computer
Science, Robotics, and Microelectronics, University of Montpellier II,
Montpellier, France, libourel@lirmm.fr

Efstratios Loizou
Department of Agriculture Products Marketing and Quality Control,
Technological Educational Institution of Western Macedonia, Terma
Kontopoulou, Florina, Greece, Lstratos@agro.auth.gr

Pramod V. Mahajan
Department of Process and Chemical Engineering, University College, Cork,
Ireland, p.mahajan@ucc.ie

Paolo Masi
Department of Food Science, University of Naples Frederico II, Naples, Italy,
paolo.masi@unina.it

Konstadinos Mattas
Department of Agriculture Economics, Aristotle University of Thessaloniki,
Thessaloniki, Greece, mattas@auth.gr

James M. McKinion
USDA-ARS-Genetics and Precision Agriculture Research Unit, Mississippi
State, Mississippi, USA, mckinion@ra.msstate.edu

Min Min
Department of Agricultural and Biological Engineering, University of Florida,
Gainesville, Florida, USA

André Miralles
Centre for Agricultural and Environmental Engineering Research, Earth
Observation and GeoInformation for Environmental and Land Development
Unit, Montpellier, France, andremiralles@teledetection.fr

Valeria Marina Monetti
Department of Water Resource Management, University of Naples Federico II,
Naples, Italy

Kelly Morgan
Soil and Water Science Department, University of Florida, Gainesville,
Florida, USA, conserve@ufl.edu

Antonio Mucherino
Center for Applied Optimization, University of Florida, Gainesville, Florida,
USA, Amucherino@ufl.edu

Fernanda A.R. Oliveira
Department of Process and Chemical Engineering, University College, Cork,
Ireland, f.oliveira@ucc.ie

Petraq Papajorgji
Center for Applied Optimization, University of Florida, Gainesville, Florida,
USA, petraq@ufl.edu

Panos M. Pardalos
Department of Industrial and System Engineering, University of Florida,
Gainesville, Florida, USA, pardalos@cao.ise.ufl.edu

François Pinet
Cemagref, Clermont Ferrand, France, Francois.pinet@cemagref.fr

Catherine Roussey
LIRIS CNRS, University of Lyon, Lyon, France,
Catherine.roussey@liris.cnrs.fr

Alessandro Santini
Division of Water Resource Management, University of Naples Federico II,
Naples, Italy, alessandro.santini@unina.it

Michel Schneider
Cemagref, Laboratory of Computer Science, Modeling, and System
Optimization, Blaise Pascal University, Clermont Ferrand, France,
Michel.schneider@cemagref.fr; Schneider@isima.fr

Michael R. Seal
ITD, Spectral Visions, Stennis Space Center, Mississippi, USA

Onur Seref
Center for Applied Optimization, University of Florida, Gainesville, Florida,
USA, seref@ufl.edu

Gerardo Severino
Division of Water Resource Management, University of Naples Federico II,
Naples, Italy, gerardo.severino@unina.it

Vincent Soulignac
Cemagref, Clermont Ferrand, France, Vincent.soulignac@cemagref.fr

Ludovic Tambour
SAS ITK, Montpellier, France, Ludovic.tambour@itkweb.com

Rohit Thummalapalli
Summer Science Training Program, University of Florida, Gainesville, Florida,
USA

Elena Torrieri
University of Naples, Federico II, Italy, Elena.torrieri@unina.it

Sibiri Traore
Agricultural Engineering and Biological Department, University of Florida,
Gainesville, Florida, USA

Sam Turner
USDA-ARS-Genetics and Precision Agriculture Research Unit, Retired,
Starkville, Mississippi, USA

Vangelis Tzouvelekas
Department of Economics, University of Crete, Rethymno, Crete, Greece,
Vangelis@econ.soc.uoc.gr

Frédéric Vigier
Cemagref, Clermont Ferrand, France, Frederic.vigier@cemagref.fr

Francesco Vuolo
Ariespace s.r.l., University of Naples Federico II, Ercolano, Italy

Jeffrey L. Willers
USDA-ARS-Genetics and Precision Agriculture Research Unit, Mississippi
State, Mississippi, USA, jeffrey.willers@ars.usda.gov

Jin Wu
Soil and Water Science Department, University of Florida, Gainesville,
Florida, USA

Vitaliy Yatsenko
Space Research Institute of National Academy of Sciences of Ukraine and
National Space Agency of Ukraine, Kiev 03187, Ukraine,
vyatsenko@gmail.com

The Model Driven Architecture Approach: A Framework for Developing Complex Agricultural Systems

Petraq Papajorgji, Ryan Clark, and Eric Jallas

Abstract Development and application of crop models is increasingly constrained by the difficulty of implementing scientific information into an efficient simulation environment. Traditionally, researchers wrote their own models and tools, but as software has become much more complex, few researchers have the means to continue using this approach. New modeling paradigms provided by the software engineering industry can be successfully used to facilitate the process of software development for crop simulation systems.

This chapter outlines a model driven architecture (MDA)-based approach to construct a crop simulation model. This new modeling paradigm is a Unified Modeling Language (UML) -based approach. A conceptual model of the problem is first constructed to depict concepts from the domain of the crop simulation and their relationships. The conceptual model is then provided with details about the role each of the concepts plays in the simulation. The multiplicity of the associations between concepts is determined, and the behavior of each of the objects representing concepts of the domain is defined. Mostly, an object's behavior in the crop simulation domain is expressed using equations. For this type of behavior, this new modeling paradigm offers a declarative way to write equations using attributes of objects participating in the conceptual diagram. For behavior that cannot be expressed through equations, a formal language is used to model behavior without the ambiguities that can be introduced by the use of natural language. Models can be validated and logical flows can be discovered before code generation.

An Extensible Markup Language (XML) representation of the conceptual model is used by an engine that generates automatically executable code in several programming environments such as Java, Enterprise Java Beans, Visual Basic, and .NET. Results obtained from this new approach are presented, and they coincide with results obtained with other approaches.

P. Papajorgji (✉)
Center for Applied Optimization, University of Florida, Gainesville, FL, USA
e-mail: petraq@ufl.edu

P.J. Papajorgji, P.M. Pardalos (eds.), *Advances in Modeling Agricultural Systems*,
DOI 10.1007/978-0-387-75181-8_1, © Springer Science+Business Media, LLC 2009

1 Introduction

The model driven architectureModel Driven Architecture (MDA) is a framework for software development defined by the Object Management Group (OMG) [14]. At the center of this approach are models; the software development process is driven by constructing models representing the software under development. The MDA approach is often referred to as a model-centric approach as it focuses on the business logic rather than on implementation technicalities of the system in a particular programming environment. This separation allows both business knowledge and technology to continue to develop without necessitating a complete rework of existing systems [20].

The MDA approach is making its advance in the software industry consistently. There are a considerable number of software companies providing MDA-based tools such as Kabira (http://kabira.com/), Accelerated Technology (http://www.acceleratedtechnology.com/), Kennedy Carter (http://www.kc.com/), and Sosy, Inc. (http://sosyinc.com/), and a more important number of companies are developing their applications using this technology. According to a recent survey organized by the well-respected Gartner, Inc. [8], given the potential savings and the linkage to requirements that MDA promises, many analysts say it is only a matter of time before MDA-like environments will be mandated by management. A study undertaken by The Middleware Company (http://www.middleware-company.com) encourages organizations wishing to improve their developer productivity to evaluate MDA-based development tools. The Middleware Company has relationships with several hundreds of thousands of developers through TheServerSide Communities and provides the most in-depth technical research on middleware technology available in the industry. Visionary Bill Gates in the Gartner symposium [7, 8] predicted that visual modeling tools will reduce software coding "by a factor of five" over the next 10 years.

The main goal of this study is to evaluate the application of this modeling paradigm in the domain of crop simulation systems. Crop simulation applications are different from business applications. Business applications have the tendency to be linear and usually do not involve a great number of iterations. Crop simulations are repetitive, and calculations are done for each time step of the simulation. There are many examples of applications using the MDA approach in the business area but none in the domain of crop simulation systems. Modeling the relationship between a client and purchases and orders is a relatively well-known process. Crop simulation systems tend to be more abstract than are business systems. Expressing the relationships between plant, soil, and weather and the processes occurring in each of these elements may not be as straightforward as modeling a client–supplier relationship. The level of calculations used in a business model is relatively simple whereas crop simulation models make heavy use of equations.

The MDA-based tool used in this study is the Oliva Nova Model Execution of Sosy, Inc. (http://sosyinc.com/). Because all MDA-based tools implement the same principles, we believe that the type of the problems to be addressed during model construction would be similar.

2 MDA and Unified Modeling Language

The OMG characterizes the MDA approach as fully specified, platform-independent models that can enable intellectual property to move away from technology-specific code, helping to insulate business applications from technology and to enable further interoperability [21, 22, 30].

One of the key architects of the MDA approach [31] states that one of the goals of MDA is that models are testable and simulatable. Thus, models developed using this new paradigm are capable of execution. In order for a model to be executed, its behavior must be represented. Behavioral modeling is essential to reach the goals of the MDA [17].

MDA uses the Unified Modeling Language (UML) to construct visual representations of models. UML is an industry standard for visualizing, specifying, constructing, and documenting the artifacts of a software-intensive system [1], and it has a set of advantages that makes it fit to be the heart of the MDA approach. First, by its nature, UML allows for developing models that are platform-independent [23, 26]. These models depict concepts from the problem domain and the relationships between them and then represent the concepts as objects provided with the appropriate data and behavior. A model specified with UML can be translated into any implementation environment. The valuable business and systems knowledge captured in models can then be leveraged, reused, shared, and implemented in any programming language [3]. A second advantage is that UML has built-in extension mechanisms that allow the creation of specialized, UML-based languages referred to as UML profiles [6]. In the case that modeling agricultural systems would require special modeling artefacts, then an agricultural UML profile would be created and plugged into the UML core system.

Modeling artifacts in UML are divided into two categories: structural models and behavioral models. Structural models include class and implementation diagrams. Behavioral models include use-case diagrams, interaction (sequence and collaboration) diagrams, and state machine diagrams. An important amount of code can be generated using class diagrams (structural models). Even before the MDA approach, many tools vendors provided code generators that produce "code skeletons" using class diagrams. Although this type of code generation is useful, it is very different from the code generation that the MDA approach offers. The source code generated using class diagrams (structural models) has no behavioral semantics. The programmer has to add code to include the business logic into the code.

The MDA approach consists of three levels of models as shown in Fig. 1. As shown in this figure, a set of transformations is needed to transform a model from the current level to the next one.

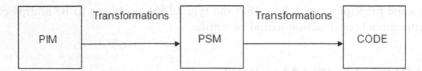

Fig. 1 Transformations are applied to a model level to obtain the next level

The approach starts with constructing a *conceptual diagram* that represents our knowledge of the problem domain expressed through concepts, abstractions, and their relationships [9, 28, 29]. Conceptual diagrams are the result of an activity referred to as *conceptual modeling*. Conceptual modeling can be defined as the process of organizing our knowledge of an application domain into hierarchical rankings or ordering of abstractions to obtain a better understanding of the phenomena under consideration [2, 33]. Conceptual diagrams have the advantage of presenting concepts and relationships in an abstract way, independent of any computing platform or programming language that may be used for their implementation. During this phase, the focus is on depicting the concepts of the system and providing them with the right data and behavior. The fact that the implementation technology may be Java, a relational database, or .NET is irrelevant at this point. Therefore, the intellectual capital invested in the model is not affected by changes in the implementation technologies. A conceptual model thus is a platform independent model (PIM).

Model construction is done visually using UML, and the participation of domain specialists in the model construction process is greatly facilitated. The MDA approach frees domain specialists from the necessity of knowing a programming language in order to be an active participant. PIMs are developed in UML, which is visual and uses plain English that can be easily understood by programmers and nonprogrammers alike [24]. A PIM is the only model that developers will have to create "by hand." Executable models will be obtained automatically by applying a set of transformations to the PIM.

After the business issues related to model construction are well-defined and presented in a PIM, then implementation matters such as the programming environment and the computing platform can be addressed. As implementation details in a specific computing environment are considered, a platform specific model (PSM) is constructed using a PIM as starting point. A PSM is a computational model that is specific to some information-formatting technology, programming language, distributed component middleware, or messaging middleware [6].

A PSM is obtained by applying a set of transformations to a PIM. A PIM could be transformed into several PSMs, in the case that different implementation technologies are selected to implement the same original PIM. Figure 2 shows an example of a PIM that will be transformed into two different PSMs: one PSM is implemented in a relational database environment and the other is implemented in a Java environment.

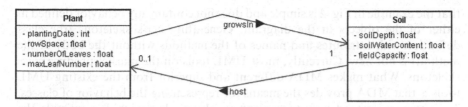

Fig. 2 A PIM representing a simple plant–soil model

The model presented in Fig. 2 does not refer to any particular computing platform or computing environment. It only shows that objects of type *Plant* and *Soil* are related to each other by the means of associations *growsIn* and *hosts*. The model says that zero or one plant can grow in a soil unit (a small area of soil used in the simulation) and that one unit of soil can host zero or one plant. Both objects have access to each other. *Plant* has a link referred to as *lnkSoil*, which allows access to data and behavior from object of type *Soil*. *Soil* has a link referred to as *lnkPlant*, which allows access to data and behavior from an object of type *Plant*. The model presented in Fig. 2 shows the type of the attributes for each of the objects. For example, *plantingDate* is of type *int*, and *fieldCapacity* is of type *float*. The presented model has the required level of detail to allow the transformations needed to obtain a precise model that includes implementation details.

If the PIM shown in Fig. 2 is to be implemented in a relational database environment, then the following transformations need to occur:

1. A table named *tbPlant* is created, and the names of its columns are *planting Date*, *rowSpace*, *numberOfLeafs*, and *maxLeafNumber*.
2. A table named *tbSoil* is created, and the names of its columns are *soilDepth*, *soilWaterContent*, and *fieldCapacity*.
3. Both tables will have a column in common, referred to as *soil–plant*, which is a foreign key (a bidirectional link) that links these two tables.

To implement the PIM presented in Fig. 2 in a Java environment, the following transformations need to occur:

1. A class referred to as *Plant* is created having for attributes *plantingDate* of type *int*, *rowSpace*, *numberOfLeafs*, and *maxLeafNumber* of type *float*.
2. A class referred to as *Soil* is created having for attributes *soilDepth*, *soilWaterContent*, and *fieldCapacity* of type *float*.
3. Class *Plant* has an attribute of type *Soil* that allows access to data and behavior defined in class *Soil*. Class *Soil* has an attribute of type *Plant* that allows for accessing data and behavior defined in class *Plant*.

PSMs are provided with enough details so that code generators can automatically translate the model in code in several programming languages. Note that the example in Fig. 2 is simple and does not contain any behavior defined in either of the objects in the diagram. Generating class skeletons (the class definition with attributes and names of the methods without the body of the method) is not new. Currently, most UML tools on the market produce class skeletons. What makes MDA different and superior from the existing UML tools is that MDA provides the means for representing the behavior of classes. To better understand how to present an object's behavior, for example the behavior of the method *calculateRate* defined in class *Soil* or *Plant*, the concept of modeling behavior needs to be introduced.

3 Modeling Behavior

Efforts for generating behavior automatically are not new in the history of software engineering. Currently, all of the integrated development environments (IDEs) provide ample support for the drag-and-drop approach. An icon representing a process is dropped on the canvas and the corresponding code is automatically generated. In this case, the icon represents a well-known process that is precoded and ready for use. Although the drag-and-drop approach facilitates enormously the process of software development, it is not a general means for modeling behavior because it is limited to cases where the behavior is known prior to model construction. As it is difficult to predict the behavior of the potential objects used in a system, this approach has not solved the problem of modeling behavior.

The problem of finding ways to express behavior has been addressed in two different ways by the researcher community. The reason for this could be found in the gap that exists between the expressive power of structural models and the potential complexity of the behavioral requirements that need to be expressed [16].

3.1 The Object Constraint Language

One line of researchers uses a formal language to address the problem of modeling behavior. There is a branch of computer science that studies how to provide precise and unambiguous descriptions of statements using formal languages. These languages are highly mathematical [36] and difficult to use by mainstream developers [35]. Efforts were undertaken to create a language that is simple and yet rigorous enough to express the behavior of objects. Thus in 1995, IBM's software engineers created the Object Constraint Language (OCL), a subset of UML that allows software developers to apply constraints to objects. These constraints are useful and provide developers with a well-defined set of rules that controls the behavior of an object. The behavior of a system can be expressed by preconditions and postconditions on operations [14]. This approach is largely inspired by ideas of the well-known modeling philosophy of "Design by Contract" [19]. Advocates of this approach do not pretend that complete code generation is possible. They state that for relatively simple operations, the body of the corresponding operation might be generated from the preconditions and postconditions, but most of the time the body of the operation must be written in PSM [14]. Furthermore, they clearly state that the dynamics of the system still cannot be fully specified in the UML–OCL combination [11, 14].

The following example shows the use of OCL to describe how to define the phenological phase of object *Plant*. The evaluation context is *Plant*, meaning that self represents object *Plant*. *phenologicalPhase* is an attribute of class *Plant*.

Plant
phenologicalPhase = **if** *self.numberOfLeaves* < *self.maximumLeafNumber*
then *vegetative*
else *reproductive*
endif

It is important to note the main characteristics of OCL. First, OCL is purely an expression language. An OCL expression does not have side effects; it does not change the status of the model. Whenever an OCL expression is evaluated, it simply delivers a value. Second, OCL is not a programming language, and therefore it is not possible to write program logic in OCL. A complete description of OCL can be found in Ref. 35. This approach has been used by a number of authors such as D'Souza and Wills [5], Cook and Daniels [4], and Walden and Nerson, [34].

3.2 The Action Language

Another group of researchers followed a different direction, the one of state machine–based models to describe behavior. They created the action language action language that allows describing actions an object performs when receiving a stimulus [18, 32]. Action languages abstract away details of the software platform so that the designer can write what is needed to be done without worrying about distribution strategies, list structure, remote procedure calls, and the like. The use of the state machines to specify behavior assumes no gap between expressive power and behavioral complexity [17]. Thus, it is possible to construct tools that express complex behavior and generate complete code from well-thought models. Because of the ability to capture arbitrarily complex behavior at a high level of abstraction, this approach seems capable of fully supporting the vision of MDA [17].

The following shows examples of statements in action language:

create object instance *newSoil* **of** *Soil*;
newSoil.name = *"sandy soil"*;

soilName = *newSoil.name;*
delete object instance *newSoil;*

The first line creates an instance of class *Soil* and assigns it to *newSoil*. Therefore, *newSoil* refers to an instance of *Soil* class. The second line assigns to attribute *soilName* a value that is the name of the soil. The third line assigns to variable *soilName* the value of the attribute *name*. The last line deletes the newly created object. Selection expressions are used to select single objects or a set of objects. The following shows examples of the use of section expressions:

select many *soils* **from instances of** *Soil;*
select many *soils* **from instances of** *Soil* **where** *selected.name* = *"sandy soil";*

The first line selects in the set of instances created from class *Soil* some of these instances. The second line selects only the instances of class *Soil* that satisfy the condition soils should belong to the category "sandy soils."

The action language approach is widely accepted by the embedded systems community, and only recently are there applications in the business information area. Among software companies that have adopted the action language approach are Accelerated Technology (http://www.acceleratedtechnology.com/), Kennedy Carter (http://www.kc.com/), and Sosy, Inc. (http://sosyinc.com/) to name a few.

4 Modeling a Crop Simulation

The crop simulation approach chosen for this study is the Kraalingen approach to modular development [13]. We choose to investigate this approach for two reasons. First, we had developed an object-oriented implementation using this approach, so this provided a basis for comparing results obtained with the new approach. Second, the Kraalingen approach is used by DSSAT-CSM crop model [12]. The experience obtained in this study may be useful in future studies when more complex crop simulation models could be considered for development using this new paradigm.

4.1 The Conceptual Model, or PIM

As previously mentioned, the MDA approach starts by constructing a conceptual model that depicts concepts from the problem domain and their relationships. The conceptual model for the Kraalingen approach is shown in Fig. 3. In this figure, object *Simulator* is the center of the model, and it plays a supervisory role. *Simulator* has relationships with *Plant*, *Soil*, and *Weather* objects. The nature of the relationship is a composition; *Simulator* plays the role of whole and *Plant*, *Soil*, and *Weather* play the role of parts. This means that object *Simulator* "owns"

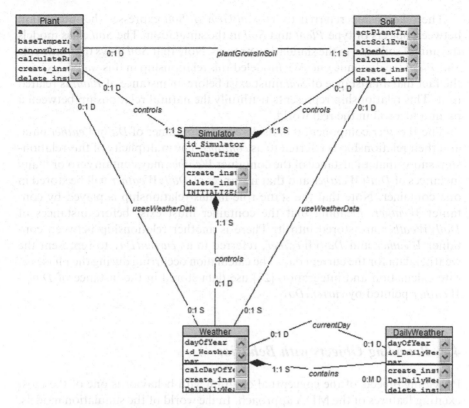

Fig. 3 Conceptual model for the Kraalingen approach

the related objects and therefore can control the manifestation of their behavior. *Simulator* is provided with behavior to send the right message at the right time to related objects to carry out the simulation process [24].

The cardinality of the association between *Simulator* and *Soil*, referred to as *controls*, shows that one instance of *Simulator* controls zero or one instance of *Soil*, and one instance of *Soil* is controlled by one instance of *Simulator*. Furthermore, besides the cardinality, the association shows that the role of *Simulator* is static, and the role of *Soil* is dynamic. Let us provide some more information on the nature (static and dynamic) of the role that classes can play in an association as it is an important concept.

The fact that *Simulator* plays a static role means that a relationship between *Simulator* and *Soil* can be established when an instance of *Soil* is created. Furthermore, the relationship can be deleted only when the instance of *Soil* is deleted. As *Simulator* has the same type of relationship with *Plant* and *Weather*, then, when an instance of *Simulator* is created, instances of *Soil*, *Plant*, and *Weather* are created and the corresponding relationships are established. The 1:1 cardinality allows *Simulator* to navigate through all the related objects.

The relationship referred to as *plantGrowsInSoil* expresses the interaction between objects of type *Plant* and *Soil* in the simulation. The *Soil* class models the unit of soil used in the simulation process. Note that *Soil* plays the static role and *Plant* the dynamic one. We modeled this relationship in this way to express the fact that an instance of *Soil* must exist before an instance of *Plant* is related to it. This relationship represents truthfully the natural relationship between a plant and a soil in the real world.

The *Weather* component is considered as a container of *DailyWeather* data, and their relationship is referred to as *contains*. The multiplicity of this relationship shows that an instance of the container *Weather* may contain zero or many instances of *DailyWeather* and that instances of *DailyWeather* will be stored in one container. Note that the static role in this relationship is played by container *Weather*, meaning that the container must exist before instances of *DailyWeather* are stored into it. There is another relationship between container *Weather* and *DailyWeather*, referred to as *currentDay* to represent the weather data for the current day. The calculation occurring during the phases of rate calculation and integration [27] use data stored in the instance of *Daily-Weather* pointed by *currentDay*.

4.2 Providing Objects with Behavior

Providing objects of the conceptual diagram with behavior is one of the most exciting features of the MDA approach. In the world of the simulation models, most of the behavior that objects should provide is expressed in the form of equations. Equations are constructed in a declarative way using attributes of objects participating in the conceptual diagram. Figure 4 shows the example of calculation of the soil attribute albedo using a declarative approach.

As shown in the figure, leaf area index data in *Plant* is needed to calculate albedo. Attributes of all objects participating in the conceptual diagram are available for use in formulas. Note that calculations may be associated with some conditions that must be satisfied before the calculations take place. To avoid errors occurring when leaf area index data is requested from a nonexistent object of type *Plant*, the condition "*EXIST(Plant)* = *true*" ensures that the calculations will take place only when an instance of *Plant* exists. In the case that the required instance does not exist, then an error will be displayed and the system halts.

In the case that the behavior of an object cannot be expressed by an equation, a formal language is provided to model behavior. The formal language used to model behavior is a type of *action language*. The behavior is referred to as *services* that an object provides. This language offers three types of services: events, transactions, and operations.

An event is an atomic unit of activity that occurs at specific instances of time. Events can create an object from a class, destroy an object, and modify

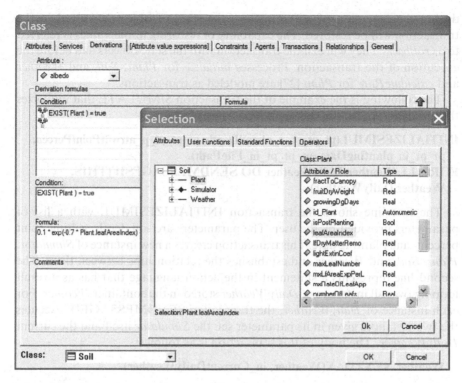

Fig. 4 Example of creating formulas in a declarative way

the life cycle of the corresponding object. Processes *calculaterDaylyNetPhotosynthesis*, *calculateDeltaLeafAreaIndex*, and *calculateOrganDryWeight* for *Plant* are modeled as events. These processes are part of the rate calculation for *Plant* [27].

Processes that are modeled as events should be simple and deal with a well-defined activity. As an example, *calculateDeltaLeafAreaIndex* is a simple process that has as a result the calculation of the value for attribute *deltaLeafAreaIndex*. An event does not consider the order in which processes should occur. Therefore, the process *calculateRate* cannot be modeled as an event. The order in which *calculaterDaylyNetPhotosynthesis*, *calculateDeltaLeafAreaIndex*, and *calculateOrganDryWeight* are executed matters. The process *calculateOrganDryWeight* uses *deltaTotalDryWeight* and the latter uses *dailyNetPhotosynthesis*. Therefore, *calculateOrganDryWeight* cannot be executed before *calculaterDaylyNetPhotosynthesis*. If processes need to be executed in a defined order, then another type of service, transactions, should be used.

A transaction is an atomic processing unit composed of one or more events or other transactions. Similar to events, transactions are used to create, destroy, and modify the life cycle of objects. The state of an object involved in a transaction cannot be queried until the transaction is finished. If an error occurs during the execution of a transaction, the state of objects involved in the transaction will be restored. The capability of restoring initial values is referred to as *rollback*. They will exit the transaction with the state objects had before the execution of the transaction. Processes *initialize* for *Plant*, *Soil*, and *Weather* and *calculateRate* for *Plant* [27] are modeled as transactions.

The following is the example of the transaction SIMULATE that initializes the *Simulator* and starts the simulation process:

INITIALIZESIMUL(pt_pt_pt_p_atrsoilDepth, pt_pt_pt_p_atrwiltPointPercen, pt_pt_in_plantingDate, pt_pt_pt_in_FilePath).
FOR ALL Weather.DailyWeather DO SENDMESSAGESIF(THIS, Weather.DailyWeather)

The first line shows the transaction INITIALIZESIMUL with a list of parameters provided by the user. The parameters are soil depth, wilting point percent, and planting date. This transaction creates a new instance of *Simulator*, *Plant*, *Soil*, and *Weather* and establishes the relationships between them. The second line represents a statement in the action language that has as a result looping over all instances of *DailyWeather* stored in the container *Weather*. For each instance of *DailyWeather*, the transaction SENDMESSAGEIF executes the two instances given in its parameter set: the *Simulator* itself and the current *DailyWeather*. The body of this transaction is as follows:

Weather.NEXTDAY(Weather, in_CurrentDailyWeather).

Soil.calculateRate(Soil).

{Plant.isPostPlanting = TRUE}
Plant.CALCULATERATE(Plant).

Soil.integrate(Soil).

{Plant.isPostPlanting = TRUE}
  Plant.integrate(Plant)

Note that for each instance of *DailyWeather*, *Plant* and *Soil* will receive messages *calculateRate* and *integrate* as defined by [13].

The third type of service is referred to as operations. An operation is similar to a transaction but does not offer rollback capabilities. Because of the similarity with transactions, operations are not used in this project.

The simulation process is controlled by the behavior of object *Plant*. A state-transition diagram is used to model the behavior of *Plant* [1]. This diagram shows the valid execution order of the services of the class and the set of possible life cycles of *Plant*. Figure 5 shows the state-transition diagram of *Plant*. The diagram has two types of elements: *states* and *transitions*. States represent the different situations through which an object of type *Plant* can pass, depending on the value of its attributes. Transitions represent executed services, events, or transactions, which produce state changes and modify the value of object's attributes.

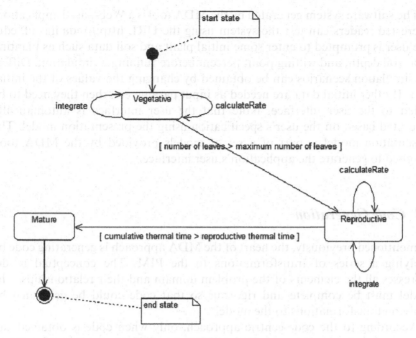

Fig. 5 State-chart diagram describing the behavior of *Plant*

According to the state-transition diagram, *Plant* will remain in the state *vegetative* and will continue to receive messages *calculateRate* and *integrate* as long as the guard condition *number of leaves > maximum number of leaves* is not satisfied. When the guard condition is satisfied, *Plant* will move to state *reproductive*. For this transition, the source state is *vegetative* and the target state is *reproductive*. *Plant* will remain in the state *reproductive* as long as the guard condition *cumulative thermal time > reproductive thermal time* is not satisfied. When the guard condition is satisfied, *Plant* will move to state *mature* and the simulation will terminate.

4.3 Data Requirements

The system uses weather data saved locally in a text file as described in Ref. 27. The process of reading weather data is not part of the model and cannot be modeled using the available artifacts. The Oliva Nova Model Execution provides access to external functionalities through *user functions*. A user function can be written in any of the languages that are generated by the MDA-based tool such as Java, .NET, and so forth. In this project, a user function referred to as *importWeatherData* is designed to provide access to weather data. The source of the weather data can be a text file, a database, or a weather station online [25].

The software system generated by the MDA tool is a Web-based application. Interested readers can test the system using the URL http://mda.ifas.ufl.edu. The user is prompted to enter some initial plant and soil data such as planting data, soil depth, and wilting point percent before running a simulation. Different simulation scenarios can be obtained by changing the values of the initial data. If other initial data are needed as input parameters, then they need to be added to the user interface. Note that the user interface is automatically generated based on the user's specification using the presentation model. The presentation model is part of a set of models provided by the MDA tool, designed to generate the application's user interface.

4.4 Code Generation

As mentioned previously, the heart of the MDA approach is generating code by applying a series of transformations to the PIM. The conceptual model expresses all the elements of the problem domain and their relationships. The model must be complete and rigorous so that code could be generated by applying transformations to the model.

According to the code-centric approach, only when code is obtained can errors be found and corrected. In the MDA approach, the process of checking the validity of the model can start while designing the model. MDA-based tools provide ample capabilities to check the correctness of the conceptual model, the behavior of each object in the model, and relationships between objects. An Extensible Markup Language (XML) file containing detailed specifications about the model is created and is used by code engines/compilers to generate code in several programming languages. Several scenarios can be considered as different parts of the system can be implemented in different languages. For example, a user can choose the C# environment for developing the user interface and a CORBA-EJB, Java-based environment for the implementation of the server. Because the conceptual model is detailed and precise, code generators can find all the needed information to translate the model into several programming environments. Besides the code representing objects of the conceptual model, code generators will provide all the wiring code that links the client and the server applications.

4.5 Results

As previously mentioned in this chapter, a Java version of the Kraalingen simulation model was developed by the authors. The validation of the Java version was done in collaboration with crop specialists and the results were compared with a FORTRAN version developed earlier [27]. Results obtained by the software developed using the MDA approach are presented in Fig. 6.

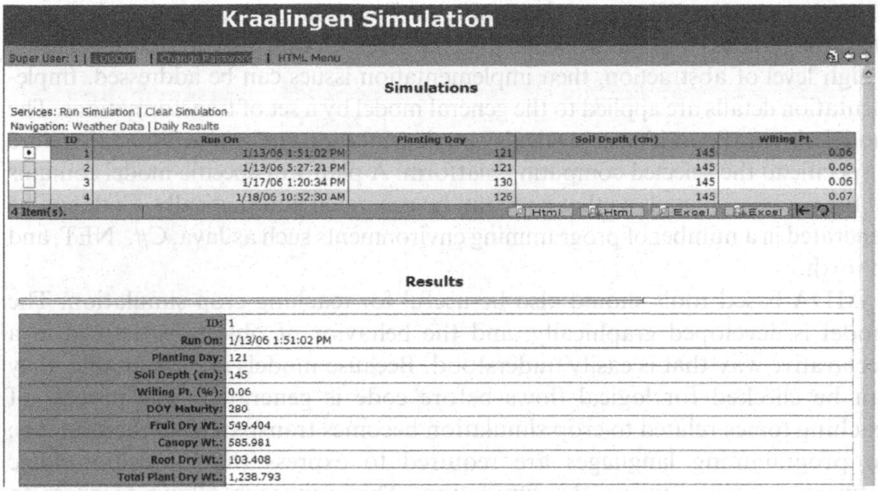

Fig. 6 Simulation results obtained by the MDA approach

As shown in Fig. 6, the system allows for creating simulation scenarios using different initial conditions. The results presented in Fig. 6 pertain to the scenario 1, which uses as initial conditions planting date = 121, soil depth = 145, and wilting point percent = 0.06. Scenario 3 uses planting date = 130, and other initial conditions are kept the same to show the impact of the planting date in simulation results. Scenario 4 uses as input values planting date = 126 and wilting point percent = 0.07. The results obtained in scenario 1 coincide with the results obtained with the Java version of the software using the same initial conditions. Thus, the quality of the obtained results is not affected by the technology used to develop the software system.

5 Conclusions

This paper describes the concept and the feasibility of using MDA-based tools for modeling software for crop simulation systems. The center of the approach is a conceptual model that expresses concepts from the domain problem and their relationships. The conceptual model is built using UML, a standard in the software industry. The model is developed visually, and the language used is simple and understandable to programmers and nonprogrammers alike. This modeling paradigm is specialist-centric, as it allows for a greater participation of specialists in model construction. The model is constructed conceptually, focusing on the business logic that is familiar to the specialists.

The conceptual model is constructed without considering any implementation or computing platform issues. Therefore, the model is platform independent; it can be implemented in different programming environments and computing platforms. Once the business logic of the problem is clarified and expressed in the model using a high level of abstraction, then implementation issues can be addressed. Implementation details are applied to the general model by a set of transformations. The model obtained considers a particular implementation environment and therefore is specific to the selected computing platform. A platform-specific model contains all the necessary details so that code can be generated automatically. Code can be generated in a number of programming environments such as Java, C#, .NET, and so forth.

MDA-based tools should also be useful for teaching crop simulation. The model is developed graphically, and the behavior of objects is defined in a declarative way that is easily understood. Because models are executable, they can be checked for logical flows before code is generated. The process of teaching topics related to crop simulation becomes transparent to the students; no programming languages are required to express the interaction logic between concepts during the simulation. This approach allows teachers to focus on the logic of the simulation without using details of programming languages to express it.

Recently, the Open Source community has shown strong interest in the MDA approach. Because specifications for the MDA approach are free of charge and can be downloaded from OMG's Web site, many projects are concurrently under development aiming to produce MDA-based tools. There are several MDA-based tools available such as ANDROMDA (http://www. andromda.org/whatisit.html) and OPENMDX (http://www.openmdx.org/) to name a few. Sun Microsystems (http://www.sun.com/) donated metadata repository modules to the NetBeans open-source development project as part of OMG's MDA effort. The NetBeans open-source project is a Java-based effort that is being positioned as a compliant solution for the MDA specification [10, 15]. Open Source MDA-based tools are an excellent resource that can be used to learn this new modeling paradigm at no cost.

Starting a project using a new paradigm is not a straightforward process. We recommend a gradual approach to this problem. First, a good mastery of UML is required to be able to present concepts of the domain and their relationships. There are quite a few good open-source UML tools that can be used for this purpose. We used ArgoUML (http://argouml.tigris.org/) for learning concepts and developing simple UML-based projects. Second, time needs to be invested in learning the language used by the tool to express behavior. Although MDA principles defined by OMG are the same, they may be implemented slightly differently by different development groups. If the MDA-based tool used can generate all the code, then the user is required to only deploy the application. If some code still needs to be implemented by hand, then the user is required to provide this code before deploying the application. Note that modeling is a rigorous discipline and modeling using an MDA-based tool becomes more abstract than usual. It takes time and effort to become comfortable using this new modeling paradigm as is always the case when a paradigm shift occurs.

References

1. Booch, G., Rumbaugh, J., Jacobson, I. 1999. The Unified Modelling Language User Guide. Addison Wesley Longman. One Jacob Way, Reading, Massachusetts 01867.
2. Carnegie Mellon University, Software Engineering Institute. 1995. The Capability Maturity Model: Guidelines for Improving the Software Process. Addison Wesley Professional. One Jacob Way, Reading, Massachusetts 01867.
3. Clark, R., Papajorgji, P. 2005. Model Driven Architecture for Software Application Generation. World Congress of Computers in Agriculture, Vila Real, Portugal.
4. Cook, S., Daniels, J. 1994. Designing Object Systems – Object-Oriented Modeling with Syntropy. Prentice Hall, Englewood Cliffs, NJ.
5. D'Souza, D.F., Wills, A.C. 1999. Objects, Components, and Frameworks with UML: The CATALYSIS Approach. Addison-Wesley, Reading, MA.
6. Frankel, S.D. 2003. Model Driven Architecture. Applying MDA to Enterprise Computing. Wiley Publishing, Inc. OMG Press, Needham, MA.
7. Gartner Symposium, March 2004. http://symwest.blog.gartner.com.
8. Gartner, Inc. http://www.gartner.com
9. Hay, D.C. 2003. Requirements Analysis: From Business Views to Architecture. Pearson Education. Pearson Education, Inc. 200 Old Tappan Road, Old Tappan, NJ 07675.
10. Jazayeri, M., Ran, A., Van Der Linden, F. 2000. Software Architecture for Product Families: Principle and Practice. Addison-Wesley, Reading, MA.
11. Jones, C. 1996. Patterns of Software Systems Failure and Success. International Thompson Computer Press, Boston, MA.
12. Jones J.W., Keating, B.A., Porter, C. 2001. Approaches to modular model development. Agricultural Systems 70, 421–443.
13. van Kraalingen. D.W.G. 1995. The FSE system for crop simulation, version 2.1. Quantitative Approaches in Systems Analysis Report no. 1. AB/DLO, PE, Wageningen.
14. Kleppe, A., Warmer, J., Bast, W. 2003. MDA Explained. The Model Driven Architecture: Practice and Promise. Addison-Wesley, Reading, MA.
15. Krill P. 2002. Sun hails open-source meta data repository. http://www.infoworld.com/articles/hn/xml/02/05/07/020507hnnetbeans.html.
16. McNeile, A. 2003. MDA: The vision with the hole. http://www.metamaxim.com
17. McNeile, A., Simons, N. 2004. Methods of Behaviour Modelling. A Commentary on Behavior Modelling Techniques for MDA. Metamaxim Ltd. 201 Harverstock Hill, Belsize Park, London NW3 4QG United Kingdom.
18. Mellor, J.S., Balcer J.M. 2002. Executable UML A foundation for Model-driven Architecture. Addison-Wesley, Reading, MA.
19. Meyer, B. 1988. Object-Oriented Software Construction. Prentice Hall, Englewood Cliffs, NJ.
20. Miller, J., Mukerj, J. (Eds). 2003. MDA Guide Version 1.0.1. Document Number omg/2003-06-01. Object Modeling Group. Needham, MA 02494, U.S.A.
21. Object Management Group (OMG). http://omg.org/.
22. OMG Model Driven Architecture: How Systems Will be Built. http//omg.org/mda/.
23. Papajorgji, P., Beck, W.B., Braga, J.L. 2004. An architecture for developing service-oriented and component-based environmental models. Ecological Modelling 179/1, 61–76.
24. Papajorgji, P., Shatar, T. 2004. Using the Unified Modelling Language to develop soil water-balance and irrigation-scheduling models. Environmental Modelling & Software 19, 451e459.
25. Papajorgji, P. 2005. A plug and play approach for developing environmental models. Environmental Modelling & Software 20, 1353e1359.
26. Papajorgji, P., Pardalos, P. 2005. Software Engineering Techniques Applied to Agricultural Systems: An Object-Oriented and UML Approach. Springer-Verlag, Berlin.

27. Porter, C.H., Braga, R., Jones, J.W. 1999. Research Report No. 99-0701. Agricultural and Biological Engineering Department, University of Florida, Gainesville, FL.
28. Rosenberg, D., Scott, K. 1999. Use Case Driven Object Modelling with UML: A Practical Approach. Addison Wesley Longman. One Jacob Way, Reading, Massachusetts 01867.
29. Sommerville, I., Sawyer, P. 1997. Requirements Engineering: A Good Practise Guide. John Wiley & Sons, Chichester, England.
30. Soley, R. 2000. Model driven Architecture. White paper draft 3.2. http://ftp.omg.org/pub/docs/omg/00-11-05.pdf.
31. Soley, R. 2002. Presentation: MDA: An introduction. http://omg.org/mda/presentations.htm.
32. Starr, L. 2002. Executable UML: How to Build Class Models. Prentice Hall, Englewood Cliffs, NJ.
33. Taivalsaari, A. 1996. On the notion of inheritance. ACM Computing Surveys 28 (3), 438–479.
34. Walden, K. Nerson, J-M. 1995. Seamless Object-Oriented Software Architecture: Analysis and Design of Reliable Systems. Prentice Hall, Englewood Cliffs, NJ.
35. Warmer, J., Kleppe, A. 1999. The Object Constraint Language Precise Modelling with UML. Addison-Wesley, Reading, MA.
36. Wordsworth, J. 1992. Software Development with Z. Addison-Wesley, Reading, MA.

A New Methodology to Automate the Transformation of GIS Models in an Iterative Development Process

André Miralles and Thérèse Libourel

Abstract In the majority of research today in areas such as evaluation of flood risks, management of organic waste as it applies to plants, and mapping ecological conditions of rivers, scientific advances are often aimed toward the development of new software or the modification of existing software. One of the particulars for software developed for agricultural or environmental fields is that this software manages geographic information. The amount of geographic information has greatly increased over the past 20 years. Geographic Information Systems (GISs) have been designed to store this information and use it to calculate indicators and to create maps to facilitate the presentation and the appropriation of the information. Often, the development of these GISs is a long and very hard process. Since the early 1970 s, in order to help project managers, software development processes have been designed and applied. These development processes have also been used for GIS developments. In this chapter, the authors present a new methodology to realize GIS more easily and more interactively. This methodology is based on model transformations, a concept introduced by the Object Management Group (OMG) in its approach called model driven architecture (MDA). When software is developed, models are often used to improve the communication between users, stakeholders, and designers. The changes of a model can be seen as a process where each action (capture of user concepts, modification of concepts, removal of concepts, etc.) transforms the model. In the MDA approach, the OMG recommends automation of these actions using model transformations. The authors have developed a complete set of model transformations that enable one to ensure the evolution of a GIS model from the analysis phase to the implementation phase.

A. Miralles (✉)
Centre for Agricultural and Environmental Engineering Research, Earth Observation and GeoInformation for Environment and Land Development Unit, Montpellier, France
e-mail: andre.miralles@teledetection.fr

P.J. Papajorgji, P.M. Pardalos (eds.), *Advances in Modeling Agricultural Systems*,   19
DOI 10.1007/978-0-387-75181-8_2, © Springer Science+Buisness Media, LLC 2009

1 Introduction

The development of a software application is becoming increasingly difficult. Since the earliest developments of software, many methodologies have been designed and used to help the project leader in developing software. Over the past 15 years, Ivar Jacobson, Grady Booch, and James Rumbaugh have been major contributors to the improvement of the methodologies used to develop software [12]. They define a *software development process* as *the set of activities needed to transform a user's requirements into a software system* (Fig. 1).

These authors have also formalized the various "ingredients" taking part in the process of developing a computer application. This model is called the 4Ps model (Fig. 2).

This model dictates that the *Result* of a *Project* is a *Product* that requires *People* in order to describe the studied domain (actors) and to manage it (analysts, designers, programmers, etc.). The realization of the *Project* is conducted in accordance with *Templates* defining, organizing, and explaining the successive steps of the development *Process*. In order to manage the development *Process*, *Tools* facilitating the expression of the needs, the modeling, the project planning, and so forth, are needed.

This description of the development process paints a set of variety of topics and issues that a project manager in charge of application development should address. To illustrate the intrinsic complexity of a development, Muller and Gaertner [21] use two metaphors reported here *in extenso*:

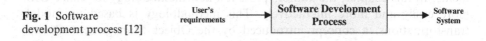

Fig. 1 Software development process [12]

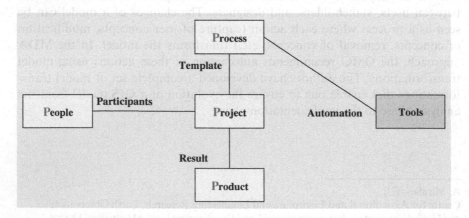

Fig. 2 The 4 Ps software development process [12]

- The first one is related to the management of the project: the development of software can be seen as the crossing of an ocean in a boat. The departure day is known; the arrival is not so well-known. In the course of the trip, it will be necessary to brave storms and to repair damage.
- The second one concerns the multidisciplinary character of necessary competencies to make a development. They write: if the computer programmer had produced furniture, he would begin by planting acorns, would cut up trees into planks, would dig the soil in search of iron ore, would manufacture ironworks to make his nails, and would finish by assembling everything to obtain a piece of furniture. . . . A mode of development which rests on the shoulders of some heroic programmers, gurus, and other magicians of software does not constitute a perennial and reproducible industrial practice.

These two metaphors perfectly illustrate the challenge with which the project leader and the programmers are confronted when they take on the realization of a data-processing application. This challenge is not entirely imaginary. The statistics of Ref. 30 give an idea of the difficulty. According to these statistics, the failure risk of the development of an application is 23%, the risk of drift is 49%, and only 28% of the developments are finished within the foreseen delay and within the projected budget. These figures are from an investigation carried out on more than 150,000 developments achieved in the United States in 2000. It is noteworthy that in this investigation, the developments aimed at creating a new application are grouped with those aimed at the evolution of an existing application. Thus, it is quite likely that the figure of 28% is overestimated, as the failure risk linked to achieving a new application is much larger than the risk linked to the evolution of an application.

2 The Software DevelopmentProcess

The creators of the Unified Modeling Language (UML) have deliberately failed to define a methodology to successfully carry out a project of development [8] in order to let each designer freely choose the most suitable method adapted to his professional environment. Generally, the designer uses methods of project leading with the aim of *increasing the satisfaction level of the customers or of the stakeholders while making the development work easier* [3] and more rationally

There are a wide variety of software development processes that can be classified into two large families:

- The so-called traditional methods (waterfall life-cycle, V life-cycle, spiral life-cycle, unified process, rapid application development, etc.) are derived most often from the methods used in industrial engineering or in civil engineering (i.e., building and public works sector) [17].
- The *agile* methods, of which the most important are extreme programming, dynamic software development method, adaptive software development, SCRUM [3, 17], and so forth. Their major characteristics are their potential for adaptation and *common sense in action* [3].

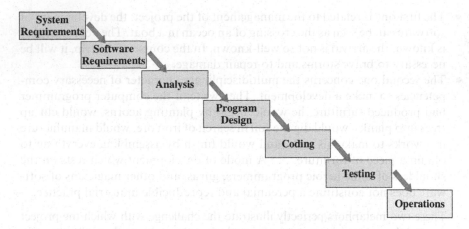

Fig. 3 Typical waterfall cycle for software development process [26]

The software development processes that were used in the 1970 s were essentially of the linear type (Fig. 3); that is to say that the analysis was conducted at the start of the project. Then, the development of the application was conducted, step after step, without any intermediate validation from users, stakeholders, or sponsors. At the end of the process, the application was presented to the stakeholders. It was not rare that the application did not correspond with the needs of the users, the stakeholders, or the sponsors. In this case, the project manager was professionally in a difficult position, especially if the duration of the development was long (several months and even 1 or 2 years).

This situation is hardly surprising because it is difficult, and even impossible, for the users or the stakeholders, *to conceive the solution in its whole* [5]. This report is all the more true when the scale of the project is substantial.

To avoid facing this type of situation, the project managers have appealed more and more to the users and the stakeholders during the development process to validate the application progress.

The experience cumulated during this type of development processes enables better formalization of the participation of the stakeholders in the development of computer applications and allows for the proposal of new methods to conduct the project. The unified process method and extreme programming method are two key methods coming from this line of thinking.

The unified process method, relying on the modeling language UML, is the synthesis of the best practices of software development over three decades in various fields[1] [12]. It assumes the adoption of the following four principles: the development process should be *use-case driven,* but it should also be *iterative and incremental, architecture-centric,* and *risk-centric* [16, 21, 25].

---

[1] Telecommunication, aeronautic, defense, transport, and so forth.

The *use-case driven* development principle has been introduced by Ivar Jacobson [11] in order to pilot application development according to the requirements of the users or the stakeholders. A use-case is *a sequence of actions, including variants, that a system (or other entity) can perform, interacting with actors of the system* [23]. This concept enables one to describe what the system should do. Once implemented, a use-case compulsorily resolves a requirement. If this is not the case, it is because the need was not properly described. Thus, it is important to describe the use-cases at the beginning of the project. In this vision, the use-cases can be used as a planning tool during the development.

The *iterative* development process (Fig. 4) has been designed to prevent the drawbacks caused by linear development. In order to do this, the system is structured into subsystems, and for each iteration, a subsystem is analyzed and implemented. Therefore, the model evolves following an *incremental* process. For Ivar Jacobson, *an increment is the result of an iteration* [12] that lasts between 2 and 4 weeks.

The principle of a development process that would be *architecture-centric* assumes that the structuring in subsystems must not be a simple description of the system under a graphic or a textual form, but that it should be materialized by a model in a *case-tool* [25].

The aim of a *risk-centric* development is to put as a priority the achievement of the systems or subsystems for which the designers have the least experience: implementation of new technologies, for instance. This principle of development enables one to take issues into account very early and to process them by anticipation.

Extreme programming [1, 4] is a method called *agile*, which recommends reducing activities that are not closely related to the production of a code, including documentation. The code is the main part of the production of the team. This method is hence often qualified as code-centric development. It is representative of the agile methods that rely on four values:

- **Communication** between the users, the stakeholders, and the designer to prevent situations described in the waterfall method.
- **Simplicity** of the code so that it is easily understandable and it is possible to integrate changes.

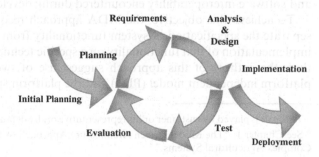

Fig. 4 Typical iteration flow in unified process [15]

- **Feedback,** which should be quick from the stakeholders and from the other members of the development team, enables the developer to have information on the quality of his development.
- **Courage** to tell things as they are and to make difficult decisions like changing a code structure or throwing it away [6].

The fulfillment of these four values is ensured by 12 practices with the aim of encouraging *quick feedback*, favoring the *incremental evolution* of the code, seeking *simplicity* in the produced code, and targeting the *code quality*.

Among these practices, that of *customer on-site*[2] is probably the most important. The aim of this practice is to fluidize the communication between the customer and the programmers by hosting the customer or his representative within the team. This practice ensures a strong reactivity and a high feedback. The main aim of this practice is to make up for the lack of detailed specifications.

The main task of the customer is the writing of the *user stories*, which will allow one to code the functionalities of the application. A second task that is just as important as the first one is the determination of the tests that the tester should implement to validate functionalities. The customer acts by fixing the priorities among the functionalities, by stating the specifications that have not been previously defined or that have remained fuzzy during the previous discussions, and so forth.

The presence of the customer in the team enables him to see the immediate result of his work of specification and to evaluate the progression of the application. This closeness also enables him to quickly assess the relevance of his specifications. If the project drifts or progress is slow, he will immediately realize it.

Actually, the practice of *customer on-site* gives a high level of interactivity to the development process, which associated with practice of the test-driven development reduces the number of bugs by a factor of 5 in some cases.

3 The Model Driven Architecture[3]

Model driven architecture (MDA) is a software design approach proposed by the Object Management Group (OMG) with the objective of improving application developments. It was conceived and formalized in 2001 to improve productivity but also to resolve problems of software portability, software integration, and software interoperability encountered during developments [14].

To achieve this objective, the MDA approach recommends that designers separate the specification of system functionality from the specification of the implementation of that functionality on a specific technology platform [18]. For that, the authors of this approach suggest use of two types of models: the platform independent model (PIM) and the platform specific model (PSM).

---

[2] This role is played by customer or his representative or, by default, by a member of the team.

[3] See Chapter 1, "The Model Driven Architecture Approach: A Framework for Developing Complex Agricultural Systems."

Fig. 5 Illustration of the
separate of concepts and
of the transformation notion

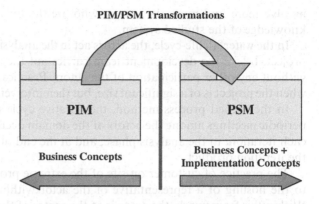

PIMs are models providing a description of the structure and functions of the system independently of platform specifications. PSMs are models defining how structure and functions of a system are implemented on a specific platform.

In fact, the MDA approach introduces a separation between concepts and specifications needed to develop software. PIMs only contain business concepts, whereas PSMs contain implementation concepts. Because all the PIM business concepts are included in PSMs, a PIM can be seen as a modified subset of a PSM [7]. Therefore, a PSM always derives from a model PIM through one or more transformations [18, 19].

Figure 5 illustrates this separation and transformation. If different platforms are used for the implementations (e.g., same standardized model implemented into different organizations), then more than one PSM may be derived from the same PIM.

The previous transformations, called PIM/PSM transformations, are not the only ones. In fact, the authors of MDA mention on the one hand the existence of PSM/PIM transformations converting a PSM into a PIM and, on the other hand, transformations whose model sources and targets are of the same standard (PIM/PIM transformations or PSM/PSM transformations).

In the process of development, PSM is not the last step as it is then necessary to project this model into a programming language. This projection is often considered as a transformation.

4 The New Interactive Development Method

4.1 The Principle of the Continuous Integration Unified Process Method

For about 40 years, the major aim of research bearing on the methods of development of computer applications has been to reduce the gap between the needs of the actors (users, clients, stakeholders, etc.) and the end product. To achieve this, the authors of the methods of development seek to associate and to involve more and more the actors, who are the only ones that have a good knowledge of the studied system.

In the waterfall life-cycle, the actors act in the analysis phase at the start of the project, before the development team carries out the application, theoretically without any other participation of the actors. Practically, the actors mostly act when the project is of a significant size, but their interventions are not formalized.

In the unified process method, the iterative cycle requires organization of periodic meetings among the actors of the domain occurring at the beginning of each iteration, in the analysis phase, and at the end of the iteration to validate the iteration product.

The practice of customer on-site of the extreme programming method leads to the hosting of a representative of the actor within the development team. Within this framework, the actor is at the center of the development.

Actually, the increased participation of the actors enables, on the one hand, improvement in the capture of knowledge and the expression of the actors' needs and, on the other hand, to have, at a more or less continuous frequency, the validation of the evolution of the application. With this type of process, the semantic side of the application is of a higher quality. The direct consequence is that the increment developed during the iteration is more stable.

Building on that report, the authors have designed a new method called the *continuous integration unified process,* which allows an increase in the interactivity between the actors and the designer.

This new method is an extension of the unified process method incorporating some practices of the extreme programming method. It is based on the following report: in the analysis phase, the actors are in a situation similar to that of the customer on-site in the extreme programming method (see Section 2) – they are at the heart of the analysis. As communication is the key value of the extreme programming method, any technique or method increasing it will result in improvement of the quality of the end application. Dialogue around a prototype is one of these techniques or methods.

It is not rare that during the development of an application, one or several prototypes are produced so that the actors have a better understanding of what the end application will be. Then, the actors implicitly validate the concepts of the field and, if it is a "dynamic prototype" [24], they validate the assumed functionalities corresponding with their requirements. A prototype is a device that fluidizes the exchanges between the actors and the designer, but it also increases the area of shared knowledge [10] called *commonness* [27]. Moreover, the implementation of the prototype accelerates learning by the actors of the modeling language used by the designers [10].

The qualities of the prototype have led the authors to formalize its use in the analysis phase, a key phase for the capture of the knowledge and the actors' requirements.

To generate a prototype requires similar development to that of the final application. In this background, the development process includes simplified analysis, design, and implementation. If all the activities to develop the prototype are done manually, the analysis will be interrupted by nonproductive slack periods that will prove to be expensive.

This exercise of analysis will quickly become tedious for the actors, and they will become demobilized and lose interest in the exercise. The result is that the analysis could be less relevant and the quality of the application could deteriorate. On the other hand, if the same activities are automated, then the slack periods do not exist, and the response of the actors to the prototype will be better than in front of a model for which all the finer points of the modeling language are not known. Then, the development process of the prototype is made according to a cycle with a very short duration, which is qualified as *rapid prototyping*.

Building on these thoughts, the definition of the new method is the following: the continuous integration unified process method superimposes, on the main cycle of the unified process method, a cycle of rapid prototyping (Fig. 6), which is provided with a process automating the evolution of the models from the analysis to the implementation.

The idea of automatic evolution of the models from the analysis up to the implementation can also be found in the concerns of the MDA community. It is obvious, from reading the fundamental texts of this approach [18, 19], that this was one of the objectives that were sought. Some authors [13, 28] describe as full MDA the complete automation of the evolution of the models. The *common warehouse metamodel* (CWM) standard [22] has been created to cover the complete cycle of design, completion, and management of the data warehouses [18].

Naturally, the challenge remains to design and implement a complete set of model transformations assuming a full MDA process. The authors reach such a

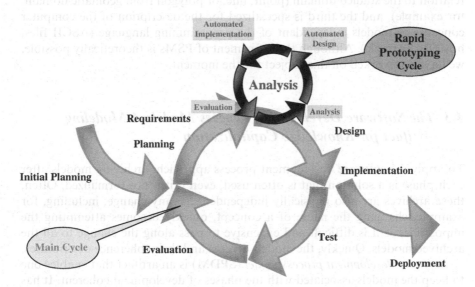

Fig. 6 Continuous integration unified process

challenge in generating automatically the structure of the database of a Geographic Information System (GIS).

Given such a set, the duration of the phases of design and implementation of the cycle of rapid prototyping is reduced to the unique time of completion of these transformations, time linked to the volume of concepts contained in the models.

4.2 The Software Development Process Approach: A Generalization of the MDA Approach

When an application is developed, one of the main preoccupations for a project manager and for the company in charge of the software development is the capitalization of knowledge and the re-use of the knowledge accumulated during development.

The capitalization of knowledge is not just the problem of separating the business concepts and implementation concepts according to the MDA vision presented in Section 3, as at each phase of the development, the type of mobilized knowledge is different. Thus, another approach involves capitalizing on the knowledge at each phase of the application development process. The *software development process approach* proposed by the authors is founded on this report [20]. Thus, in this new approach, a model is associated at each one of the phases.

In fact, this approach generalizes the MDA approach by refining the PIMs into three types of models: the analysis model, the preliminary design model, and the advanced design model. The first one is used to analyze the system with the actors, the second one is dedicated to the concepts coming from a domain in relation to the studied domain (point, line, or polygon from geomatic domain, for example), and the third is specialized for the description of the computer concepts or models independent of the programming language (ASCII files, host, for example). Although the refinement of PSMs is theoretically possible, we have not worked on this subject for the moment.

4.3 The Software Development Process Model: A Modeling Artifact for Knowledge Capitalization

To apply the software development process approach, archiving models after each phase is a solution that is often used, even if it is not formalized. Often, these archives are also physically independent, so any change, including, for example, changing the name of a concept, quickly becomes attempting the impossible, as it is difficult and expensive to pass along the change to all the archived models. Quickly, the models diverge and their coherence deteriorates.

Software development process model (SPDM) is an artifact that enables one to keep the models associated with the phases of development coherent. It has been conceived and implemented by Miralles [20] into case-tools. This modeling

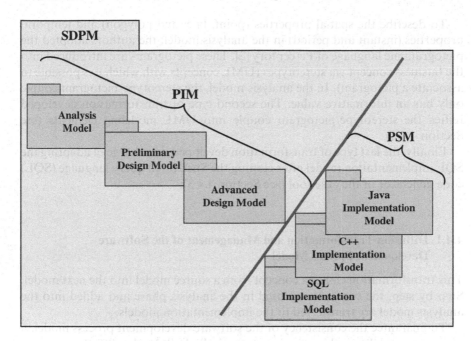

Fig. 7 The software development process model

artifact contains the different models associated with the phases of the software development process. In the vision of the authors, the software development process model is the MODEL of the application under development.

Figure 7 shows the software development process model for the development of software following the *two-track unified process* method [25], a method derived from the unified process method. This figure also shows that the PIM/PSM separation introduced by the MDA approach occurs when the project moves from the advanced design phase to the implementation phase.

4.4 The Complete Set of Transformations Enabling a Full MDA Process for Databases

To realize a full MDA[4] process for GIS, the first set of transformations implemented into the case-tool is in charge of diffusing the captured business concepts from the analysis model to the implementation models (see Section 4.4.1).

---

[4] Normally, a full MDA process must include the model verification and the model compilation. Currently, the model verification is not made but it is one of the future subjects of research. The model compilation is held by code generators (C++ and C#, Java, Corba, and SQL) proposed by case-tool.

To describe the spatial properties (point, line, and polygon) and temporal properties (instant and period) in the analysis model, the authors adopted the pictogrammic language of Perceptory [2]. These pictograms are introduced into the business concept via stereotypes (UML concepts with which it is possible to associate a pictogram). In the analysis model, the stereotype/pictogram couple only has an informative value. The second type of transformation developed reifies the stereotype/pictogram couple into UML modeling elements (see Section 4.4.2).

Finally, the last type of transformation developed is in charge of adapting the SQL implementation model after cloning the Structured query language (SQL) code generator of the case-tool (see Section 4.4.3).

4.4.1 Diffusion Transformation and Management of the Software Development Process Model

This transformation clones a concept from a source model into the next model. Step by step, the concepts captured in the analysis phase and added into the analysis model are transferred to the implementation models.

To guarantee the consistency of the software development process model, a *cloning traceability architecture* is automatically built by the *diffusion transformation*. After cloning, this transformation establishes an individual cloning traceability link between each one of the source concepts and the cloned concepts. Figure 8 illustrates the cloning traceability architecture.

In an iterative development process, the *diffusion transformation* adds, with every iteration, a new clone of the same source into the following model. To avoid this problem, when an individual cloning traceability link exists, the *diffusion transformation* does not clone the concepts but only carries out one update of the clone.

4.4.2 The GIS Transformations

The GIS Design Pattern Generation Transformation

The spatial and temporal concepts have stable relationships that are completely known. They constitute recurrent minimodels having the main property of design patterns[5]: recurrence [9]. It is this property that led authors to call these minimodels *design patterns*. These GIS design patterns do not have the

---

[5] A design pattern systematically names, motivates, and explains a general design that addresses a recurring design problem in object-oriented systems. It describes the problem, the solution, when to apply the solution, and its consequences. It also gives implementation hints and examples. The solution is a general arrangement of objects and classes that solve the problem. The solution is customized and implemented to solve the problem in a particular context [9].

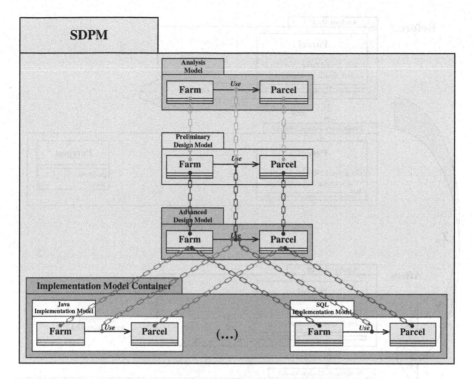

Fig. 8 Example of the cloning traceability architecture

same statutes as the design patterns described in Ref. 9, but they are fundamental design patterns in the geomatic domain. Figure 9 shows an example of
design pattern of the GIS domain. The set of these patterns is called the *GIS
design pattern.*

   Given that the design patterns are always identical, they can be automatically
generated with a case-tool without any difficulty. The *GIS design pattern generation transformation* is the transformation in charge of generating the set of
GIS design patterns.

The Pictogram Translation Transformation

Once the GIS design patterns have been created, the business and the spatial
or temporal concepts represented by the pictogram are totally disassociated
(Fig. 10, "Before"). The goal of the *pictogram translation transformation* T_p

Fig. 9 Example of a GIS
design pattern

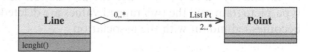

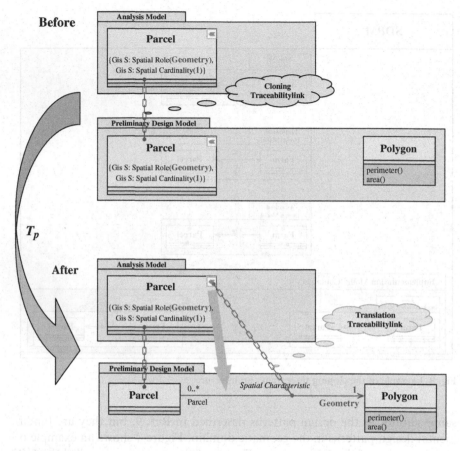

Fig. 10 The pictogram translation transformation T_P

(Fig. 10) is to automatically establish a relationship between the *Parcel* and *Polygon* concepts. This transformation creates an association, called *spatial characteristic*.

During the capture of the pictogram, two tagged values are added to the business concept to specify the role of the spatial concept ({*Gis S: Spatial Role(Geometry)*}) and its cardinality ({*Gis S: Spatial Cardinality(1)*}). By default, this role and this cardinality have the values *Geometry* and *1*, respectively, but the designer can subsequently modify them. In this association, the entity name has been allocated to its role, *Parcel* in this example, and its cardinality value is 0.1. Once the association has been created, the stereotype/pictogram and the two tagged values are deleted because this information becomes redundant with the association.

To ensure traceability, the transformation T_p creates a traceability link, called *translation traceability link*, between the pictogram of the business entity of the analysis model and the *spatial characteristic* association.

4.4.3 The SQL Transformation

To achieve a full MDA process, the SQL transformation T_{SQL} has been conceived and implemented. It is applied on the SQL implementation model. The objective of this transformation is to adapt the SQL implementation model after cloning (Fig. 11, "Before") to the SQL code generator of the case-tool. To do this, it adds SQL concepts, such as persistence and key primary (Fig. 11, "After"), to the business concepts. These SQL concepts are not systematically added to all the business concepts but only to a certain number of them.

Persistence is an "SQL" property that should be added to all concepts that should be converted into tables (Fig. 11, "After"). Considering that the "son" concepts involved in a hierarchy of business concepts inherit properties of "father" concepts, the persistence property should be put in the "root" concept of the hierarchy.

Although the primary key properties are as essential as the concepts involved in a relationship of association, of aggregation, or of composition, the transformation T_{SQL} systematically adds the primary key property on all the persistent classes (Fig. 11, "After"). Just like for persistence, the "son" concepts of a hierarchy of concepts inherit the primary key of the "root" concept.

Annotated with SQL concepts, the SQL code generator can be applied on the SQL implementation model to produce the SQL code for creating the database.

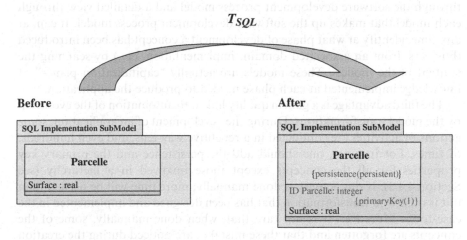

Fig. 11 The SQL transformation T_{SQL}

5 Conclusions

The continuous integration unified process method proposed above presents a number of advantages. Below, we clarify three important advantages.

The first advantage is to have split research of excellence into the subcategories *semantic excellence* and *technical excellence*. This dichotomy introduced by the continuous integration unified process method has been obtained by superimposing a cycle of rapid prototyping in the analysis phase onto the main cycle (Fig. 6). Successive prototypes completed during this cycle of rapid prototyping are devices that form the topics of discussion and criticisms from the actors. Progressively, these prototypes tend toward "the ideal application" and the actors of the field need this. To reach this ideal state, the actors describe the business concepts, and their description becomes finer as one proceeds through the iterations of the cycle of rapid prototyping. The semantic excellence is then reached, and the model coming from the phase of analysis in the main cycle can be stabilized from the semantic point of view. In the phase of design and implementation of the main cycle, the development team has the leisure to address and to solve the technical sides linked to the production of the binary code. During this main cycle, the aim is toward technical excellence.

The second advantage is the capitalization of knowledge. The generalization of the idea to separate the business concepts from those of implementation, an idea suggested by the MDA approach, has led to the design of the software development process approach, which associates a model with each of the phases of the development cycle of an application. Thus, the software development process model, an artifact that reifies the software development process approach, groups together all the models that are associated with the development phases. Thus, the project manager has both a global view of development through the software development process model and a detailed view through each model that makes up the software development process model. It can, at any time, identify at what phase of development a concept has been introduced (business, from an associated domain, implementation, etc.) by scanning the content of the models. These models are actually "capitalization planes" of knowledge implemented at each phase needed to produce the application.

The third advantage is a gain in quality linked to automation of the evolution by the model transformations. During the development of an application, some actions or activities are conducted in a repetitive way tens, and even hundreds, of times. For instance, one should add the persistence and the primary key properties for all the concepts except those involved in a hierarchy (see Section 4.4.3). If this activity is done manually, more time will be needed than if it is done by a transformation that has been designed and implemented in the case-tools. Moreover, it is not rare that, when done manually, some of the concepts are forgotten and that these mistakes are noticed during the creation of the database. In this case, all the implementation process should be resumed. A designer, who may be poorly experienced in SQL language, may also add this information to all the concepts of a hierarchy even though this is not necessary. In this case, the coherence of the model deteriorates. In the software development process approach, nothing forbids the designer to add these SQL properties in the design or analysis phase. The capitalization is then affected. The implementation of the transformation T_{SQL} on the SQL implementation model avoids this problem. The quality of the SQL implementation model is better and the productivity is increased. It is the same for all the other transformations described in Section 4.4. These thoughts are corroborated by the study conducted by The Middleware Company [29]. This consulting business has been in charge of conducting a productivity analysis between two teams with an equivalent competency level: the first one was to develop an application in a traditional way, and the second one had to create the same application according to the MDA approach. The team working according to the MDA approach completed the application with a time savings of 35% and a gain in quality as the team did not have to correct development bugs. The analysts of the consulting business attribute this gain in quality to the automated transformations of the models.

References

1. Beck K. 2000. eXtreme Programming Explained – Embrace Change. Addison-Wesley. 190 pp.
2. Bédard Y, Larrivée S, Proulx M-J, Nadeau M. 2004. Modeling Geospatial Databases with Plug-ins for Visual Languages: A Pragmatic Approach and the Impacts of 16 Years of Research and Experimentations on Perceptory. Presented at ER Workshops 2004 CoMoGIS, Shanghai, China.
3. Bénard J-L. 2001. Méthodes agiles (1) – Panorama. Développeur Référence. http://www. devreference.net/devrefv205.pdf. Last access: September 2004.
4. Bénard J-L, Bossavit L, Médina R, Williams D. 2002. Gestion de projet eXtreme Programming. Eyrolles. 298 pp.
5. Booch G, Rumbaugh J, Jacobson I. 2000. Guide de l'utilisateur UML. Eyrolles. 500 pp.
6. Cros T. 2001. La conception dans l'eXtreme Programming. Développeur Référence. http://www.devreference.net/devrefv201.pdf. Last access: September 2004.
7. Desfray P. 1994. Object Engineering – The Fourth Dimension. Addison-Wesley. 342 pp.
8. Fayet E. 2002. Forum Utilisateurs Rational – Le discours de la méthode. Développeur Référence. http://www.devreference.net/devrefv220.pdf. Last access: September 2004.
9. Gamma E, Helm R, Johnson R, Vlissides J. 2001. Design patterns – Elements of Reusable Object-Oriented Software. Addison-Wesley Professional. 416 pp.
10. Guimond L-E. 2005. Conception d'un environnement de découverte des besoins pour le développement de solutions SOLAP. Thèse. Université Laval, Québec. 124 pp.
11. Jacobson I. 2003. Use Cases – Yesterday, Today, and Tomorrow. http://www.ivarjacob son.com/html/content/publications_papers.html; http://www.ivarjacob son.com/publi cations/uc/UseCases_TheRationalEdge_Mar2003.pdf. Last access: August 2005.
12. Jacobson I, Booch G, Rumbaugh J. 1999. The Unified Software Development Process. Addison-Wesley. 463 pp.
13. Kleppe A. 2004. Interview with Anneke Kleppe. Code Generation Network. http://www. codegeneration.net/tiki-read_article.php articleId=21. Last access: August 2006.

14. Kleppe A, Warmer J, Bast W. 2003. MDA Explained: The Model Driven Architecture—Practice and Promise. Addison-Wesley Professional. 170 pp.
15. Kruchten PB. 1999. The Rational Unified Process: An Introduction. Addison-Wesley Professional. 336 pp.
16. Larman C. 2002. Applying UML and Patterns: An Introduction to Object-Oriented Analysis and Design and the Unified Process. Prentice Hall PTR. 627 pp.
17. Larman C. 2002. UML et les Design Patterns. CampusPress. 672 pp.
18. Miller J, Mukerji J. 2001. Model Driven Architecture (MDA). OMG. http://www.omg.org/cgi-bin/apps/doc? 07-01.pdf. Last access: September 2004.
19. Miller J, Mukerji J. 2003. MDA Guide Version 1.0.1. OMG. http://www.omg.org/cgi-bin/doc? -01. Last access: May 2006.
20. Miralles A. 2006. Ingénierie des modèles pour les applications environnementales. Thèse de doctorat. Université Montpellier II, Montpellier. http://www.teledetection.fr/ingenierie-des-modeles-pour-les-applications-environnementales-3.html. 322 pp.
21. Muller P-A, Gaertner N. 2000. Modélisation objet avec UML. Eyrolles. 520 pp.
22. OMG. 2001. Common Warehouse Metamodel – Version 1.0. OMG. http://www.omg.org/cgi-bin/doc ?ad/2001-02-01. Last access: June 2004.
23. OMG. 2003. Unified Modeling Language – Specification – Version 1.5. http://www.omg.org/cgi-bin/apps/doc? formal/03-03-01.pdf. 736 pp.
24. Région Wallonne. 2004. Le prototypage: Définition et objectifs. Portail Wallonie.
25. Roques P, Vallée F. 2002. UML en Action – De l'analyse des besoins à la conception en Java. Eyrolles. 388 pp.
26. Royce WW. 1970. Managing the Development of Large Software Systems. Presented at IEEE Westcon, Monterey, CA.
27. Schramm WL. 1954. How communication works. In: The Process and Effects of Communication. University of Illinois Press. pp. 3–26.
28. Softeam. 2005. Formation sur les Modèles Objet et UML.
29. The Middleware Company. 2003. Model Driven Development for J2EE Utilizing a Model Driven Architecture (MDA) Approach – Productivity Analysis.

Application of a Model Transformation Paradigm in Agriculture: A Simple Environmental System Case Study

André Miralles and Thérèse Libourel

Abstract In this chapter, the authors use the methodology presented in Chapter 2 to develop a system that manages the spreading of organic waste on agricultural parcels. The proposed method uses a process of iterative and incremental development. Two complete iterations of the development process are presented starting from the analysis model and ending with the code produced by the case-tools Structured query language (SQL) code generator. The first iteration deals with the description of territory objects and the second one deals with the business objects used in the context of the spreading of organic waste. As a result of transformations applied, models are enriched with new concepts and, therefore, are more complex. The growing complexity of the model may negatively affect an actor's understanding, which may become an impediment by slowing down the analysis phase. The authors show how the software development process model, a modeling artifact associated with the continuous integration unified process method, avoids the apparent complexity of the model and improves productivity.

1 Introduction

The main purpose of developing an application is to convert the requirements of the concerned actors (users, clients, stakeholders, etc.) into a software application. To attain this objective –shown in Fig.1– a method for conducting projects is used. Section 2 of Chapter 2 presents different methods for conducting a project.

To overcome this challenge, the developed application should not only correspond with the actors' requirements but should also satisfy them. Experience has shown that it is not easy to be successful in this aim. Often, there is a gap between

A. Miralles (✉)
Centre for Agricultural and Environmental Engineering Research,
Earth Observation and GeoInformation for Environment and Land Development Unit,
Montpellier, France
e-mail: andre.miralles@teledetection.fr

P.J. Papajorgji, P.M. Pardalos (eds.), *Advances in Modeling Agricultural Systems*,
DOI 10.1007/978-0-387-75181-8_3, © Springer Science+Business Media, LLC 2009

Fig. 1 Software development process [9]

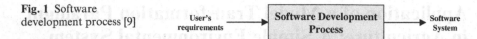

User's requirements → **Software Development Process** → Software System

the requirements expressed initially by the actors and those actually covered by the developed application.

This gap between the requirements that the application actually satisfies and the requirements expressed at the beginning of the project is the cause of most disagreements between the client and the company developing the application. To avoid this very situation, the companies that develop software applications have long adopted software development processes[1] for monitoring the *project's progress* [4] and for *cost and time control* [17].

At the very beginning, the first software applications were developed without any method. The first method to have been formalized and described is the one based on the waterfall cycle [18]. Since then, several development methods have been designed and implemented in different software development projects. In these past 10 years, a significant portion of computer-related research has focused on a multitude of new methods [1, 5, 7, 8, 9, 11, 13, 19] because application-development companies and project leaders have realized the impact that the method used for a project can have on its success.

In the development process, the designer of an application should understand the business concepts of the actors to be able to reproduce as faithfully as possible the properties and the behavior of these concepts in the software components. A good understanding of the concepts is fundamental because this will determine whether development will succeed or fail. If the properties and behavior of the business concepts are not correctly reproduced, the developed application will have a different behavior from the one expected by the actors. To begin with, they will feel disoriented by thinking that they are the cause of the unexpected functioning. Later on, when they realize that they are not responsible for the unexpected functioning, it is likely that they will feel disappointed, even upset, that the application does not meet their expectations.

The understanding of concepts is part of a development phase that is most often called the *analysis phase*. Some methods may give this phase a different name, but the activity conducted during the phase corresponds exactly with analysis. For example, extreme programming calls it the *exploration* phase, during which the client describes the application's features in the form of *user stories*.

By whatever name it may be known, every method of development contains this appropriation phase. This phase always brings together the designer of the application and one or more actors of the concerned domain in analysis sessions. It cannot be otherwise unless one is actor and designer simultaneously, a situation that arises frequently in scientific circles.

---

[1] A software development process is the set of activities needed to transform a user's requirements into a software system.

As detailed in Chapter 2, the hours or days spent on the analysis is fundamentally important, because how well the designer understands and captures the business concepts will depend directly on it. Nevertheless, the number of hours or days spent is not the only important criterion for the appropriation phase. Experience has shown that the frequency of meetings between designers and actors is also important. When the analysis was conducted *in toto* at the beginning of the cycle (waterfall cycle, spiral cycle, RAD[2] method, etc.), the gap between the expressed requirements and those delivered by the application was often large. It is this observation that led to the design and adoption of iterative methods (UP,[3] FDD,[4] etc.). These latter all have the common characteristics of development cycles (analysis, design, implementation, validation) whose durations are fixed in advance, varying from 2 to 4 weeks. These cycles are called iterations. At each iteration, a new "brick" of the application is produced. During an iteration, the actors of the concerned domain participate not only during the analysis but also during the validation (i.e., both at the beginning and at the end of the iteration). This increased participation on the part of the actors allows them to see the application evolve in quasi–real time and to correct as soon as possible any deviation in the development. These new methods are often called *user-centric development.* With its *on-site customer* practice, the extreme programming method has pushed this idea to the limit. This practice involves either including a representative of the client within the development team or to have a team member plays the client's role. In this way, the frequency of the actors' participation becomes infinite. This practice is probably the one with the best communication between the designer and the actors and is the most notable feature of the extreme programming method.

2 The Continuous Integration Unified Process

The observations and the theoretical fundamentals that have led the authors to design and implement software artifacts for implementing the continuous integration unified process method are presented in Chapter 2 and, in much more detail, in [14]. Nevertheless, for reasons of completeness, a brief recap of the broad ideas that form the basis of this method is presented here.

It is an extension of the unified process method [9]. Its specialty is in superimposing a rapid-prototyping cycle on the main cycle during the analysis phase (Fig.2). The objective of this rapid-prototyping cycle is to improve the exchange between actors and designer.

Building a prototype is the same as building an application during the main cycle. The main difference resides in the fact that the prototype has to be built

---

[2] RAD: rapid application development [13, 22].

[3] UP: unified process [9, 11].

[4] FDD: feature driven development [3].

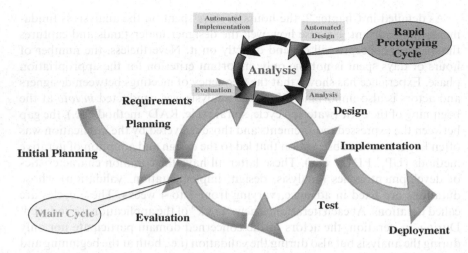

Fig. 2 Continuous integration unified process

automatically via model transformations, otherwise the actors present will soon become bored and lose interest in the ongoing analysis. It is this automated process that Anneke Kleppe calls full MDA[5] [10].

3 Transformations of the Continuous Integration Unified Process in Action

The aim of this section is show how the continuous integration unified process method can be used during the design of an information system.

At the very beginning of the development cycle, all the team has is a very short description of the system, often less than one page long [9]. The goal of the initial planning phase (Fig.2) is to impart "consistency" to the system. To do this, the domain's sponsors should *define the ultimate goal of the system* [6, 12, 15] and *demarcate the boundaries of the concerned domain* [4, 6, 12].

Wastewater from various human activities is now increasingly being processed at water treatment plants. These facilities improve the water quality before discharging it into rivers and streams. The major downside to such a system is the sludge produced. This sludge is heavy with organic wastes and has to be disposed of. One of the solutions currently proposed is to spread this organic waste on agricultural parcels as a replacement in part for nitrogenous fertilizer because of its high nitrogen content.

---

[5] Model driven architecture (MDA) approach, recommended by the Object Management Group [16].

The *ultimate goal of the system* that we shall present here is to improve the management of organic waste spread on agricultural parcels.

To simplify the case study,[6] the *boundaries of the concerned domain* were limited to parcels belonging to agricultural enterprises, the organic waste to be spread, and the qualitative study of the soil by regular analysis.

This case study is part of the Système d'Information pour la Gestion des Epandages de Matières Organiques (SIGEMO) project [21]. The model in Fig.3 is extracted from a general SIGEMO model and consists of concepts relevant to the case study. This model is also used in Chapter 4. It is this model that we shall construct step by step by applying the continuous integration unified process method.

Of course, the modeling process will implement the software development process approach and its reification, the software development process model (SDPM), an artifact for knowledge capitalization. The process will also implement the pictogramic language for expressing spatial and temporal properties of business concepts [2, 14] as well as all the transformations that are necessary to automatically evolve a model from analysis up to implementation. The SQL implementation model thus obtained will then be projected into SQL code.

To simplify the description, we will present here just two iterations: the first will be dedicated to the description of territory objects, and the second will be devoted to the business objects used in the context of the spreading of organic waste.

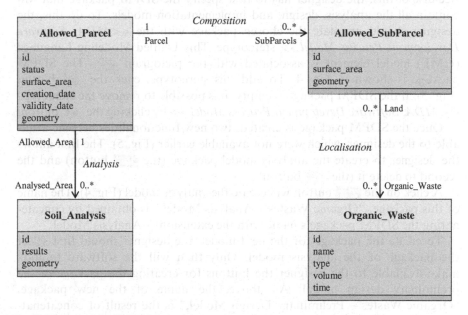

Fig. 3 Extract of the SIGEMO model

---

[6] This decision is based on the fact that models become complex very rapidly and hence difficult to read.

As mentioned above, the main interest in the continuous integration unified process method lies in its use of the rapid-prototyping cycle during the analysis phase. Because of this, we shall focus our explanations on this development phase. We recall that during this phase, the concerned domain's actors have to participate in the system analysis and, as such, will be present during the different analysis sessions.

3.1 Construction of the Software Development Process Model

We have decided to devote a section to describing the construction of the software development process model so as to simplify subsequent explanations. This is possible, on the one hand, because its construction can take place at the same time as the model's transformations and, on the other hand, because its construction always takes place during the first iteration. Once created, its structure does not change unless, during the development, the need arises to use a new programming language. In that case, a new implementation model becomes necessary.

As said by Muller and Gaertner [15], the models are reified within packages. Because of this, the designer has to first specify the SDPM package that will contain all the analysis, design, and implementation models. To do this, the designer should annotate the chosen package with the <<MDA: Software Development Process Model>> stereotype. This Unified Modeling Language (UML) model element is associated with the pictogram ⬤. The SDPM package is shown in Fig. 4. To add this stereotype, click the ⬤ button (Fig. 5). If the SDPM package is empty, it is possible to remove the stereotype <<MDA: Software Development Process Model>> by clicking the ✖ button.

Once the SDPM package is created, two new functionalities become available to the designer, which were not available earlier (Fig. 5). The first allows the designer to create the analysis model package (the ▦ button) and the second to delete it (the ✖ button).

A click on the ▦ button will create the analysis model (Fig. 4). The name of this package, "Organic Wastes – Analysis Model," is obtained by concatenating the SDPM package's name with the extension "- Analysis Model."

To create the package for the next model, the designer should first select the package of the analysis model. Only then will the software toolbox make available to the designer the buttons for creating the package of the preliminary design model. As above, the name of the new package, "Organic Wastes – Preliminary Design Model," is the result of concatenating the SDPM package's name with the extension "- Preliminary Design Model."

Step by step, by selecting the last package created, the designer can thus construct the architecture of the SDPM, as shown in Fig. 5.

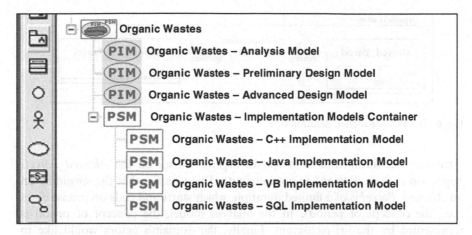

Fig. 4 Example of the structure of software development process model

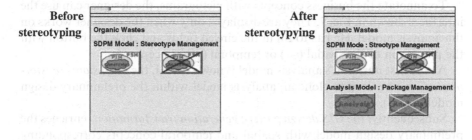

Fig. 5 Case-tool interface before and after stereotyping the SDPM package

In the example that follows, only the SQL implementation model will be created as the aim is to show the full MDA process that was designed and implemented.

3.2 First Iteration

During the first analysis session, the actors and the designer will identify and describe the domain's relevant objects. Here, the important real-world objects are *Allowed_Parcels* and *Allowed_SubParcels*. The latter are geographical partitions of the former; that is, an *Allowed_Parcel* is composed of a set of *Allowed_SubParcels*. The cardinalities 1 and 0..* respectively of the classes *Allowed_Parcel* and *Allowed_SubParcel* translate this geographic structure (Fig. 6).

The concept of spreading brings with it the concept of surface area. This leads the designer to introduce the concepts of geometry and area. In the analysis model, the geometry is represented by the ▣ pictogram [2] and the area by the *surface_area*

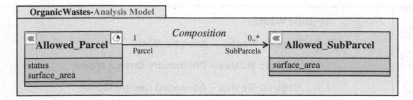

Fig. 6 The analysis model: iteration one

attribute. The authorization to spread organic wastes on the *Allowed_Parcels* begins on a given date and is not indefinite. This signifies that the spreading on an *Allowed_Parcel* is of a limited duration, which starts and ends on precise dates (i.e., the concept of period). In the analysis model, the concept of period is represented by the ⓐ pictogram. Finally, the domain's actors would like to know if the *Allowed_Parcel* can always be used (or not) to spread organic wastes. The *status* attribute shows and stores this information.

To annotate the business concepts with pictograms, the designer can use the interfaces shown in Fig. 7. They are displayed only when the designer works on the analysis model. He has to fill in the circled fields and click the button with the pictogram of the spatial (◀) or temporal (ⓐ) concept.

As the first iteration's analysis model is now created, the *Diffusion Transformation*[7] can be used to clone an analysis model within the preliminary design model (Fig. 8).

Subsequently, the *GIS design pattern generation transformation*[7] enriches the preliminary design model with spatial and temporal concepts corresponding with the pictograms used in the model, *Polygon* and *Time_Period* in our case. The state of the model at this stage can be seen in Fig. 9.[8] The business concepts,

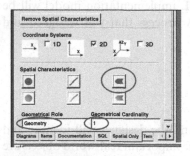
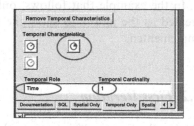

Fig. 7 Example of spatial and temporal user interfaces

---

[7] See Chapter 2.

[8] For the sake of simplification, the complete *GIS design patterns* are not shown on this figure, nor on the following ones. Only the spatial and temporal concepts corresponding with the pictograms used during the modeling have been added. For additional information on the *GIS design patterns*, refer to [14].

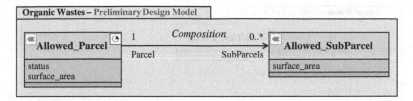

Fig. 8 The preliminary design model: iteration one

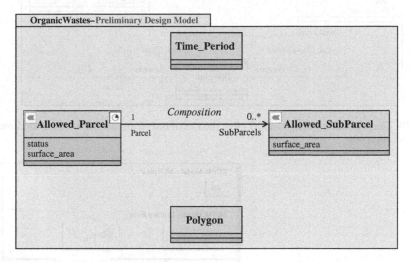

Fig. 9 The preliminary design model after the enrichment by the *design pattern generation transformation*: iteration one

Allowed_Parcel and *Allowed_SubParcel*, do not yet have any relationship with the spatial and temporal concepts that were added by the *GIS design pattern generation transformation*. This transformation applies on the preliminary design model and modifies only it. It does not alter the analysis model in any way.

The following stage consists of concretizing the relationships between the business concepts and the *Polygon* and *Time_Period* concepts. This is the job of the *pictogram translation transformation*.[7] This transformation links the business concepts annotated with pictograms to the corresponding spatial and temporal concepts. To do this, it creates *Spatial_Characteristic* and *Temporal_-Characteristic* associations. It also defines the roles (*Geometry* or *Period*) and the cardinalities (1) of the spatial and temporal concepts. It also specifies the roles (*Allowed_Parcel* or *Allowed_SubParcel*) and the cardinalities (0..*) of the business concepts. This latest evolution of the preliminary design model is shown in Fig. 10.

The *GIS design pattern generation* and *pictogram translation transformations* are triggered when the designer clicks the � and ◉ buttons, respectively (Fig. 11). This user interface is only displayed by the case-tools when the designer is working with the preliminary design model.

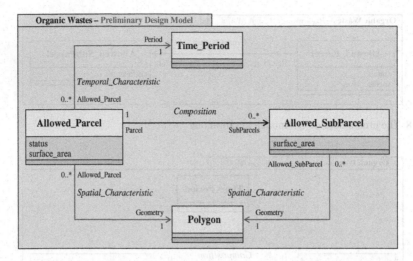

Fig. 10 The preliminary design model after the run of the *pictogram translation transformation*: iteration one

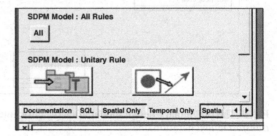

Fig. 11 User interface for
GIS transformations

To continue the full MDA process, the preliminary design model is in its turn cloned into the advanced design model. We shall not linger on this latter model because, as of now, we have not identified and implemented any transformation on it.

Once again, the *diffusion transformation* is executed to clone the advanced design model into the SQL implementation model. At this stage, it is identical to that in Fig.10.

To be able to finalize the full MDA process, this last model has to be complemented with information specific to the SQL language and interpretable by the *SQLDesigner* code generator of the case-tools. The *SQL transformation*[7] was designed and implemented to assume this role. The added information is the *persistence* and an attribute. The attribute is introduced to play the role of *Primary Key*. Its name is made up of that of the class preceded by the "ID_" character string. A tagged value *{primaryKey(1)}* confers upon it this status (Fig.12). For all practical purposes, the *SQL transformation* automates the actions that would have had to be undertaken manually.

The SQL implementation model is now ready to be projected into SQL code. The projection rules are implemented in the SQLDesigner generator developed

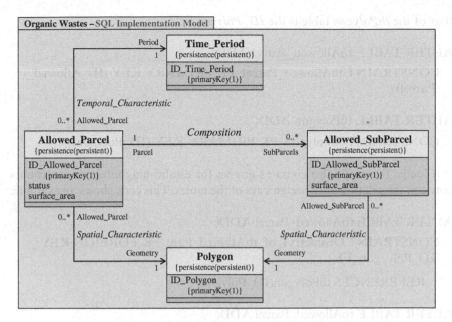

Fig 12 The SQL implementation model after the run of the *SQL transformation*: iteration one

by the Softeam company. They are documented in [20]. This generator produces a SQL script that contains all the queries that allow tables to be created in a database management system (DBMS)[9]. We provide below the code for creating the *tbAllowed_Parcel* and *tbPolygon* tables.

```
CREATE TABLE tbAllowed_Parcel(
    ID_Allowed_Parcel INTEGER NOT NULL ,
    status INTEGER ,
    surface_area FLOAT ,
    ID_Polygon_FK INTEGER NOT NULL ,
    ID_Time_Period_FK INTEGER NOT NULL );
```

```
CREATE TABLE tbPolygon(
    ID_Polygon INTEGER NOT NULL );
```

The SQLDesigner generator also produces queries for indicating which is the table's field that plays the role of primary key. As shown by the code below, the primary key of the *tbAllowed_Parcel* table is the *ID_Allowed_Parcel* field and

---

[9] To facilitate the comprehension of the codes used, the table names have been prefixed by "tb" and those of foreign keys suffixed by "_FK."

that of the *tbPolygon* table is the *ID_Polygon* field.

ALTER TABLE tbAllowed_Parcel ADD(

CONSTRAINT tbAllowed_Parcel_PK PRIMARY KEY (ID_Allowed_ Parcel));

ALTER TABLE tbPolygon ADD(

CONSTRAINT tbPolygon_PK PRIMARY KEY (ID_Polygon));

Finally, the script also contains queries for establishing integrity constraints between the primary and foreign keys of the tables. This code shows an example:

ALTER TABLE tbAllowed_Parcel ADD(

CONSTRAINT Geometry1_of_tbAllowed_Parc_FK FOREIGN KEY (ID_Poly- gon_FK)

REFERENCES tbPolygon(ID_Polygon));

ALTER TABLE tbAllowed_Parcel ADD(

CONSTRAINT Period_of_tbAllowed_Parcel_FK FOREIGN KEY (ID_Time_Pe- riod_FK)

REFERENCES tbTime_Period(ID_Time_Period));

Loaded into a DBMS, the script creates database tables with their primary and foreign keys and the integrity constraints linking these two key types. The creation of the tables in the DBMS ends the first iteration.

Even though the database is complete, it should on no account be populated with data because, at the end of the first iteration, the analysis is far from over. Several more iterations will be necessary to stabilize the initial analysis model. In the following section, we present the major stages of the second iteration.

3.3 Second Iteration

After having described the territory that will be subject to spreading, the business concepts specific to the domain are identified and added to the analysis model of Fig. 6. The main concept is *Organic_Waste*, which is spread on a given date, whence the presence of the ⊘ pictogram. To be able to monitor the proper spreading of the organic waste, the rules often require recording of name, type, and volume spread. These three items of information are characteristics, which are included as attributes in the *Organic_Waste* class. With these roles and its cardinalities, the *Localization* association allows the listing of the products spread on an *Allowed_SubParcel*.

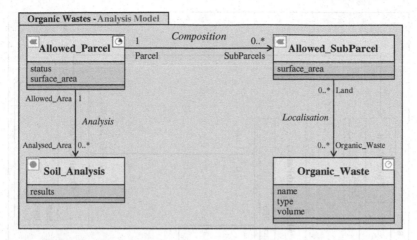

Fig. 13 The analysis model: iteration wo

The rules also require that *Soil_Analysis* be conducted to determine the composition of the *Organic_Waste* spread. The spatial position of the soil sampling is an important property that the information system has to manage. This is the reason for the ▣ pictogram. The *Analysis* association establishes a relationship between the *Soil_Analysis* and the parcels from which samples have been collected.

The analysis model in Fig. 6 is now enriched by two new concepts (Fig. 13) that have to be diffused and processed in the design and implementation models of the software development process model to modify the database of the first iteration.

After the diffusion of the new concepts to the preliminary design model, the *pictogram translation transformation* establishes the *Spatial_Characteristic* and *Temporal_Characteristic* relationships with the *Point* and *Time_Point* concepts. These spatial and temporal concepts corresponding with the ▣ and ▣ pictograms were masked in the previous iteration to simplify the different models. The model resulting from this transformation is shown in Fig. 14. It is easy to see that the complexity of the model is increasing rapidly.

As earlier, the new concepts of the preliminary design model are diffused first and foremost to the advanced design model.

Because no transformation is available on this model, the new concepts are once again diffused to the SQL implementation model. The *SQL transformation* adds the persistence and the attributes that will play the role of the primary key in the classes that do not have them. The result of these transformations is shown in Fig. 15.

The *SQLDesigner* code generator is now ready to use the contents of the SQL implementation model to produce a SQL script containing the queries for generating the new database. This script consists of 33 queries, some of which we have extracted below to show the final result.

We have reproduced the code that creates the *tbAllowed_SubParcel* and *tbOrganic_Waste* tables and that implements the *Localization* association,

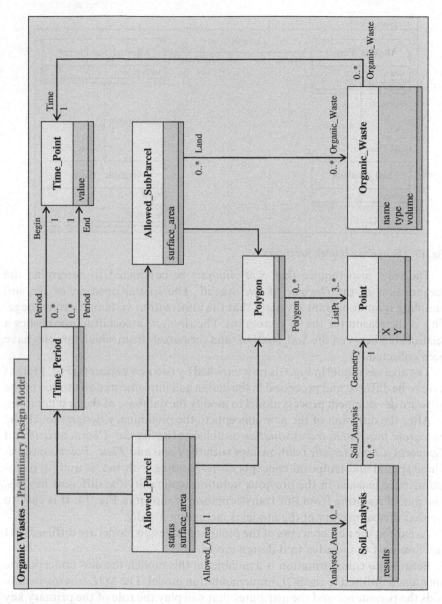

Fig. 14 The preliminary design model: iteration two

which has the distinctiveness of being a N,N association (Fig. 6). This association typology necessitates the creation of the *Localization* table.

CREATE TABLE tbAllowed_SubParcel(

  ID_Allowed_SubParcel INTEGER NOT NULL ,

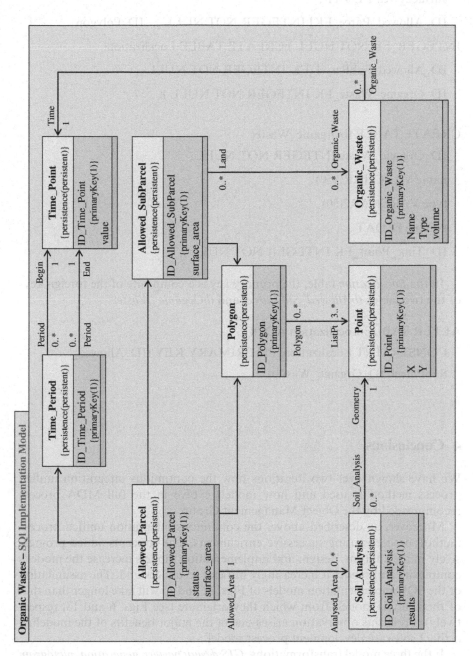

Fig. 15 The SQL implementation model: iteration two

surface_area FLOAT ,

ID_Allowed_Parce_FK1 INTEGER NOT NULL ,   ID_Polygon

INTEGER_FK NOT NULL);CREATE TABLE Localization(

ID_Allowed_SubParcel_FK INTEGER NOT NULL ,

ID_Organic_Waste_FK INTEGER NOT NULL);

CREATE TABLE tbOrganic_Waste(

ID_Organic_Waste INTEGER NOT NULL ,

name VARCHAR2(50) ,

type VARCHAR2(50) ,

volume FLOAT ,

ID_Time_Point_FK INTEGER NOT NULL);

In the *Localization* table, the primary key is a composite of the foreign keys of the two tables *tbAllowed_SubParcel* and *tbOrganic_Waste*.

ALTER TABLE Localization ADD(

CONSTRAINT Localization_PK PRIMARY KEY (ID_Allowed_

SubParcel, ID_Organic_Waste));

4 Conclusions

We have shown over two iterations how the continuous integration unified process method is used and how models evolve in the full MDA process recommended by the Object Management Group.

Moreover, as described above, the continuous integration unified process method proceeds using successive enrichments. The concepts added progressively in the analysis, design, and implementation models increase the model's complexity and render it increasingly difficult to understand. The assimilation of the SQL implementation models of Figs. 12 and 15 will take longer than that of the analysis models from which they originate (see Figs. 6 and 13, respectively). From this observation arises one of the major benefits of the modeling artifact software development process model.

If the three model transformations, *GIS design pattern generation, pictogram translation transformations,* and *SQL transformation*, had all been applied to the same model, the analysis model of Fig. 6 would not exist at the time of the first iteration. The actors would have been confronted by the model in Fig.12, which is much more complex than the model of Fig. 6. They would have been disturbed or even upset by a model enriched by a whole host of concepts whose significance they would not have necessarily understood.

With the software development process model, this potential inconvenience disappears. In fact, during the course of an iteration, the analysis model remains unchanged; it is the design and implementation models that are enriched. At the beginning of the next iteration, the actors thus find the same analysis model. They therefore feel comfortable and do not have to make an extra effort to understand the significance of concepts added during the iteration because they do not see them. This leads to a direct gain in productivity during the analysis.

References

1. Beck K. 2000. eXtreme Programming Explained – Embrace Change. Addison-Wesley. 190 pp.
2. Bédard Y, Larrivée S, Proulx M-J, Nadeau M. 2004. Modeling Geospatial Databases with Plug-ins for Visual Languages: A Pragmatic Approach and the Impacts of 16 Years of Research and Experimentations on Perceptory. Presented at ER Workshops 2004 CoMoGIS, Shanghai, China.
3. Bénard J-L. 2002. Méthodes agiles (6) – Feature Drive Development. Développeur Référence. http://www.devreference.net/devrefv210.pdf. Last access: September 2004.
4. Bénard J-L. 2002. Méthodes agiles (7) – Unified Process. Développeur Référence. http://www.devreference.net/devrefv212.pdf. Last access: September 2004.
5. Boehm BW. 1988. A Spiral Model of Software Development and Enhancement. IEEE Computer, Vol. 21(5) pp. 61–72.
6. Booch G, Rumbaugh J, Jacobson I. 2000. Guide de l'utilisateur UML. Eyrolles. 500 pp.
7. Fowler M, Highsmith J. 2001. The Agile Manifesto. Software Development magazine. http://agilemanifesto.org/Last access: 5 July 2008.
8. Highsmith J. 2002. Agile Software Development Ecosystems. Addison-Wesley Professional. 448 pp.
9. Jacobson I, Booch G, Rumbaugh J. 1999. The Unified Software Development Process. Addison-Wesley. 463 pp.
10. Kleppe A. 2004. Interview with Anneke Kleppe. Code Generation Network. http://www.codegeneration.net/tiki-read_article.php?articleId = 21. Last access: August 2006.
11. Kruchten PB. 1999. The Rational Unified Process: An Introduction: Addison-Wesley Professional. 336 pp.
12. Larman C. 2002. Applying UML and Patterns: An Introduction to Object-Oriented Analysis and Design and the Unified Process. Prentice Hall PTR. 627 pp.
13. Martin J. 1991. Rapid Application Development. Macmillan Publishing. 788 pp.
14. Miralles A. 2006. Ingénierie des modèles pour les applications environnementales. Thèse de doctorat. Université Montpellier II, Montpellier. http://www.teledetection.fr/ingenierie-des-modeles-pour-les-applications-environnementales-3.html. 322 pp.
15. Muller P-A, Gaertner N. 2000. Modélisation objet avec UML. Eyrolles. 520 pp.
16. OMG. Object Management Group home page. http://www.omg.org/. Last access: October 2004.
17. Roques P, Vallée F. 2002. UML en Action – De l'analyse des besoins à la conception en Java. Eyrolles. 388 pp.
18. Royce WW. 1970. Managing the Development of Large Software Systems. Presented at IEEE Westcon, Monterey, CA.

19. Schwaber K, Beedle M. 2001. Agile Software Development with Scrum. Prentice Hall. 158 pp.
20. Softeam. 2003. Objecteering/UML – Objecteering/SQL Designer User guide – Version 5. 2.2. 236 pp.
21. Soulignac V, Gibold F, Pinet F, Vigier F. 2005. Spreading Matter Management in France within Sigemo. Presented at 5th European Conference for Information Technologies in Agriculture (EFITA 2005), Vila Real, Portugal.
22. Vickoff J-P. 2000. Méthode RAD – Éléments fondamentaux. http://mapage.noos.fr/rad/radmetho.pdf. 32 pp.

Constraints Modeling in Agricultural Databases

François Pinet, Magali Duboisset, Birgit Demuth, Michel Schneider,
Vincent Soulignac, and François Barnabé

Abstract The size of agricultural databases continues to increase, and sources of information are growing more and more diversified. This is especially the case for databases dedicated to the traceability of agricultural practices. Some data are directly collected from the field using embedded devices; other data are entered by means of different computer-based applications. Once stored in the same database, all this information must be consistent to guarantee the quality of the data. This consistency issue is becoming a new challenge for agricultural databases, especially when complex data are stored (for instance, georeferenced information). To achieve consistency in a database, a precise, formal specification of the integrity constraints is needed. Indeed, database designers and administrators need a language that facilitates the conceptual modeling of this type of constraint. In this chapter, we introduce the Object Constraint Language (OCL), using the example of an agricultural database for organic waste management. The example of a tool supporting OCL (the Dresden OCL Toolkit) and an overview of a spatial extension of the language will be also presented.

1 Introduction

Sustainable development directly concerns agricultural activity. In such a context, a response to people's needs involves traceability of the activities at all levels, from the level of the farm to that of the state. Decisions taken must be based on reliable indicators that require the monitoring of agricultural practices and consequently the recording and the exchange of information related to these practices. To achieve this goal, ever larger environmental and agricultural databases will be developed.

Often, certain data may be directly collected from the field using an embedded device; other data may be imported from other databases or entered by means of different computer-based applications. Once stored in the same

F. Pinet (✉)
Cemagref, Clermont Ferrand, France
e-mail: francois.pinet@cemagref.fr

P.J. Papajorgji, P.M. Pardalos (eds.), *Advances in Modeling Agricultural Systems*,
DOI 10.1007/978-0-387-75181-8_4, © Springer Science+Business Media, LLC 2009

database, all this information must be consistent, that is, all the data must satisfy rules defined by experts; for instance, organic waste cannot be spread in a lake. Respecting these rules helps to guarantee the quality of data. By checking these rules, only consistent information will be stored in the database.

The insertion of inconsistent data may lead to the production of inaccurate environmental indicators. Integrity constraints are well-known techniques to guarantee the consistency of the data. According to Miliauskaitė and Nemuraitė [14], integrity constraints are logical formulae that are dependent on a problem domain, and they must be held to be true for all meaningful states of information systems. The modeling of integrity constraints in a conceptual data model may be viewed as a representation of a set of business rules for the information system. In the context of environmental and agricultural databases, the satisfaction of these constraints will tend to guarantee the consistency and quality of data and consequently the production of reliable indicators.

The work of Demuth et al. [4, 5] proposes to make use of the Object Constraint Language (OCL) [15] to model database integrity constraints. The goal of this chapter is to introduce the basic concepts of this language by modeling the integrity constraints of an agricultural database for organic waste management. An example of a tool supporting OCL (the Dresden OCL Toolkit [2, 3]) and an overview of a spatial extension of the language will also be presented.

2 The Object Constraint Language

The OCL provides a standard to define constraints based on a Unified Modeling Language (UML) model formally and with precision. OCL is textual and integrates several concepts that come from traditional object-oriented languages. In the context of databases, an important advantage of OCL is due the fact that constraints are expressed at a conceptual level. OCL is used to specify invariants (i.e., conditions that "must be true for all instances of a class at any time") [20].

OCL was first developed by a group of scientists at IBM around 1995 during a business modeling project. It was influenced by Syntropy, an object-oriented modeling language that relies to a large extent on mathematical concepts [1]. OCL is now part of the UML standard supported by the Object Management Group, and its role is important in the model driven architecture (MDA) approach [12]. A growing number of system designers use this constraint language as a complement to their class diagram models. OCL is suitable for information system engineers and is especially dedicated to formal constraint specification without side effects.

This section illustrates OCL by examples based on the spreading information system described in Ref. 21. This system has been developed to monitor and analyze the spreading of organic wastes on agricultural land in France. A main functionality of the tool allows users to define the planning of the agricultural spreading. The different types of wastes permitted on each agricultural area can be specified and stored in a database (Fig. 1).

Fig. 1 Different types of wastes permitted on each agricultural area can be specified and stored in a database

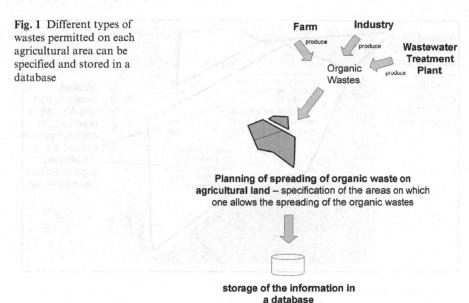

Planning of spreading of organic waste on agricultural land – specification of the areas on which one allows the spreading of the organic wastes

storage of the information in a database

Figure 2 presents a small part of the database model of the information system. In Fig. 2, the `Allowed_Parcel` class models the area on which the regulation allows the agricultural spreading of organic wastes. Each allowed parcel can be broken down into several subparts (`Allowed_Subpart` class); for each subpart, a specific list of organic wastes can be defined (see example of

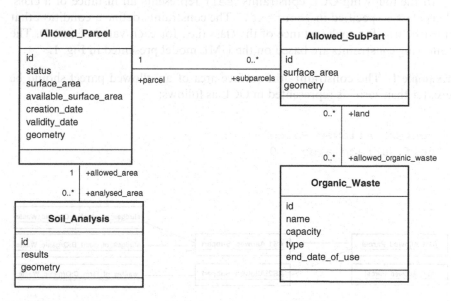

Fig. 2 Part of the database conceptual schema (represented in UML)

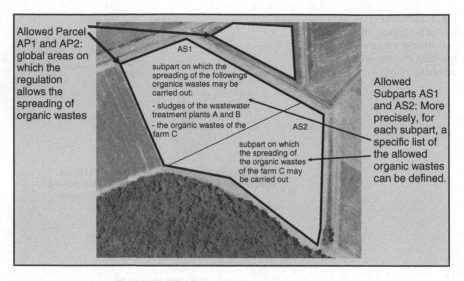

Fig. 3 Spatial example

Fig. 3). This list corresponds with an enumeration of all the types of materials (Organic_Waste class) that can be spread inside the subpart. The relationship between the subparts and the organic wastes is represented in Fig. 1 by the association "land ... allowed_organic_waste." Soil analyses are also carried out. An object diagram corresponding with Fig. 3 is presented in Fig. 4.

In the following OCL constraints, self represents an instance of a class. This class is specified in "context." The constraints define a condition that must be true for each instance of the class (i.e., for each value of self). The following constraints are based on the UML model presented in Fig. 1.

Example 1. The constraint "the surface area of an allowed parcel should be greater than zero" is represented in OCL as follows:

```
context Allowed_Parcel inv:
self.surface_area > 0
```

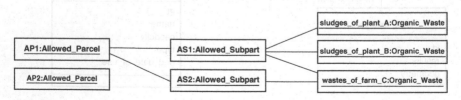

Fig. 4 Object diagram

In the constraint definition, `self` represents an instance of the "context" (i.e., an allowed parcel). Thus the constraint defines a condition that must be true for each instance of `Allowed_Parcel`; `self.surface_area` is an attribute of `self`.

Example 2. In practice, for each allowed parcel, the surface area available for spreading is not necessarily exactly equal to the total surface area. The following OCL constraint expresses that the total surface area of an allowed parcel should be greater than or equal to its surface area available for spreading:

```
context Allowed_Parcel inv:
self.surface_area >= self.available_surface_area
```

In the previous constraint definition, `self` represents an allowed parcel; `self.surface_area` and `self.available_surface_area` are attributes of `self`.

More complex constraints can be built by using navigations along the associations between classes. Starting from an object, we can navigate through an association to refer to other objects and their attributes. The next constraint illustrates the use of this technique in OCL; it presents a navigation through the "subparcels ... parcel" association, starting from allowed subpart objects to allowed parcel objects.

Example 3. The surface area of an allowed subpart should be lower than or equal to that of the associated allowed parcel:

```
context Allowed_Subpart inv:
self.surface_area <= self.parcel.surface_area
```

For instance, the surface areas of AS1 or AS2 should be lower than or equal to the surface area of AP1 in Fig. 4. A complete expression in natural language of the constraint of Example 3 is "the surface area of an allowed subpart `self` is lower or equal to the surface area of the allowed parcel of `self`."

By navigating from the class `Allowed_Subpart` to the class `Allowed_Parcel`, the expression `self.parcel` returns the allowed parcel linked with the subpart `self` by the "subparcels ... parcel" association. The expression `self.parcel.surface_area` returns the surface area of the allowed parcel.

Universal and existential quantifiers are denoted in OCL by `forAll` and `exists`. These operations are used on collections of objects. The logical implication can be expressed by `implies`. The next expression exemplifies the operations `implies` and `forAll`.

Example 4. An organic waste should not be associated with a new allowed parcel if the end date of use of the organic material has expired:

```
context Organic_Waste inv:
self.land->forAll (as |
( not(self.end_date_of_use = 0) ) implies
    self.end_date_of_use
        > as.parcel.creation_date )
```

As presented in this constraint, it is possible to navigate between Organic_ Waste and Allowed_Parcel by using the relationships "allowed_ organic_waste ... land" and "subparcels ... parcel." The context class is Organic_Waste, and self.land returns the collection of the allowed subparts associated with the organic waste self. For example, by starting from the object wastes_of_farm_C (=self), this navigation returns a collection composed of AS1 and AS2 (Fig. 4). For each allowed subpart as in this collection, if the end date of use of self has been entered in the database, then this date must be greater than the creation date of the allowed parcel of as. Note that a numeric representation of dates is used in this example.

OCL provides a platform-independent and generic method to model constraints. It can be interpreted by code engines/compilers to generate code automatically [11]. For instance, the Dresden OCL Toolkit developed by the Technische Universität Dresden can generate SQL code (Structured Query Language) from OCL constraints [3, 4, 5]. The produced code can be used to check if a database verifies the constraints or to forbid inserting data that do not verify a constraint. The next section describes the global architecture of this code engine supporting OCL.

3 Example of a Tool Supporting OCL: The Dresden OCL Toolkit

The Dresden OCL Toolkit is a software platform for OCL tool support that is designed for openness and modularity and is provided as open source. The goal of the platform is, for one thing, to enable practical experiments with various variants of OCL tool support, and then, to allow tool builders and users to integrate and adapt the existing OCL tools into their own environments.

The toolkit has been developed in two versions. The older one supports OCL 1.x. The more recent version supports OCL 2.0 [15]. The definition of the OCL 2.0 language version was a step from a pure constraint language (as presented in Section 2) to a query and constraint language [12]. OCL queries can be used to construct any model expressions just as you can construct any tables using SQL queries. This is helpful in writing model transformations that are needed in model driven software development. The architecture of the Dresden OCL2 Toolkit is metamodel-based (Fig. 5). This means that all models have a metamodel, and metamodels are managed as MOF (Meta Object Facility) instances [16].

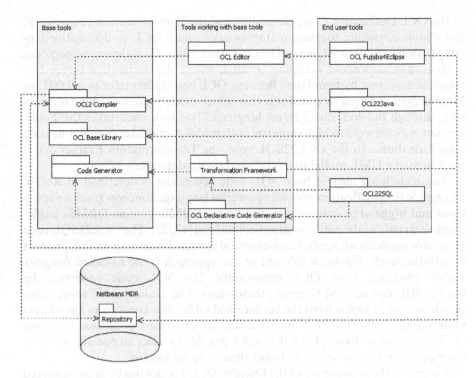

Fig. 5 Architecture of the Dresden OCL2 Toolkit

Figure 5 shows the most used Dresden OCL tools. We divide these tools into three categories: base tools, tools working with base tools, and end-user tools.

The most important base tool is the *OCL2 compiler*, which parses an OCL string and generates the *abstract syntax model (ASM)* of the OCL expression. An ASM represents a UML/OCL model or an MOF/OCL metamodel and is stored in the repository. The *OCL Base Library* provides Java code for all predefined OCL types such as *Integer* and *Collection*. The *Code Generator* is an abstraction of multiple general template engines (such as StringTemplate) helpful for re-use by the *Transformation Framework* and the *OCL Declarative Code Generator*.

Figure 5 presents the modular structure of the existing tools by dependencies. Tools working with base tools and end-user tools are dependent on base tools. For example, the UML tool *Fujaba4Eclipse* integrates our OCL tools, and the *OCL editor* Eclipse plug-in calls the OCL2 compiler. The *OCL22Java* tool generates Java code out of OCL expressions and also instruments Java programs with the generated code.

From the end-user point of view, there is still another code generator, the *OCL22SQL* tool, generating SQL code for checking the database that implements the UML model. Strictly speaking, the OCL22SQL tool is an application of the OCL Declarative Code Generator and the Transformation Framework. Both tools constitute together a framework to map OCL to declarative languages such as SQL or XQuery. The *OCL Declarative Code Generator* provides a language-independent mapping schema, where the declarative target language is described by templates. Because OCL constraints refer to a UML or MOF model, we need to transform this source model, too, and make it accessible through the declarative target language. This task undertakes the *Transformation Framework*, which supports both model-to-model and model-to-code transformations. In the OCL22SQL case, the Transformation Framework is used to map a UML model to a relational (SQL) database schema.

The underlying idea of both SQL code generators is described in Ref. 4. Though SQL specification provides powerful language concepts (such as assertions and triggers) to enforce integrity by the DBMS, current DBMSs implement them only at the entry or intermediate level [13, 22]. The OCL22SQL code generator supports an application-enforced approach realizing an independent constraint check. The basic element of our approach is the so-called *Integrity Views* generated from OCL constraints. The SQL code generated by OCL22SQL can be used to create these views. The evaluation of these views provides a set of tuples from the constrained tables that constitutes the objects violating the constraint. For instance, the integrity view generated from the OCL constraint of Example 1 will make it possible to select all parcels that have their attribute `surface_area` lower than or equal to zero.

Currently, the architecture of the Dresden OCL Toolkit has been reengineered again to support any MOF-based metamodels and, besides Netbeans MDR, further repositories such as EMF (the Eclipse Modeling Framework). The repository independence allows for a broader reuse of the Dresden OCL tools, especially by the large Eclipse community. The support of any metamodels enormously broadens the application of OCL. Thus, for example, OCL expressions cannot only be specified on UML or MOF models, but can also be used for Extensible Markup Language (XML) and Domain Specific Languages (DSLs).

4 Extending OCL for Spatial Objects

Environmental events occur in time and space, and often data used by agricultural systems are georeferenced. As an example, the databases developed to monitor and analyze the spreading of organic wastes [21] make considerable use of georeferenced data. It is very important to model spatial constraints that model the different spatial configurations of the objects accurately. The purpose of this chapter is to show how it is possible to extend OCL to model this type of constraint.

The chapter will present the integration of spatial functions based on Egenhofer's relationships into OCL. The extended language will be called "Spatial OCL" [6]. The eight Egenhofer binary relationships presented in Fig. 6 have been actively studied [8, 9]; they constitute the basis of Oracle Spatial SQL [17].

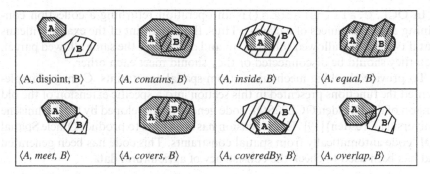

Fig. 6 Eight Egenhofer binary relationships between two simple regions

The general syntax of the proposed OCL spatial functions is

A.Egenhofer_topological_relation(*B*) : *Boolean*

Thus, Egenhofer_topological_relation can be: disjoint, contains, inside, equal, meet, covers, coveredBy, overlap. *A* and *B* are the parameters of the operations (i.e., the two simple geometries to compare). These operations return true or false depending on whether the topological relation between *A* and *B* is true or false. Only the relationships between regions are considered in this chapter. The following example of OCL constraint illustrates the use of the proposed functions.

Example 5. An allowed subpart is spatially inside or equal to or covered by its allowed parcel:

```
context Allowed_Subpart inv:
self.geometry.inside(self.parcel.geometry) or
self.geometry.equal(self.parcel.geometry) or
self.geometry.coveredBy(self.parcel.geometry)
```

For instance, in Fig. 4, AS1 and AS2 are covered by AP1.

Example 6. All allowed subparts of the same allowed parcels should be disconnected or should meet each other:

```
context Allowed_Subpart inv:
Allowed_Subpart.allInstances()->forAll(as |
( not(self.id = as.id) and
  self.parcel.id = as.parcel.id )
    implies self.geometry.disjoint(as.geometry) or
            self.geometry.meet(self.geometry) )
```

In OCL, C.allInstances() is an operation returning a collection containing all the instances of a class C. Thus, the constraint of the example means that if two distinct allowed subparts as and self have the same allowed parcel, then they should be disconnected or they should meet each other.

To provide checking mechanisms from spatial constraints, Cemagref implemented the functions presented in this section into a specific extension of the old version of the Dresden OCL22SQL code generator (developed by the Technische Universität Dresden) [19]. This extension has been used to produce Oracle Spatial SQL code automatically from spatial constraints. This code has been generated and has been used to check the consistency of agricultural data.

5 Conclusions

OCL is a standard and conceptual language [15]. It is used in model driven software development to describe the behavior of an object formally and to generate the corresponding code automatically [12]. Furthermore, OCL integrates notations close to a spoken language to express constraints.

OCL provides a platform-independent and generic method to model constraints. As presented in Section 3, it can be interpreted by code engines/compilers to generate code automatically. Indeed, some tools allow for the production of integrity checking mechanisms in different languages from specifications of constraints expressed in OCL [3, 11].

In database management systems, OCL can be used to model integrity constraints. OCL provides a formal language to define these constraints precisely. In the context of environmental and agricultural databases, the satisfaction of these constraints will tend to guarantee the consistency and the quality of data and consequently the production of reliable indicators.

Concerning the modeling of spatial constraints, several recent papers deal with the spatial extensions of OCL [6, 7, 10, 18, 19]. The proposed languages could be viewed as DSLs [2].

References

1. Cook, S., Daniels, J.: Designing Object Systems—Object Oriented Modeling with Syntropy. Prentice-Hall: New York (1994).
2. Demuth, B.: The Dresden OCL Toolkit and the Business Rules Approach. European Business Rules Conference (EBRC2005), Amsterdam, <http://st.inf.tu-dresden.de/files/papers/EBRC2005_Demuth.pdf> (2005).
3. Demuth, B., Loecher, S., Zschaler, S.: Structure of the Dresden OCL Toolkit. In: 2nd International Fujaba Days "MDA with UML and Rule-based Object Manipulation." Darmstadt, Germany, September 15–17 (2004).
4. Demuth, B., Hußmann, H., Loecher, S.: OCL as a Specification Language for Business Rules in Database Applications. Lecture Notes in Computer Science vol. 2185, 104–117. Springer: New York (2001).

5. Demuth, B., Hußmann, H.: Using UML/OCL Constraints for Relational Database Design. Lecture Notes in Computer Science vol. 1723, 598–613. Springer: New York (1999).
6. Duboisset, M, Pinet, F., Kang, M.A., Schneider, M.: Precise Modeling and Verification of Topological Integrity Constraints in Spatial Databases: From an Expressive Power Study to Code Generation Principles. Lecture Notes in Computer Science vol. 3716, 465–482. Springer: New York (2005).
7. Duboisset, M., Pinet, F., Kang, M.A., Schneider, M.: Integrating the Calculus-Based Method into OCL: Study of Expressiveness and Code Generation. DEXA Workshops, 502–506 (2005).
8. Egenhofer, M., Franzosa, R.: Point-Set Topological Spatial Relations. International Journal of Geographical Information Systems 5 (2) 161–174 (1991).
9. Egenhofer, M., Herring, J.: Categorizing Binary Topological Relationships between Regions, Lines, and Points in Geographic Databases. Technical report. Department of Surveying Engineering, University of Maine, Orono, ME, 28 p. (1992).
10. Hasenohr, P., Pinet, F.: Modeling of a Spatial DSS Template in Support to the Common Agricultural Policy. Journal of Decision Systems 15 (2) 181–196 (2006).
11. Klasse Objecten: OCL Tools and Services Web site, <http://www.klasse.nl/ocl> (2005).
12. Kleppe, A., Warmer, J.: Object Constraint Language, the Getting your Models Ready for MDA. Addison-Wesley: Reading, MA (2003).
13. Melton, J., Simon, A.: Understanding the New SQL: A Complete Guide. Morgan Kaufmann, San Francisco (1993).
14. Miliauskaitė, E., Nemuraitė., L.: Representation of Integrity Constraints in Conceptual Models. Information Technology and Control, vol. 34 (4) 355–365 (2005).
15. OMG: OCL 2.0 specification. OMG specification, 185 p. (2005).
16. OMG: OMG MOF Web site, <http://www.omg.org/mof > (2007).
17. Oracle Corp.: Oracle Spatial. User's Guide and Reference. Oracle documentation (2005).
18. Pinet, F., Kang, M.A., Vigier, F.: Spatial Constraint Modelling with a GIS Extension of UML and OCL: Application to Agricultural Information Systems. Lecture Notes in Computer Science vol. 3511, 160–178. Springer: New York (2005).
19. Pinet, F., Duboisset, M., Soulignac, V.: Using UML and OCL to Maintain the Consistency of Spatial Data in Environmental Information Systems. Environmental Modelling and Software 22 (8) 1217–1220 (2007).
20. Schmid, B., Warmer, J., Clark, T.: Object Modeling with the OCL: The Rationale Behind the Object Constraint Language. Springer: New York, 281 p. (2002).
21. Soulignac, V., Gibold, F., Pinet, F., Vigier, F.: Spreading Matter Management in France within Sigemo. In: Proceedings of the 5th European Conference for Information Technologies in Agriculture (EFITA 2005), Vila Real, Portugal, July 25–28, 8 p. (2005).
22. Türker, C., Gertz, M.: Semantic Integrity Support in SQL:1999 and Commercial (Object-) Relational Database Management Systems. The VLDB Journal 10 (4) 241–269 (2001).

Design of a Model-Driven Web Decision Support System in Agriculture: From Scientific Models to the Final Software

Ludovic Tambour, Vianney Houlès, Laurence Cohen-Jonathan,
Valérie Auffray, Pierre Escande, and Eric Jallas

Abstract This chapter aims at introducing a new type of design of decision support systems (DSSs). The DSS presented here is a software based on client–server technology that enables great accessibility by the Web. Its conception flow has been established to be generic and not explicitly problem-oriented. In this way, once the first DSS is built, the creation of other DSSs will be easy and time-saving. The creation of the DSS requires the collaboration of different experts such as agronomists, computer specialists, and interface experts. Their communication is improved by the use of the formal language Unified Modeling Language (UML) throughout the process of software design. The relevance of the DSS comes from its use of scientific mechanistic models adapted to the users' needs and from a flexible architecture that allows easy software maintenance. The chapter is structured as follows: after the introduction, the second section will explain in detail the methods used to build the scientific models that describe the biological system. The third section describes the methods for the validation and implementation of those models, and the fourth section deals with the transcription of the models into software components processable in the DSS. Finally, the last section of this chapter describes the architecture of the client–server application.

1 Introduction

1.1 General Points

Modern agriculture has multiples stakes in economics, environment, society, health, ethics, and even geopolitics. Farmers must integrate more and more information with increasing production constraints. Decision support systems (DSSs) have been developed with the intention of providing farmers with relevant information for diagnosis assistance or more generally to facilitate

L. Tambour (✉)
SAS ITK, Montpellier, France
e-mail: ludovic.tambour@itkweb.com

P.J. Papajorgji, P.M. Pardalos (eds.), *Advances in Modeling Agricultural Systems*,
DOI 10.1007/978-0-387-75181-8_5, © Springer Science+Business Media, LLC 2009

strategic or operational decision-making in an inaccurate and/or uncertain environment.

These systems are particularly useful for pest control; indeed, these tools can provide risk indicators through the use of models running with meteorological data, or more basically, with a description of land history (preceding crop, soil management, etc). This enables the farmer to make a more accurate diagnosis and anticipate treatments or change his strategy. For instance, when the DSS foresees a low pest pressure, the farmer may favor environmental characteristics instead of efficiency when planning his phytosanitary program.

In France, the Plant Protection Service (Service de la Protection des Végétaux) has been working on disease prediction models since the 1980s and has included them in DSS since the beginning of the 1990s. More than 30 pests (insects or diseases) have been studied, and 21 models are currently used in France [1, 2]. Since then, many other DSSs have appeared. They have benefited from technological advances in data acquisition and treatment and from the Internet for the diffusion and update of information. Many agriculture-oriented institutes are involved in the development of DSSs such as research centers (INRA,[1] CIRAD[2]), technical institutes, cooperatives, and agropharmaceutical companies. The aims of DSSs are varied and may concern, for instance, variety choice (Culti-LIS), weed control (Decid'herb [3]), disease control (Sépale +), or nitrogen fertilization (Ramsès).

Despite this variety of offerings, DSSs are not currently used by farmers. The cooperative In Vivo, one of the major DSS providers, covers only 1.45 million hectares with its DSS for fertilization and plant protection [4] among 29.5 million hectares of agricultural lands in France. To be profitable, a DSS must be simultaneously reliable (scientifically validated) and user-friendly for the farmers.

With regard to plant phytosanitary protection, major effort is still required to adapt DSSs to practical needs of farmers. Special attention should be paid to the following three points:

1. *Integration of plant sensitivity*. Currently, most DSSs aimed at plant protection provide a pest risk indicator that depends on weather data but does not take into account the host plant. It would be more interesting to provide a risk indicator that considers plant sensitivity, for instance, by modeling the phenological stage as well as age and surface of the different organs.
2. *Integration of the major diseases for a specific crop*. Fungicides are indeed seldom specific to only one disease; each treatment often controls two diseases or more. Moreover, farmers strive, whenever it is possible, to group together the treatments to limit workload and fee. An efficient DSS should

---

[1] Institut National de Recherche Agronomique (French National Institute for Agricultural Research).

[2] Centre de coopération Internationale en Recherche Agronomique pour le Développement (French Agricultural Research Center for International Development).

therefore take all the major diseases of a crop into account, yet most current DSSs have been built for only one disease. The aggregation of different DSS outputs to define a global protection strategy is therefore difficult.

3. *Modeling the effects of the applied treatments.* Most models simulate disease evolution without any control. They thus provide useful information for the first treatment but they cannot indicate if other treatments are necessary afterwards.

The next point depicts the different steps and know-how involved in the design of a generic DSS that tries to avoid the drawbacks described above.

1.2 Generic Design of Decision Support Systems

The LOUISA project (Layers of UML for Integrated Systems in Agriculture) launched by the ITK[3] company is a general design for DSSs in agriculture with the following characteristics:

1. It integrates scientific models simulating the entire biological system. This requires the description of not only each element composing the system but also the interactions between those elements. The relevance of the information simulated by the DSS depends directly on the quality of the models.
2. The computing environment is flexible in order to integrate new formalisms, for example when agricultural knowledge improves.
3. The outputs of the models must be useful and easily accessible for the users, including farmers and farming advisers.

Therefore, the proposed DSS is a Web-based system. The Web-service is preferred over the stand-alone model because it enables an easy management of the diffusion of the tool, its update, and communication with linked databases such as those providing meteorological data.

The design of a DSS involves the following different specialists:

1. Agronomists who build the scientific model by conceptualizing the system and then creating a prototype.
2. Computer specialists who translate the scientific model into a processable model. This model needs to be easy to implement in the final DSS and must be simultaneously composed of independent components reusable in other DSSs.
3. Web interface specialists who define the functionalities of the DSS and design an appropriate and user-friendly interface considering the needs of users.

---

[3] See http://www.itkweb.com.

Those specialists have their own language, needs, aims, and tools. In such a multidisciplinary team, communication is the keystone of success. In our case, UML [5] is used by all specialists to facilitate this communication.

The developed DSS greatly depends on the agronomic knowledge and technologies. As these elements can evolve quickly, attention has been paid to flexibility throughout the design of the DSS.

The different stages of a DSS construction are depicted in the flowchart of Fig. 1.

The first phase is to establish the document of requirements specification, which defines the technical characteristics and functionalities of the complete DSS. Then two main objects are performed: the model and the interface. The model is first studied from a scientific point of view. A prototype is created to verify that the outputs are consistent. Once the model is validated, it is transcribed by computer specialists who are interested in the practical integration of the model into the DSS. The design of the interface can be done directly from the requirements specification. Thus, model and interface can be designed simultaneously and evenly matched into the DSS.

The next section describes in detail these phases on the basis of a practical case of DSS. For each phase, we will present our experience and progress, the encountered difficulties, and will justify our choices.

1.3 Development of DSS Software for Phytosanitary Plant Protection

This document illustrates a practical use of the generic DSS design LOUISA. The example taken here is a DSS designed for phytosanitary plant protection against diseases, although other aspects of the crop management could have been chosen as well.

The aim of this DSS is to enable farmers to choose their phytosanitary program with the support of risk indicators illustrating parasite pressure and short-term predictions of contamination risks.

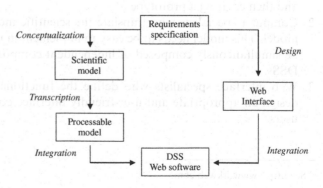

Fig. 1 Flowchart of the construction of the DSS

These indicators are obtained by means of scientific model simulating the behavior of the entire biological system. This system is composed of three submodels:

- Host plant
- Disease
- Phytosanitary products (pesticides).

The three submodels interact with each other and are driven by meteorological data including daily measures and short-term predictions. The accuracy of the model could be improved using an hour time-step for weather data, however we preferred the daily time-step as these data are far more easily obtained by farmers nowadays.

The main functionalities of the application are synthesized in the UML use-case diagram in Fig. 2.

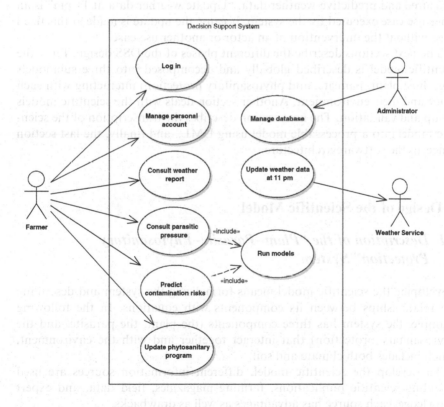

Fig. 2 UML Use-case diagram of the functionalities of a DSS intended for phytosanitary plant protection

This diagram depicts the different actors of the systems, human and non-human, and their interactions with the system. The main actor within the software is the farmer. The functionalities related to this user are

- Connection and disconnection to a server
- Personal account management including registration of land characteristics and cropping practices
- Access to weather data (daily records and predictions)
- Access to current parasitic pressure thanks to simulated indicators
- Access to contamination risks predictions thanks to simulations using predictive weather data
- Update of the phytosanitary program according to the DSS suggestions, taking into account simulated risk indicators and phytosanitary products regulations.

The second major actor is the administrator, who must manage the database.

The weather service is the last actor and, more precisely, a non-human secondary actor. The DSS must be connected to the weather service to get measured and predictive weather data. "Update weather data at 11 pm" is an inner use-case executed by the system, because the update is made at this fixed hour without the intervention of an actor or another use-case.

The next sections describe the different phases of the DSS design. First, the scientific model is described globally and decomposed into three submodels (e.g., host plant, parasite, and phytosanitary protection) interacting with each other and their environment. Another section deals with the scientific models set-up and validation. The next section describes the transcription of the scientific model into a processable model using UML, and finally, the last section concerns the software architecture.

2 Design of the Scientific Model

2.1 Description of the "Plant–Parasite–Phytosanitary Protection" System

Developing the scientific model means formalizing the system and describing the relationships between its components with equations. In the following example, the system has three components (the plant, the parasite, and the phytosanitary protection) that interact together and with the environment, which includes both climate and soil.

To develop the scientific model, different information sources are used including scientific publications, farming magazines, field data, and expert knowledge; each source has advantages as well as drawbacks.

In general, scientific publications describe a particular phenomenon accurately thanks to models, in the best case, or at least with a qualitative description.

Concerning our example, it is possible to find models describing the influence of temperature and humidity on disease development in the scientific literature but it is not sufficient for describing the entire system. For instance, no model is available that treats the influence of plant growth on phytosanitary protection. Moreover, the final aim of published models is more often the understanding of biological phenomena rather than their use in DSSs. Before integrating a published model in a DSS , it must be verified that the inputs are consistent with the DSS specification requirements. Moreover, published models may not be generic but only valid for specific experimental conditions.

Farming magazines provide general information that cannot directly be used to construct models but may help in formalizing and expressing hypotheses. For instance, advice concerning pesticides spray frequency can be useful information with respect to the phytosanitary protection model and its interaction with the crop and the climate. This information source is also helpful for keeping in touch with the concerns of farmers.

In the case when accumulated information is not sufficient for describing a system, it is recommended to interact with experts and to use field data.

The conception of the scientific model is not a simple bibliographic search and compilation. Published models are not prefabricated bricks that can be assembled to build the DSS. To be integrated in a coherent system, models must first be adapted with specific regard to their accuracy in relation with the DSS needs. Moreover, lack of information has to be filled in with new formalisms coming from hypotheses that will be tested and validated. Even if specific published models are useful, it is important to consider their collaboration for modeling the entire system. The scientific model is just an element that must serve the final aim of the DSS, taking into account technical constraints and specific concerns of farmers.

For a DSS designed for phytosanitary protection, a few judicious risk indicators have been selected:

- Parasitic pressure
- Phytosanitary protection level
- Plant sensitivity that may depend on specific organs and their age.

These risk indicators must be part of the outputs generated by the model. The UML class diagram in Fig. 3 expresses the relationships between concepts/entities and the outside user.

The system has five components, which are the submodels plant, parasite, and phytosanitary protection added to the soil and the climate. In this conceptual view, climate has an influence on the four other components. For instance, rain refills the soil with water and washes off phytosanitary products. The plant is the medium for the parasite's development. The phytosanitary protection destroys the parasite but its efficiency may decrease because of rain and plant growth (the latter phenomenon is called "dilution" in the diagram). The soil provides the plant with water and nutrients and modifies the

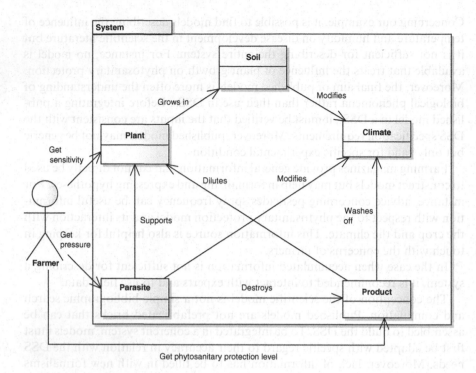

Fig. 3 UML class diagram of the whole system and its components

microclimate. For example, a soil with high water retention can increase humidity and thus favor the development of some parasites.

For the moment, soil behavior is not explicitly modeled. As a first approach, the plant grows without soil-limiting factors (i.e., neither water stress nor nitrogen stress, for example). In the same way, the soil influence on the parasite's development (through action on microlimate) was not modelled, because of the scarcity of knowledge on the issue.

It could, of course, be possible to add more relationships in this diagram. For example, the influence of the parasite's development on plant growth or the effect of the plant on microclimate could be modeled. For an initial approach, we decided to model only the most influential phenomena regarding the three selected risk indicators. Once a prototype is built, the outputs of the model can be compared with observed data. In case of discrepancies due to excessive simplification, the model can be made more complex by adding more relationships.

2.2 The Plant Model

Plant models can be divided into three groups according to their abstraction level:

- *Big leaf models.* In these models, the plant is considered as a homogeneous structure. Organs, such as leaves, are not modeled individually but aggregated into one large organ. This is for instance the case for the "CERES-maize" [6] and "STICS" [7] models.
- *Topological models.* These models describe the structure and the relative position of organs with relationships such as organ X bears organ Y. Each organ is described individually regarding its behavior and its attributes. For example, the rice model "EcoMersitem" [8] and grapevine model [9] can be cited.
- *Architectural models.* These models are even more accurate than topological models by adding information about organs geometry and spatial position. An example of this kind of model is "COTONS" [10].

An output of the model concerns plant sensitivity to a particular parasite. This sensitivity depends on the plant's phenological stage as well as on the age and surface of susceptible organs. For this reason, a topological model seems to be the most appropriate for the DSS. Indeed, a big leaf model cannot simulate properly the age and surface of each organ (e.g., leaves), and all the details provided by an architectural model do not seem useful to simulate the selected indicators. Modeling at the organ scale is also convenient for flexibility. For example, if knowledge concerning fruit growth improves and gives birth to a new model, it is possible to integrate it without rewriting the entire plant model.

The topological model can describe the plant's structure. However, it is not sufficient for the DSS needs, as plant evolution modeling involves other submodels (Fig. 4):

- *Phenological model.* This submodel aims to determine the major plant phenological stages such as growth start, flowering, fruit setting, and so forth. The phenological model drives the appearance of organs type (e.g., fruits appear with fruit setting), and the equations describing organs evolution may change according to phenological stage.

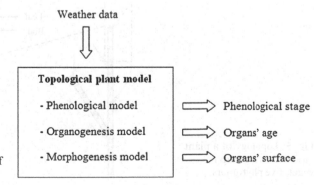

Fig. 4 Inputs and outputs of the plant model

- *Organogenesis model*. This submodel determines the apparition of new organs. It adds dynamics to the plant model by describing the evolution of plant topology with time.
- *Morphogenesis model*. This model simulates the growth of each organ individually.

The chosen topological model considers the above-ground part of a plant as a group of phytomers (Fig. 5). A phytomer is a cell cluster that will evolve into organs having the same age, and it can be considered as the basic plant element. Organs coming from a phytomer can vary with plants or within a plant. For example, a plant can have vegetative phytomers composed of a node, an internode, a leaf, and a bud and fruiting phytomers composed of a node, an internode, a leaf, a flower, and a bud.

A plant's structure can thus be modeled in terms of axes (e.g., branches for trees and tillers for *graminaceae*). Each axis is composed of a series of phytomers, and each phytomer may bear another axis as its bud itself can evolve into a new phytomer. Assuming this organization, describing a plant consists of determining the composition of its phytomers and defining the rules driving their evolution. For instance, the following questions have to be answered: where are the fruiting phytomers? In what conditions does a bud develop into a new phytomer? The rules describing the plant's structure can be determined with support of a statistical analysis using Markov chain models [11, 12, 13].

The evolution of an individual organ is influenced by its surroundings and is narrowly linked with the rest of the plant [14], meaning that the organogenesis and morphogenesis models have to be regulated by the status of the whole plant. Regulations are taken into account through parameters included in the equations describing organogenesis and morphogenesis.

If there is no limiting factor, phytomer emission (organogenesis) and organ growth (morphogenesis) are only driven by thermal time (timescale depending on temperature), although the equations can vary with the type of axis. For instance, equations describing the phytomer emission rhythm (also called

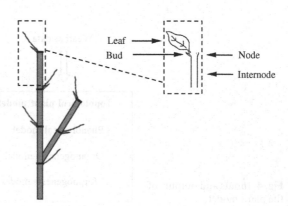

Fig. 5 Topology of a plant modeled as a group of vegetative phytomers

plastochrone) and the final size of leaves can be different between a phytomer constituting the main axis and another phytomer constituting a secondary axis.

Phenology has a major impact on the plant and on its organs. For example, during the vegetative stage, only vegetative phytomers appear, whereas during the reproductive stage, both vegetative and fruiting phytomers can appear. Phytomer emission rhythm and vegetative organ growth can also be slowed down from fruit setting on because of preferential carbohydrates allocation to fruits.

Water stress and nitrogen deficiency can also be factors slowing down phytomer emission and organ growth. Stresses are generally modeled at the plant scale.

2.3 Parasite Model

The risk indicators selected for the DSS concern phytosanitary protection, plant sensitivity, and parasitic pressure. The latter element is the most important factor to decide phytosanitary products application, which is why the parasite submodel interacts with all the other submodels. As the DSS is not designed to deal with yield prediction, the effects of the parasite on the plant are not modeled. Thus, it is also assumed that there is no retrospective effect on the parasite. In reality, as parasites weaken or destroy parts of the plant that represent their support, their own development should also be affected, but this phenomenon is voluntary neglected.

The following description of the parasite model concerns diseases in general resulting from bacteria, viruses, or fungi. Parasites transform into several biological stages during their development. For instance, fungi have two principal forms: sexual and asexual. Other forms can be added, specialized in propagation, development, or survival during critical periods (winter for temperate zones, dry period for tropical zones). The evolution from one form to another is modeled more or less accurately depending on the chosen formalism.

Three main kinds of models are used to simulate disease evolution:

- Epidemiologic models
- Mechanistic models
- Intermediate models based on the parasite biological forms.

Epidemiologic models [15] describe disease progression by means of a single equation with the following type:

$$\frac{dX}{dt} = kX(t) \cdot (1 - X(t)),$$

where X stands for the modeled variable that is often the surface colonized by the disease, t stands for the time, and k is the disease amplification factor. This differential equation is solved by approximating the temporal derivative term by finite differences method resulting in:

$$\Delta X = k \cdot \Delta t \cdot X(t_n) \cdot (1 - X(t_n)),$$

with $\Delta X = X(t_{n+1}) - X(t_n)$ standing for the variation of X during $\Delta t = t_{n+1} - t_n$ corresponding with the model time-step.

Factor k is difficult to determine because it varies with climate and plant evolution. In epidemiologic models, disease progression is modeled globally without distinguishing the different parasite forms.

Mechanistic models describe the different physical, chemical, and biological processes involved in disease development and simulate accurately the parasite's biological cycle (e.g., [16, 17]). This kind of model requires a meticulous description of plant structure, microclimate, and the parasite's biological processes. Such a complexity can be useful for research purposes but it requires too many inputs to be integrated in the DSS. Moreover, increasing complexity often leads to less reliability and may result in erroneous outputs [18]. These models are also specific to a particular parasite and are hardly adaptable to other parasites.

An intermediate complexity model consists of representing each parasite's biological form as a stock containing a quantity of items. The evolution of items from one form to another is modeled but not necessarily in a mechanistic way. For instance, to pass from an ungerminated to a germinated fungus spore, an empirical relationship calculating the germination rate according to climate can be used instead of the description of the physical and chemical reactions involved. The number of items in a stock can also increase by intern multiplication without implicating the other stocks. This intermediate complexity is easier to understand and set up compared with mechanistic models and more factors can be explicitly integrated compared with epidemiologic models. Furthermore, this type of model is more generic. Indeed, the number of stocks and equations driving the transitions can be adapted to model another disease. For all the above-mentioned reasons, this kind of model has been chosen.

Most diseases start with an initiation phase followed by an amplification phase corresponding with epidemic cycles. Figure 6 depicts this general representation of diseases with the Stella formalism [19]. Rectangles on the left-hand side (stages 1 and 2) illustrate the stocks during the initiation phase; rectangles forming a cycle (stages 3 to 6) illustrate the amplification phase. The movement of items from one stage to the next is controlled by a flow. This flow can depend on factors related to weather conditions, the plant, or the disease itself. This dependency is depicted by a circle named "Factor" on the figure. Each stock can also be affected by mortality, depicted by a flow named "Mort." for mortality. There is no stock following mortality so that items concerned by mortality are destroyed. Of course, Fig. 6 has only illustrative value, and the number of stocks depicted will depend on the considered disease.

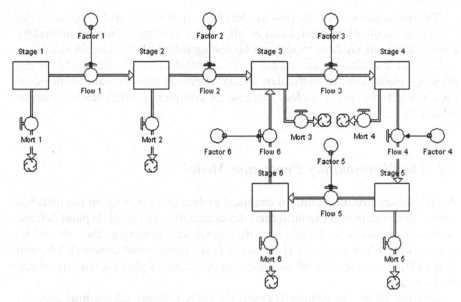

Fig. 6 A virtual parasite's biological
(according to the Stella formalism [19])

Disease development can also be considered from the plant point of view. In this case, stocks would represent the different status of plant (e.g., healthy, contaminated in latency, contaminated and infectious, contaminated and no longer infectious), while from the parasite's point of view, stocks symbolize the different biological forms such as larval and adult forms for insects. Items contained in these stocks can be surface units for the plant and individuals for the parasite. The choice of this point of view will depend on knowledge availability.

It is then necessary to determine the different factors influencing disease development.

Of course, phytosanitary protection has a major influence, although *a priori* all crop management techniques can positively or negatively influence disease development. For example, pruning modifies susceptible organs available for contamination. Therefore, the parasite submodel must use outputs from the plant and the phytosanitary protection submodels as inputs.

Weather data must also be used as inputs, because the climate influences the disease development. It is thus necessary to analyze *a priori* the weather data in terms of availability, accuracy, and costs. A simple model using few inputs should be preferred compared with a more complex model using inputs hardly available or inaccurately measured [18].

The influence of crop location characteristics on disease development can be modeled by an impact on microclimate. For instance, the water absorption and retention characteristics of the soil influence the humidity in the plant vicinity. Another example is the land slope that influences the plant's microclimate and the radiations intercepted by the plant.

This point shows that the pest model choice depends on its biological cycle, its action on the plant, and, most of all, on the existing scientific knowledge. Even if different kinds of model can be distinguished, there is no fixed methodology for building a pest model. It is essential to adapt the model to the project's global aim. In particular, special attention must be paid to the accuracy needed because a useless increase of complexity often leads to lower reliability.

2.4 The Phytosanitary Protection Model

Phytosanitary products contain chemical molecules operating on the metabolism of parasites in order to limit their development or to stimulate plant defense processes. As these molecules are often toxic and polluting, they should be applied only when necessary (Integrated Pest Management concept). The aim of the DSS is precisely the adaptation of applications of phytosanitary products to the needs.

Existing DSSs are usually focused on pests without taking into account phytosanitary protection. Yet the necessity of a phytosanitary product application depends not only on the parasitic pressure but also on the current plant protection level. It would be useful to determine an application remanence, taking into account the plant's development, climate (including rain and radiation), and characteristics of phytosanitary products.

The phytosanitary protection model has to simulate a protection level that is one of the three selected risk indicators for the DSS. This protection level takes into account:

- Active period of products corresponding with the duration of effectiveness of molecules.
- Plant growth: A product applied at time t will be less efficient at $t + \Delta t$ because of the plant's volume and surface increase. The effect of the plant growth depends on product properties.
- Rain wash-off.
- Spraying quality depending on equipment and weather conditions, in particular wind and humidity.

There are three types of phytosanitary products:

- *Protective products.* They have a surface action and protect only the sprayed surfaces.
- *Eradicant products.* They can penetrate plant tissues and are thus protected from rain wash-off. Their penetration speed depends on the characteristics of products and on weather conditions.
- *Systemic products.* They are capable of moving throughout a plant using the vascular system and thus they can protect organs created after spraying. They are very useful when the crop has an intense growth phase.

Each product operates on specific stages of a parasite's cycle. For instance, in the case of fungicides, protective products usually block spore germination: using the formalism depicted in Fig. 6, this action can be modeled by decreasing the flow linking the stocks "ungerminated spore" and "germinated spores."

To model plant phytosanitary protection, each product has to be characterized in terms of biological action on parasites, active period, rain wash-off sensitivity, and efficacy decrease due to crop growth. Thanks to this submodel, the DSS can provide a graph simulating the decrease of plant protection with time.

3 The Scientific Model 's Set Up and Validation

3.1 Principle

Once the formalisms are chosen, the scientific model needs calibration. This operation consists of determining the parameters values that minimize deviations between simulations and observations.

The parameters values can be specific to particular situations such as the geographic zone, the cultivated variety, or, of course, the parasite. In reality, these situation-dependent parameters express variations of factors that are not explicitly modeled and their value must be adapted according to the case [20]. For example, rather than modeling the influence of soil on parasite development, a set of parameters can be proposed for each type of soil.

A model can have an important number of parameters but they do not have the same influence. Some parameters will have a low impact on the model; it is thus unnecessary to spend energy to optimize these parameters. On the contrary, other parameters are essential and have a large impact on simulations. A widely used method for determining the influence of parameters is called *sensitivity analysis* [21, 22].

Finally, the model's performance has to be analyzed in terms of prediction quality and accuracy regarding to the final user needs.

The following section briefly presents the methods for sensitivity analysis, parameter estimation, and model assessment. All of these methods are worth implementation during model design. For more information on this subject, an exhaustive description of good agronomic modeling practices has been previously published [23].

The methods described below require high data quantity. Quality and quantity of data directly impact calibration quality, model accuracy assessment, and consequently model relevance. To be exploitable, a data set must be composed of variables corresponding with model inputs, observations corresponding with outputs, and, if possible, intermediate variables corresponding with variables calculated by the model during the simulation process. The latter variables may refine the diagnosis and may detect the origin of discrepancies. In the DSS, for example, leaf area is not an output shown to the farmer, but it is important to control that it is correctly modeled because it influences parasite growth.

3.2 Methods Used for Sensitivity Analysis, Calibration, and Validation

Two types of sensitivity analysis are distinguished: local methods and global methods [24].

Local methods explore the model's behavior in the vicinity of an input parameter set. They are used when an approximate value of input parameters is known.

Global methods explore all the input parameters' possible values constituting the parameter space. They often use stochastic methods based on random number selection inside the parameter space (Monte Carlo methods). They are preferably used if there is little prior information on input parameter values and give an accurate idea of importance of parameters. A major drawback of some global methods is that the effect of a parameter can be hidden if the model is strongly nonlinear. Some methods are model-independent [21] and cope with nonlinearity and interaction between factors (e.g., the FAST method [21]). Global methods generally require a large computing time.

Local and global methods provide sensitivity indicators allowing classification of parameters according to their influence on the model, but the obtained results depend highly on the explored space.

Parameter estimation consists of finding the parameters values minimizing discrepancies between model outputs and observations. The cost function is a metric of that discrepancy and is to be minimized. In the famous least-square method, for instance, the cost function is the sum of squared errors, but many other cost functions exist that are adapted to different purposes [20]. Observed data must cover various situations and be numerous enough, with the ideal number depending on the number of parameters to be estimated. Many algorithms can be used for parameter estimation such as the Gauss–Newton method, simulated annealing method, or genetic algorithm. The choice of both cost function and optimization algorithm is not easy. It can be interesting to test several couples but the most important point is data quality.

After model calibration, the next stage is the model performance assessment. Concerning a DSS, two complementary analyses should be made [25]. The first one classically compares model outputs with observations. It may detect bias or dispersion and assesses calibration quality. The data used here must be different from those used for calibration. The second analysis concerns recommendations performed by the model and requires an adapted observed data set. For instance, if the DSS recommends phytosanitary spraying dates, it would be useful to have experimental data sets assessing phytosanitary protection with different spraying dates. The model's ability to perform recommendations is validated if the recommended dates are close to the observed spraying dates leading to the best phytosanitary protection.

Once the model's reliability is validated, it is interesting to compare the DSS recommendations with a classic phytosanitary spraying program. The aim of this phase is the assessment of the DSS benefits, for example, in terms of cost, plant protection efficiency, or environmental impact.

3.3 The Choice of Modeling and Validation Tools

The design of scientific models requires appropriate modeling tools. In our experience, an appropriate tool should have the following capabilities:

- To enable quick and easy modeling of agronomic concepts
- To be sufficiently flexible to easily modify the model when changing assumptions
- To be executable in order to check the modeling assumptions by simulation
- To provide a flexible environment facilitating result validation (e.g., graphics, analysis functions of results).

Many tools and languages are available for model set up. The following is a historic classification of these tools. For each type, the advantages and drawbacks are discussed.

The first supports used for modeling were classic procedural languages or object-oriented languages (e.g., C [26], C++ [27], JAVA [28], Delphi [29]). These languages are still intensively used, even if they are not well adapted to model conception and development. They are indeed too verbose to be quickly implemented and modified. At least 50% of a model written with these languages is destined to purely computing aspects (e.g., variables' declaration, tables' size management), decreasing the attention that is paid by the agronomist in his modeling task. Moreover, as those languages do not supply a validation environment, validation methods must be computerized.

To avoid those computing problems, new languages appeared, intended for numerical computing (MATLAB [30], R [31], Scilab [32]), and formal calculus such as Mathematica [33], Maple [34]). These languages are similar to programming languages but they suppress pure programming aspects. For instance, variables are not declared, table size is automatically managed, and simple functions allow table manipulation. Interpretive languages also supply a wide range of mathematical functions assisting modeling and validation. These latter two tasks are concise and also easy to set up and modify. Validation is also facilitated by command lines coupled with the execution environment. After a simulation, all the simulated variables are available from the workspace and, by means of command lines, the user can rapidly visualize, manipulate, and compare the results. In return, this conciseness makes the code difficult to understand and to maintain. Moreover, as these languages are interpretive, their execution is slower. For example, a well-written MATLAB program is at least 10 times slower than a C++ program, and this factor can reach 1,000 if the MATLAB program is written without taking care of executive efficiency.

If numerical computing languages are intended for modeling, most modeling agronomists are resistant to the use of textual languages as modeling support. To overcome this problem, graphical modeling environments have been created (Simulink [30], Stella [19], ModelMaker [35]). By this graphical approach, a model can be decomposed into submodels (describing a process) linked to each other by wires symbolizing dependencies. This graphical modeling environment is easier to handle for a neophyte and further favors the understanding of the model's general concepts inside the multidisciplinary modeling team or with external partners.

On the contrary, these graphical environments are not flexible to change, especially when models are complex or are decomposed into fine granularity components. Another drawback is their limited expressive power. Indeed, available tools impose construction rules that may not be adapted to agronomic concepts. That is, for instance, the case when modeling conditional aspects or states (in the UML sense of state) or for dynamic construction/destruction of components (e.g., simulation of appearance of new organs). Moreover, graphical modeling environments generally execute simulations far slower than do compiled or interpretive programs.

Agronomists also use spreadsheet programs (e.g., Excel [36]) for modeling. These programs allow immediate visualization of numerical results, but they are not adapted for complex models due to their restriction to spreadsheets, and their equations edition is not sufficiently clear. However, they are still very convenient for the quick modeling of simple phenomena.

To develop our scientific model, we chose to use MATLAB because of its flexibility. This flexibility involves the first model description, its assessment, and its possible modifications (change flexibility). Prior to the MATLAB model, a preliminary graphical description was generated with Stella. This software can also simulate, but we only use the graphical functionality to study how the model can be decomposed into submodels of approachable complexity and to depict the phenomena that need studying. Stella graphical aspect also favors communication between people involved in the DSS project.

4 Software Architecture of the Scientific Model

The MATLAB model aims at checking the reliability of scientific concepts prior to their implementation into the final software. Even though MATLAB provides the environment needed for modeling and analysis, this language is not suitable for the complex DSS software because of the following problems:

- *Integration.* A MATLAB model cannot be executed out of the MATLAB environment. Thus, it is difficult to link a MATLAB model with other software components such as database, Web services, or functional components written in another language.
- *Performance.* MATLAB executes slowly, yet response time is decisive for software acceptance by users.
- *Robustness.* MATLAB is devoted to numerical computing but not to programming. It is thus not robust enough for software design.

Therefore, the MATLAB scientific model has to be transcribed into a model using a programming language that we call the processable model. For our DSS, we have chosen the JAVA [28] language because it is well adapted to Internet applications programming. The transcription from a scientific MATLAB model to a processable JAVA model cannot be done directly. Whereas the scientific model aims at simulating reliable outputs, the JAVA model must produce a model with the same functionality that is able to be integrated with other software components and allow quick execution and flexibility regarding change. This latter point is essential because the scientific model continues evolving, and the processable model must be able to go along with this evolution. Using design patterns [37] is a good practice for achieving flexibility. In software engineering, a design pattern formally describes (via UML diagrams) a standard solution to a general recurrent computing problem.

This fourth section deals with the processable model design from the MATLAB model. This design phase uses the UML formalism as specification support, independently from the programming language. Consequently, the work presented here can be generalized to any object-oriented programming language.

4.1 Class Diagram of the Plant–Parasite–Phytosanitary Protection System

The system is studied with a modular approach [38], allowing the decomposition of a complex modeling problem into subelements having a more controllable complexity. The subelements, called components or classes, are autonomous and can communicate with each other. A component, for instance, can be a model of the system (e.g., plant, parasite, phytosanitary protection) or a part of these models (leaves or fruits). Components contain the objects' characteristics constituted by attributes and methods. Attributes correspond with the physical elements to model, and methods correspond with the physical processes.

We will now define the relationships between the classes of the studied system. As the soil influence is not considered at the present time (cf. Section 2.1), it is not depicted in the diagrams in order to lighten the figures.

Explanations given here have a generic value and can be applied to any system decomposed into classes.

As recommended previously in [39], each class diagram element implements an interface depicted by circles as shown in Fig. 7. The use of interfaces creates a "plug and play" architecture where each component is interchangeable without changing the whole implementation; components to interchange only have to share the same interface. For instance, if the modeling agronomist prefers using a big leaf plant model instead of the topological model first chosen, the computer scientist only replaces the "PlantTopo" class by a "PlantBigLeaf" class without modifying the other system components and connections. As a result, the system is very flexible. Moreover, interface creation only requires minor additional implementation time, which is not significant compared with the time saved to implement modifications. For this reason, the plug and play architecture is systematically used for our complex systems implementation.

Figure 8 depicts the general class diagram with the introduction of a simulation controller. The simulation controller aims at controlling the model execution flow and the tasks scheduling. The "singleton" design pattern [37] is used to instantiate the controller so as to ensure its uniqueness. Indeed, if several instances of the controller command the same system, a conflict may be induced.

The control job can be decomposed as follows using pseudo code:

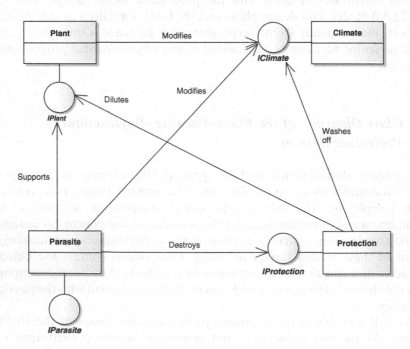

Fig. 7 General class diagram: "plug and play" architecture

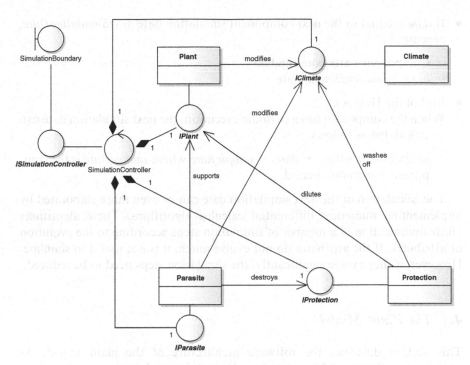

Fig. 8 General class diagram: inclusion of the simulation controller

- Creation of the different components and realization of the components' connections
- Components initialization
- For *date* going from the first simulation date until the last date:
  - Run the execution of the simulation step corresponding with the simulation date *date* for each component
  - Determine the next simulation date *date*
- End of the loop
- Components finalization (e.g., save the simulation results).

The creation of concrete class instances is delegated to a factory and follows the factory design pattern [37]. According to the simulation query, this design pattern creates different simulation scenarios (including the phytosanitary protection or not, instantiating the crop "wheat" or "maize," for example).

A multifrequencies management is added to the modular approach [38] so that components can run with different time-steps. For example, plant growth can be modeled at a daily time-step, whereas the light interception component runs with an hourly time-step in order to take into account the solar angle variations during the day. To take multifrequency into consideration, the execution of a component's simulation step is unlocked in two stages:

- If *date* is equal to the next component simulation date *nextSimulationDate*, execute:

  - Component's attributes update
  - *nextSimulationDate* update

- End of the IF block.
  When the component has a periodic execution, the next simulation date can be calculated as follows:

  - *nextSimulationDate* = *date* + *sampleRate* where *sampleRate* is the component's execution period.

The calculation of the next simulation date can be even more elaborated by implementing numerical differential calculus algorithms. These algorithms allow optimization of a number of simulation steps according to the evolution of attributes. If the attributes do not evolve much, it is not useful to simulate. However, if they evolve significantly, the simulation steps need to be reduced.

4.2 The Plant Model

This section describes the software architecture of the plant model. As explained in Section 2.2, a plant can be seen as a phytomers set having a determined topology. The plant's structure evolves with time (organogenesis) and each organ, through the effect of the environment, evolves individually (morphogenesis) but in close connection with the entire plant's status. The evolution of an individual organ can be reduced due to global constraints operating at the plant scale, such as water stress or carbohydrates limitation. Organogenesis and morphogenesis rules can be modified according to the plant's phenological stage.

For these reasons, the software architecture of the plant model must take into account the following elements:

- A topological structure evolving with time
- The individual evolution of organs
- The modification of individual organ evolution according to global factors (water stress, carbohydrates balance)
- The modification of the evolution rules of organs at phenological key stages.

Concerning the topology, a plant can be implemented as a tree structure in the computer science sense (Fig. 9) with a node of the tree structure representing the abstract class "Phytomer. " The composite relation in Fig. 9 depicts the structure's recursion. This relation means that a phytomer (called the *parent phytomer*) can bear other phytomers (called the *child phytomers*), which subsequently can bear other phytomers on their turn; child phytomers will constitute the axis (branch/tiller) held by the parent phytomer. A phytomer may not hold

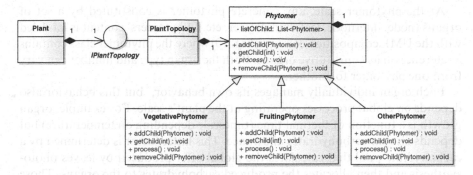

Fig. 9 Class diagram of the plant's topology

another phytomer (which is the case when the phytomer's bud is dormant). In this case, the phytomer can be seen as the tree structure's terminal element.

The abstract class "Phytomer" provides the methods for construction and destruction of phytomers and for the topological structure evolution. For a plant, the destruction concerns not only a phytomer and its descent but also the younger siblings. Indeed, when a phytomer is cut, both a part of the main axis constituted by sibling phytomers and the secondary axes developed from buds are destroyed. Each concrete class modeling a phytomer (e.g., Vegetative Phytomer, Fruiting Phytomer, Other Phytomer depicted in Fig. 9) inherits from the abstract class Phytomer. The inheritance link means that these concrete classes inherit the construction/destruction methods although they define their own individual growth and evolution methods.

The exact concrete object type is decided during execution according to the rules defined by the modeling agronomists. It is thus possible to model all kinds of a plant's structure, no matter the complexity. This architecture is therefore very flexible because it can be used to model different plants or varieties of the same plant with different structures.

In return, it is the computer scientist's responsibility to prevent absurd structures from being modeled (e.g., a phytomer cannot bear another phytomer if its bud is dormant). Indeed, with a tree structure, these errors cannot be checked during the compilation and are only detected during the execution.

Formal verification tools can aid in error detection. Plant construction rules can indeed be formalized as grammar in the language theory sense. Postsimulation, a formal verification tool can thus be used to check the plant's structure consistency of the grammar.

Recursion also facilitates the complex structures management. For example, applying a global plant growth method consists of using the phytomer's own growth method for each phytomer and then to invoke the growth methods of the child phytomers. The use of a tree structure can become complex when the execution order is different from the one naturally imposed by recursion. This is, for instance, the case if the execution order follows the phytomers creation order. This problem can be bypassed by using an external iterator.

At the phytomer scale, any concrete phytomer is constituted by a set of organs (node, internode, leaf, bud, flower, etc.). Phytomers' content is modeled with the UML composition concept (Fig. 10) where the phytomer class contains a reference on its constitutive organs. Both the organ type and number can vary from one phytomer to another.

Each organ individually manages its own behavior, but this behavior also depends on global processes operating at the plant's scale. For example, organ growth is a function of thermal time (timescale depending on temperature) but depends also on carbohydrates allowance. This latter factor is determined by a carbohydrate pool that calculates the global offer produced by leaves photosynthesis and then allocates the produced carbohydrates to the organs. Those global processes are naturally modeled as objects composing the "Plant" class. The question is therefore how to allow an information or service exchange between classes managing global and local processes (Fig. 11).

A solution to enable global–local relationships in a tree structure is the use of the "visitor" design pattern [37]. By definition, this design pattern adds new functionalities to a composite objects set without modifying the structure itself.

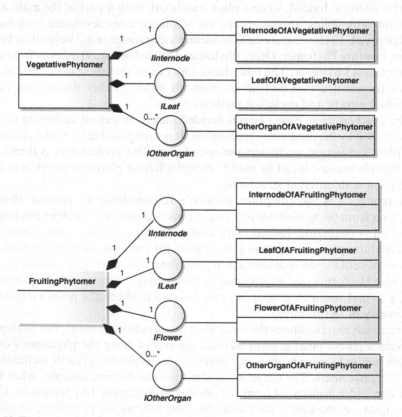

Fig. 10 Two examples of a phytomer's composition

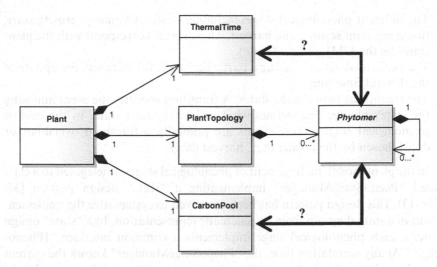

Fig. 11 Each phytomer individually manages its own behavior, but this behavior also depends on global processes operating at the plant's scale. The problem is here: how to establish relationships between local and global processes?

Nevertheless, this solution has not been chosen because it breaks the encapsulation principle. The chosen solution consists of passing a reference to the global object into an argument of organs processes' methods. Therefore, each organ manages its processes taking into account the global object (it may also modify this object's attributes state) and transmits it to its child.

As explained in Section 2.2, the plant passes through different phenological stages and its behavior as well as that of its organs can change radically from one stage to another. The evolution from a stage to the following is generally modeled in an abrupt way as a response to a discrete event (for instance, if a defined thermal time sum reaches a threshold). Consequently, a state diagram is well adapted to model the plant's phenology (Fig. 12). A state diagram enables the modeling of the different states of a system and the events ruling the passing from one state to another. For the plant, the state diagram describes:

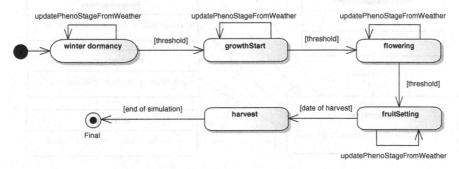

Fig. 12 Example of a state diagram concerning the plant's phenological stages

- The different phenological stages including winter dormancy, growth start, flowering, fruit setting, and harvest. These stages correspond with the plant states (in the UML sense of state).
- The performed actions during a particular stage; for instance, the update of the thermal time sum.
- The transitions between the states: A transition specifies the event imposing the state's change and defines what will be the next state. In the case of phenological stages, these events are primarily a threshold overshoot or dates chosen by the farmer (e.g., harvest date).

In the plant model, management of phenological stages is delegated to a class called "PhenoStageManager" implementing a "state" design pattern [37] (Fig. 13). This design pattern has been conceived to systematize the implementation of a state diagram from its schematic representation. In a "state" design pattern, each phenological stage implements a common interface "IPhenoStage." At any simulation time, the "PhenoStageManager" knows the current phenological stage by the "currentStage" attribute and can ask for the update of this phenological stage by executing the object referenced by "currentStage." If this execution causes a state change, the manager is informed by the "current-Stage" attribute modification. As the phenological stages implement a common interface, the "PhenoStageManager" does not need to know what stage is concretely executed. This "state" design pattern is thus flexible as regards

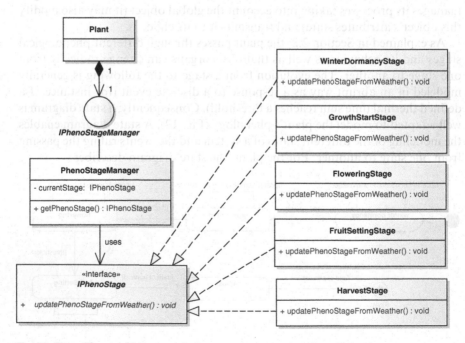

Fig. 13 Class diagram of the state design pattern concerning phenology

changes because it is independent from the phenological stages type and it allows integration of new stages and removal of others.

4.3 The Parasite Model

As explained in Section 2.3, each parasitic item follows different developmental stages. The developmental stages of parasites could be modeled like the plant's phenological stages by using a "state" design pattern. Yet this solution is not adapted to simulation of parasites because of the great number of "item" objects whose management has an important impact on simulation time and memory space.

The chosen solution models the disease as a stocks set where each stock simulates a quantity of items having the same developmental stage. This conceptual view is similar to the "flyweight" design pattern [37]. In this approach (Fig. 14), each object corresponds with a particular developmental stage, and the relationships between objects model the migration of a given number of items from one stage to another. Items migration can be seen as an assembly line where, once operated by a stage, a certain number of items migrate to the next stage.

Considering the development, stage behavior as a stock can be generalized to any other developmental stages or type of disease. Noting a developmental stage N, N − 1 the previous stage, and N + 1 the next, this behavior can be divided in three methods:

- Items migration from stage N − 1 to stage N, causing a decrease of N − 1 stage population and an increase of N stage population.
- Population increase by multiplication (for instance, in the case of mycelium increase).
- Population decrease due to mortality.

For each method, the number of items either added or subtracted depends on climate, phytosanitary protection, and plant sensitivity.

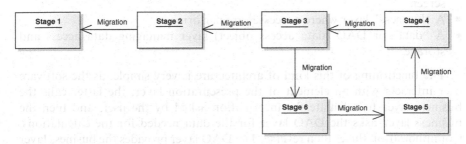

Fig. 14 Object diagram of the parasite model

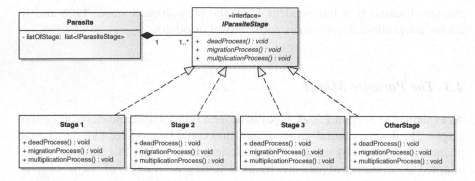

Fig. 15 Class diagram of the parasite model

The concrete classes modeling the developmental stages therefore share a common interface (Fig. 15). This interface has both the advantage of being generic to every developmental stage and flexible by allowing any developmental stages sequence.

5 The Application's Architecture

This section presents the application's software architecture and the different technologies used to ensure its cohesion and function.

5.1 The Three-Tier Architecture and the Design Pattern "Strategy"

As for many multiplatform Web developments nowadays, the three-tier architecture is chosen for the DSS software. In this kind of architecture, the generated application is divided into three separate layers:

- A "presentation" layer corresponding with what the software shows on screen
- A "business" layer where processes are performed
- A "data" or DAO (data access object) layer managing data access and storage.

The functioning of this kind of architecture is very simple, as the software user interacts with an element of the presentation layer, the latter calls the business layer to calculate the information asked by the user, and then the business layer asks the DAO layer for the data needed for the calculations. Communication fluxes then reverse. The DAO layer provides the business layer with the requested data, the business layer performs the calculations or more generally the asked service and transmits it to the presentation layer that finally displays the results. The "normal" operating mode thus induces to a round trip through the three layers. The presentation layer should never have direct access to the DAO layer and vice versa.

This organization can be easily adapted to the hardware architecture distributed on a network. Terminals on user-side present the information (corresponding with the presentation layer), whereas remote computers on server-side build this information (business layer) and manage data access (DAO layer).

The use of a software architecture based on this three-tier architecture type offers several advantages.

First of all, computer workload is dispatched. For instance, the client's PC on the user side only supports the presentation layer, whereas servers support the business and data layers.

Another advantage is the possible use of multiplatforms. Computers on both client and server sides can run under different operating systems and interpret various programming languages, which is particularly adapted to the Web-applications context.

Moreover, the organization in layers can improve coupling quality and control between these modules. As the three-tier architecture distinguishes the presentation, the business, and the data modules, the code implementing these elements is also organized in three separate software layers. The coupling control induced by the three-tier architecture helps conserve the natural independence of the different layers.

Coupling control also facilitates work management inside the developing team. The computer specialists do not have to know all the software development aspects and can therefore be specialized in one layer (interface, business, or data), without being preoccupied by the other layers' implementation details. Besides, dialogue between developers is more oriented on functionalities and less on purely technical problems concerning implementation. Communication between computer specialists is focused on the services to be provided by each layer and on the global system's coherence regarding the requirements specification. Therefore, the organization and communication within the development team are improved, according to the interaction between the different software layers of the architecture.

This three-tier architecture is very close to the design pattern "strategy" as these two conception models aim at improving coupling quality between separate software agents. The design pattern strategy uses interfaces allowing one:

- to expose the services provided by software components without showing their implementation. Therefore, a component knows that another component is able to provide a certain service because this latter is exposed by its interface, but it does not know how this component is implemented to provide this service.
- to choose, when launching the application, the implementations to use for each software component by the mean of a configuration file. The whole

application is thus totally flexible. Moreover, an application can evolve by adding a new implementation in a component without constraints regarding the other components because they are all independent.

Thus, using the design pattern strategy inside a three-tier architecture provides interesting advantages [39]. The architecture is easy to maintain because, as each software module has an independent code, interleaving is minimum. This architecture is also globally reusable with respect to its components. Each component is upgradeable and specialized with little interleaving plug-in providing services. A component can thus be used in another application needing its services. Furthermore, this architecture is flexible by means of configuration files and can evolve easily, as components can be added or updated without impact on the rest of the system.

The three-tier architecture is developed with Spring,[4] which is an open source technical framework J2EE (Java 2 Platform, Enterprise Edition) simplifying implementation [40] and providing a complete and secure solution. Spring provides a lightweight container application for the objects used by the application with the Java Bean standard and enables working on simple POJOs (Plain Old Java Objects). These objects do not need to implement interfaces linked to the technical environment. Therefore, the generated code is easy to maintain and to reuse. In addition, Spring enables three-tier architecture programming in accordance with the design pattern strategy. On each layer, interfaces are used to configure the implementation of objects called JAVA beans in a totally simplified manner. This configuration includes object connection using a programming technique called "dependency injection" that will be explained later. During the application activation, objects implementation and dependencies between objects are chosen by means of Extensible Markup Language (XML) configuration files. The application is therefore flexible and upgradeable. During the development, Spring also facilitates the integration of Hibernate, a technology managing data persistence, inside the three-tier architecture.

5.2 The Three-Tier Architecture Layers and the Technologies Used

5.2.1 The Presentation Layer and Client–Server Communication

In our application, the presentation layer is relatively particular, as it is totally supported on the client side. It is designed with the Flash[5] software that also produces interactive graphical layouts facilitating interactions with potential users.

As the technologies and platforms usable for the architecture are various, Web services are chosen to enable information exchanges and more generally

---

[4] See http://www.springframework.org/.
[5] See http://www.adobe.com/fr/products/flash/.

client–server communications. Data flow is standardized using the XML markup language with the use of Simple Object Access Protocol (SOAP) and Web Services Description Language (WSDL), for example.

Web services have the major advantage of facilitating the building of architectures distributed on heterogeneous systems. Their conception is very close to the design pattern strategy and the three-tier architecture described above. Web services operate by exposing the JAVA beans methods of the business layer to the remote presentation layer. According to its needs, the latter layer can thus ask for services, considered as a Web service. For example, when the user interacts with the interface and asks for a plant growth simulation, the presentation layer asks to use the "SimulatePlantGrowth" method exposed by a bean of the business layer. Web services are therefore very interesting because clients of a service only need to know the URL where the Web service is exposed and also be able to read the XML WSDL file describing the technical specifications. Thus, the client can exactly know the operations supported by the Web service and the parameters needed for the operation.

The Xfire[6] environment has been chosen to expose the beans' methods of the business layer as Web services. It is an open source SOAP environment for JAVA from the CodeHaus community. It can be easily integrated to the Spring environment and allows the generation of WSDL files describing beans' available services.

5.2.2 The Business Layer and the *Dependency Injection* Design Pattern

The business layer exposes business logic–related services through interfaces. These services can be the computation of plant growth or disease development for a particular situation, for example, characterized by climate, variety, and crop practices. To access the data needed for calculation, the business layer calls methods exposed by beans belonging to the DAO layer. The business layer is thus linked with the DAO layer but, as explained in a previous paragraph, the Spring environment preserves the layers' independence.

The Spring configuration files allow the instantiation of the beans used by the business layer to provide services as a singleton. According to the singleton design pattern, the objects only exist in a single copy.

The configuration code also allows dependency injection between objects, no matter their location. This corresponds with the *dependency injection* design pattern [41], also called *inversion of control*. The two concepts of dependency injection and AOP (aspect-oriented programming) constitute the Spring environment's core. Among other things, AOP allows transaction management. These notions are illustrated with the following example, corresponding with an XML Spring configuration file extract:

---

[6] See http://xfire.codehaus.org.

```
<?xml version = "1.0" encoding = "UTF-8"? >
<beans xmlns: = "http://www.springframework.org/schema/beans"
        xmlns:xsi = "http://www.w3.org/2001/XMLSchema-instance"
        xsi:schemaLocation = "
http://www.springframework.org/schema/beans
http://www.springframework.org/schema/beans/spring-beans-2.0.xsd" " >

        <bean id = "simulation"
                class = "com.services.simulation.impl.SimulationMP1ServiceImpl" >
<property name = "plantDAO" ref = "plantDAO" / >
                <property name = "searchService" ref = "searchService" / >
        </bean >

        <bean id = "searchService"
                class = "com.services.search.impl.SearchServiceImpl" >
                <property name = "plantDAO" ref = " plantDAO " / >
        </bean >

        [...]
```

The above code shows a bean in charge of the service called "simulation" (<bean id = "simulation"), logically located in the business layer. This bean has a reference to a bean located in the DAO layer (ref = "plantDAO") and another one referred to a bean located in its own layer (ref = "searchService"). The class implementing the bean "simulation" is class = "[...] SimulationMP1Service Impl">. This class has attributes referencing (meaning allowing access to) "plantDAO" and "searchService" also instantiated by Spring. To provide the object simulation with these two references, a dependency injection is needed. These dependencies are simply defined as properties (<property name = "plant-DAO") of the simulation bean "simulation." To have access to these properties, meaning injecting references to the objects "plantDAO" and "searchService," Spring uses getters/setters of the "simulation" object. Spring is also able to generate all the declared dependencies between the application objects. In concrete terms, the dependencies do not have to be hard coded any more and are externalized in easy-to-modify XML files. Therefore, to modify interfaces' implementation and dependencies between objects, it is not necessary to modify and recompile the classes' code.

5.2.3 The DAO Layer and Hibernate

Like the business layer, the DAO layer contains singleton beans instantiated by Spring. In this layer, an interface is defined for each object, such as PlantDAO, LeafDAO, ObservationDAO, and for each interface; N implementations are available such as PlantDAOImplA, PlantDAOImplB, LeafDAOImplA, Leaf-DAOImplB, LeafDAOImplC, and so forth.

In the architecture, the DAO objects aim at accessing a relational database management system (RDBMS), but they could access any type of system providing or storing data. This layer is decoupled from the technology used for data persistence, so that architecture is more generic and upgradeable. Persistence of JAVA/J2EE objects to a relational database is in charge of the Hibernate[7] environment. Indeed, this environment facilitates separation, or, more precisely, improves its quality [42, 43]. Spring is planned to integrate Hibernate and offers native classes that assist this integration. XML files called *mapping files* are used to establish a correspondence between JAVA beans and database. The link between objects and their backup in the database is called *object relational mapping* (ORM). A class, for example "Plant" in the object sense, corresponds with a table of the database in the relational sense. Each attribute of this class such as size or age corresponds with a column of this table. Each object is individually identified with an attribute "id" called the identifier, also recorded in a column of the table. The use of Spring and Hibernate allows the achievement of operations such as CRUD (Create, Read, Update, Delete) in a very simple way inside the database, although more complex queries can also be built. Regarding optimization, it is important to write queries that will exactly answer to the business layer objects' expectations that are usually accurate. These queries can be decoupled from the DAO classes' code. It is possible to store them in mapping files, which favors maintenance operations.

6 Conclusions

This chapter reviews the different design phases of a DSS intended for plant protection. The final product presented here is a Web software aiming at a large distribution among farmers. The creation of this software was shown to be complex and involved a multidisciplinary team. The first work consisted of identifying the different tasks in accordance with the requirement specifications. Four main tasks appeared: the design of the scientific model, the processable model, the interface, and the software architecture, respectively. As the specialists involved in this project belong to different professional domains, particular attention has been paid to the aspect of communication. The latter has been made simpler by adapted modeling tools such as UML. Concerning the scientific model, a major characteristic is the use of mechanistic models to simulate the biological system. This system is composed of modules, corresponding with the major actors of the system, evolving in interaction with each other. The behavior of each module, such as the plant, is precisely described in relation with its environment by means of equations. This aspect constitutes the basis of the

---

[7] See http://www.hibernate.org/.

tool relevance because the recommendations given by the model are scientifically based and reflect the system's behavior in the specified environmental conditions. Concerning the processable model and the software architecture, particular attention was paid to flexibility and evolutionary capacity so that applications can be easily upgraded or adapted to other systems.

This work is the beginning of an important agronomic modeling project. Even if the example taken concerns only the phytosanitary plant protection, the creation process is generic and can be applied to other DSSs. Moreover, the acquired experience will allow a quick design and set up of future DSSs. The potential use of this DSS is large. Indeed, new models concerning other pests and other phytosanitary products can be easily introduced in the current tool as new modules or new functionalities, for example. The integration of other disease models concerning the same crop opens new work prospects for the DSS improvement. This will indeed enable a more accurate assessment of crop damage and subsequently of phytonsanitary protection profitability. Moreover, it will be possible to take into account the competition between diseases. Furthermore, these tools also provide a pedagogical purpose aiming at good farming practices. For instance, concerning phytosanitary protection, the farmer can adapt his application program according to simple risk indicators that are easily understandable but are still scientifically based.

In the current context, environmental protection is an absolute priority, and concerning agriculture, it is essential to aim at more environment-friendly farming practices in the short-term future. This change is urgent because agriculture is going to be at the center of major stakes such as global human population growth and massive biofuel use, requiring huge production. All these conjunctural factors show that a rapid change in farming practices is necessary. The software presented in this chapter is designed to help this evolution. Indeed, the developed DSS allows a better diagnosis and short-term predictions of the current situation, and, furthermore, it can forecast the consequences of different scenarios tested by means of simulations. This system can also easily integrate new data or constraints to propose recommendations, always with a scientific justification. An increased use of these kinds of tools can allow a significant reduction of chemical products in agriculture without detriment to yield or production quality.

References

1. Rouzet J., Pueyo C., "Modèles de prévision et conseil phytosanitaire. Bilan des modèles en France, aperçu américain et perspectives";, Phytoma, 591: 32–36, 2006.
2. Decoin M., "OAD vus par la SdQPV, du côté des modèles", Phytoma, 603: 24–25, 2007.
3. Munier-Jolain N.M., Savois V., Kubiak P., Maillet-Mezeray, J. Jouy L., Quere L., "Decid'Herb : un logiciel d'aide au choix d'une méthode de lutte contre les mauvaises herbes pour une agriculture respectueuse de l'environnement", Proceeding AFPP – 19ième conférence du Columa – journées internationales sur la lutte contre les mauvaises herbes, Dijon, Décembre 2004.
4. http://www.invivo-group.com.

5. Booch G., Rumbaugh J., Jacobson I., "The Unified Modeling Language User Guide", Addison-Wesley, 1999.
6. Jones C.A., Kiniry J.R., ";CERES-maize, a Simulation Model of Maize Growth and Development", A&M University Press, 1986.
7. Brisson N., Mary B., Ripoche D., Jeuffroy M.H., Ruget F., Nicoullaud B., Gate P., De-vienne-Barret F., Antonioletti R., Durr C., Richard G., Beaudoin N., Recous S., Tayot X., Plenet D., Cellier P., Machet J.M., Meynard J.M., Delecolle R., "STICS: a Generic Model for the Simulation of Crops and Their Water and Nitrogen Balances. I. Theory and Parame-terization Applied to Wheat and Corn", Agronomie, 18: 311–346, 1998.
8. Luquet D., Dingkuhn M., Kim H.K., Tambour L., Clément-Vidal A., " EcoMeristem, a Model of Morphogenesis and Competition Among Sinks in Rice : 1. Concept, Validation and Sensitivity Analysis ", Functional Plant Biology, 33(4): 309–323, 2006.
9. Louarn G., "Analyse et modélisation de l'organogenèse et de l'architecture du Rameau de la vigne (Vitis vinifera L.)", thesis PhD, école nationale supérieure agronomique de Montpellier, 2005.
10. Jallas E., Martin P., Sequeira R., Turner S., Crétenet M., Gérardeaux E., "Virtual COTONS®, the Firstborn of the Next Generation of Simulation Model", NLAI 1834, pp. 235–245, Springer, 2000.
11. Costes E., Guedon T., "Modelling the Sylleptic Branching on One-Year-Old Trunks of Apple Cultivars", Journal of the American Society of Horticultural Science, 122: 53–62, 1997.
12. Seleznyova A.N., Thorp T.G., Barnett A.M., Costes E., "Quantitative Analysis of Shoot Development and Branching Patterns in Actinidia", Annals of Botany, 89: 471–482, 2002.
13. Buhlmann P., Wyner A.J., "Variable Length Markov Chains", The Annals of Statistics, 27: 480–513, 1999.
14. Nozeran R., "Réflexions sur les enchaînements de fonctionnement au cours du cycle des végétaux supérieurs", Bull. Soc. Bot. Fr., 125: 263–280, 1978.
15. Vanderplank J.E., "Plant Diseases: Epidemics and Control", Academic Press, 1963.
16. Fleming R.A., "Development of a simple mechanistic model of cereal rust progress." Phytopathology, 73(2): 308–312, 1983.
17. Keeling M.J., Rohani P., "Estimating spatial coupling in epidemiological systems: a mechanistic approach", Ecology Letters, 5(1): 20–29, 2002.
18. Passioura J.B., "Simulation models: snake oil, education, or engineering? " Agronomy Journal, 88: 690–694, 1996.
19. http://www.hps-inc.com/edu/stella/stella.htm.
20. Makowski D., Hillier J., Wallach D., Andrieu B., Jeuffroy M.H., "Parameter estimation for crop models" in: "Working with Dynamic Crop Models – Evaluation, Analysis, Parameterization, and Applications", pp. 101–140, Wallach D., Makowski D., and Jones J.W. (eds), Elsevier, 2006.
21. Saltelli A., Chan K., Scott E.M., "Sensitivity Analysis". Wiley, 2000.
22. Saltelli A., Tarantola S., Campolongo F., Ratto F., "Sensitivity Analysis in Practice", Wiley, 2004.
23. Wallach D., Makowski D., and Jones J.W, "Working with Dynamic Crop Models – Evaluation, Analysis, Parameterization, and Applications", Wallach D., Makowski D., and Jones J.W. (eds), Elsevier, 2006.
24. Monod H., Naud C., Makowski D., "Uncertainty and sensitivity analysis for crop models", in: "Working with Dynamic Crop Models – Evaluation, Analysis, Parameter-ization, and Applications", pp. 55–96, Wallach D., Makowski D., and Jones J.W. (eds), Elsevier, 2006.
25. Wallach D., "Evaluating crop models" in: "Working with Dynamic Crop Models – Evaluation, Analysis, Parameterization, and Applications", pp. 11–50, Wallach D., Makowski D., and Jones J.W. (eds), Elsevier, 2006.
26. Delannoy C., "Programmer en langage C", Eyrolles, 1997.

102   L. Tambour et al.

27. Delannoy C., "Programmer en langage C++", Eyrolles, 1998.
28. Niemeyer P., Knudsen J., "Introduction à JAVA", O'Reilly, 2002.
29. Martin M., "Programmeur Delphi 2005", CampusPress, 2005.
30. http://www.mathworks.com.
31. http://www.r-project.org.
32. http://www.scilab.org.
33. http://www.wolfram.com.
34. http://www.maplesoft.com.
35. http://www.modelmakertools.com.
36. Cier P., Dorin R., "Excel 2000 en pratique", Dunod, 1999.
37. Gamma E., Helm R., Johnson R., Vlissides J., "Design Patterns Elements of Reusable Object-Oriented Software", Addison-Wesley, 1995.
38. Porter C.H., Braga R., Jones J.W., "An Approach for Modular Crop Model Development", Agricultural and Biological Engineering Department, University of Florida, Gainesville, Florida, available at http://www.icasa.net/modular/downloads.html, 1999.
39. Papajorgji P., Pardalos P., "Software Engineering Techniques Applied to Agricultural Systems, an Object-Oriented and UML Approach", Springer, 2005.
40. Johnson R., Hoeller J. "Expert One-on-One J2EE Development without EJB ", Wiley Publishing, 2004.
41. Fowler M.,"Inversion of Control Containers and the Dependency Injection pattern", http://martinfowler.com/articles/injection.html, 2004.
42. King G., Bauer C., "Hibernate in action", Manning Publications, 2004.
43. Salvatori O., Patricio A, "Hibernate 3.0", Eyrolles, 2005.

How2QnD: Design and Construction of a Game-Style, Environmental Simulation Engine and Interface Using UML, XML, and Java

Gregory A. Kiker and Rohit Thummalapalli

Abstract Within wicked environmental challenges, problems that exist in the nexus of environmental science and environmental values, neatly and elegantly optimized solutions are difficult to find and rarely accepted by stakeholders. Different role players must explore the challenge adaptively and through viewpoints to contribute to their understanding of the situation and to learn about the dynamics and values of other relevant stakeholders. The *Questions and Decisions* (QnD) system (Kiker, G.A., et al., *Springer Science*, 2006, *11*, 151–186) was created to provide an effective and efficient tool to integrate ecosystem, management, economics, and sociopolitical factors into a user-friendly game/model framework. QnD is written in object-oriented Java and can be deployed in stand-alone or Web-based (browser-accessed) modes. The QnD model links spatial components within geographic information system (GIS) files to the abiotic (climatic) and biotic interactions that exist in an environmental system. QnD can be used in a rigorous modeling role to mimic system elements obtained from scientific data or it can be used to create a "cartoon" style depiction of the system to promote greater learning and discussion from decision participants. Elephant and vegetation dynamics in Africa provide an excellent example of a wicked environmental challenge as conservation objectives and societal values (both local and international) often have conflicting goals concerning appropriate elephant densities and population control options in protected areas. In attempting to capture many dynamic aspects of elephant–vegetation relationships, previous models depicting the savanna ecosystem of the Kruger National Park (KNP), South Africa, can become quite complex and demanding in terms of detailed parameter inputs. Therefore, the purpose of this modeling project was to create a simplified, management-focused, visual simulation of the KNP in order to chart future elephant, tree, and grass scenarios. QnD:EleSim has been designed to spatially simulate elephant–vegetation dynamics in 195 areas at 10-km resolution at a monthly time-step. As the effects of elephant populations on the tree–grass equilibrium of the savanna are documented, future management decisions can be advised after analysis of potential scenarios.

G.A. Kiker (✉)
Department of Agricultural and Biological Engineering, University of Florida,
Gainesville, FL, USA

P.J. Papajorgji, P.M. Pardalos (eds.), *Advances in Modeling Agricultural Systems*,
DOI 10.1007/978-0-387-75181-8_6, © Springer Science+Buisness Media, LLC 2009

1 Introduction

As a matter of recent history, environmental decision-makers are increasingly facing a set of problems that appear to have no easy solution. These "wicked" problems [28] exist "at the intersection of science and values" ([36] p. 2) and defy neat, optimized, numerical solutions. These wicked problems require an integration of scientific information, uncertainty estimation, and social/cultural valuation for environmental decision-making.

In reaction to such intellectually and emotionally complex challenges, decision makers and scientists have increasingly turned to the use of computationally complex systems models that attempt to incorporate multiple system dynamics at very fine spatial and temporal resolutions to match the complexity challenge with computational "shock and awe." However, simple, pragmatic models that require fewer parameters than complex models can be surprisingly useful in ecological studies [14, 31]. This simple-model approach was useful in highlighting selected management issues within the river ecosystems [33], where a suite of simple models at multiple scales of time and space were used to assist scientists and managers.

The concept of managing environmental systems as being a game involving different role players and options has revealed important general patterns of system behavior [6]. The Non-Point simulation program described a simple model of ecosystem management from the perspective of selected role players. This model served to show the interaction between fast and slow variables (multiple timescales) and illustrated the point that continual learning was fundamental for adaptive and resilient systems.

Often, management decisions must be made in the absence of adequate data, which is where modeling becomes a useful management tool. Thus a model's development may be driven by the objectives of the management program rather than by the available data [31]. Scenario modeling is a useful tool for envisioning future situations in an unknown future [31]. Models help to expose gaps in data and understanding and help to screen policy options, especially under conditions where time is limited and systems are sensitive [33].

1.1 Conceptual Background: Learning Through Games

Increasingly, the effort of building a model and its associated execution for exploring system/management dynamics are seen within an *adaptive learning* context. Aldritch [1] describes three fundamental and intersecting elements of successful educational simulation: Simulation elements, game elements, and pedagogy. Simulation elements systematically represent reality into computer science–based structures (i.e., objects, stochastic elements, and temporal/spatial databases) and allow concepts to be judged, altered, and reformed in an iterative fashion to provide discovery and experimentation within curriculum.

Gaming elements provide the recognizable and entertaining aspects to content although their interaction with simulation components must be systematically managed to support the sustainability of educational content. Aldritch [1] points out the challenges of scale representation of gaming elements, which is also mirrored in the environmental simulation research [13, 26, 34, 35]. Pedagogical elements provide the practical educational management of the simulation and gaming content into specific and monitored outcomes. The interaction of these three elements is adaptive and thus is managed both at the curriculum design (strategic) level as well as within the classroom (tactical) level.

1.2 QnD: A Game-Style Simulation for Adaptive Learning and Decision Making

The *Questions and Decisions* (QnD) model system is a problem-exploration tool that increases understanding of potential ecosystem behaviors and management options of a particular socioecological system [15, 16, 17, 18]. QnD is written in object-oriented Java and can be deployed in stand-alone or Web-based (browser-accessed) modes [27]. The QnD model links spatial components within Geographic Information System (GIS) files to the abiotic (climatic) and biotic interactions that exist in an environmental system. The QnD system is divided into two primary elements: the *SimulationEngine* and the *GameView* as shown in Fig. 1. Through the user-friendly graphical interface (GameView), stakeholders can "play" their system by manipulating institutional and ecological components of interest. Results generated by the SimulationEngine element differ from the combinations of various environmental and economic drivers and the player's response to them via management options implemented in different spatial areas and over different time periods. As various scenarios are

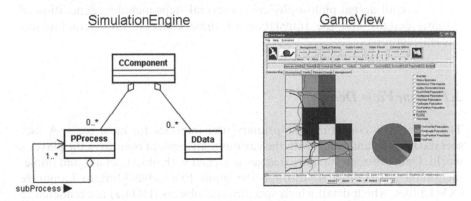

Fig. 1 QnD's two primary elements: the SimulationEngine and GameView (adapted from Ref. 17)

played, the interactions between institutional and ecological parameters are exposed, and future possibilities of the system can be envisioned.

The QnD system not only helps stakeholders deepen their understanding of a particular system's components and dynamics but also acts as a device to bring cohesion to a stakeholder community. Throughout its iterative design process, a QnD system engages stakeholders to accomplish multiple tasks: facilitating initial agreement on key forces and themes, broadening understandings of others' interests, developing scenarios and analyzing various policy options. QnD was created as a technical tool that is complemented by and works in tandem with an iterative exploration process [17]. The system's design process was intended to be compatible with several social science methodologies, such as Soft Systems Modeling (SSM), scenario planning [32], cognitive mapping or mental modeling [24].

As the fundamental philosophy and background of the QnD system was reviewed and documented in [17] and [18], the primary objective of this chapter is to provide a technical design overview in Unified Modeling Language (UML) and illustrate these designs with an application toward ongoing modeling efforts to simulate elephant population and savanna vegetation dynamics in the Kruger National Park, South Africa. Thus, the chapter is divided into two sections: (1) a technical design overview that covers the object design of the GameView and SimulationEngine elements as well as the Use Case designs and (2) a specific application of the design elements using the elephant–vegetation version of the model.

2 QnD Design Overview: Designing from Ideas to a Playable Game

This section provides an overview of the object design using UML. The entire QnD system is coded in the Java language and is a combination of original code and open source libraries/application programming interfaces (APIs) [27].

The overall design philosophy covers several steps including generation of systems designs and the translation of these concepts into actual object implementations.

2.1 GameView Design

The GameView constitutes the primary user interface for most users. A user sees data results and reacts with the various management options in the player's world. QnD utilizes a standardized game format with object details, and implementation are configured through the input Extensible Markup Language (XML) files, which detail which specific data objects (DData) are rendered in several graphic forms. DData objects that are spatially explicit can be rendered into both collective maps (selected by radio buttons) or line charts.

Fig. 2 Basic layout of
GameView elements

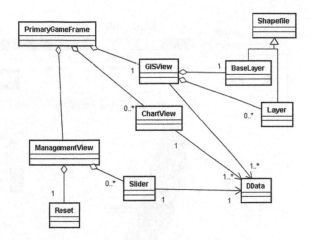

Figure 2 highlights the basic object design. GameView is made of a map viewer (GeoToolsLite API: http://geotools.codehaus.org/), scrolling time series charts (Chart2D API: http://chart2d.sourceforge.net/index.php), warning lights, and management selection widgets. By presenting the outputs in a selectable form, the QnD system allows users to choose how they want to see their output, including the following output options as listed below:

- GIS maps that are updated on each simulated time step.
- Mouse-activated charts and text for individual spatial areas (pie charts and text line descriptions).
- Warning lights that change at user-selected critical levels.
- Scrolling time-series charts (listed on user-defined, tabbed pages).
- User-defined, text output files in comma-separated format.

Figure 3 illustrates the basic parts of the QnD GameView interface for the QnD:EleSim example (detailed in later sections). A variety of maps can be viewed by selecting the desired radio button. Time series charts are accessed by selecting the tabbed panes above the map. As the user simulates each 1-month or 6-month time step, each of the various graphic objects is refreshed with new data values. When the reset (white flag) button is selected, all values return to their original settings and the game is ready for another session.

2.2 Simulation Engine Design

QnD's Simulation Engine is created through deployment of three primary elements (illustrated in Fig. 4): *component, process,* and *data* objects [17, 18]. For clarification, within QnD designs and labeling, a **"C"** prefixes Components, a **"P"** prefixes Processes, and a **"D"** prefixes Data objects. *CComponent* objects

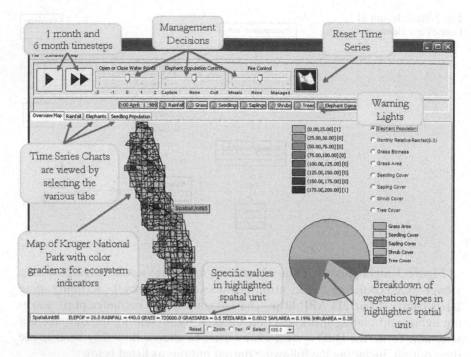

Fig. 3 An instantiation of the GameView element from QnD:EleSim

form the basic items of interest within a simulation. *PProcess* objects provide the action and changes from one state to another. *DData* objects provide the necessary description of various attributes. These same objects have analogies in nouns, verbs, and adjective/adverbs. Through various systems-envisioning methodologies such as Soft Systems Methodology [7] or Mental Modeling [24], configurations of interacting system elements can be envisioned and rendered in

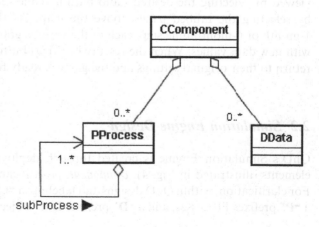

Fig. 4 Primary elements of the QnD simulation engine are components, processes, and data

rich pictures and then into UML designs. The construction of various design diagrams in UML can be converted into XML-based input files for efficient instantiation as Java objects.

Spatially explicit areas and non–spatially explicit areas are specifically represented through two primary CComponent objects, CSpatialArea and CHabitat, as shown in Fig. 5. A CSpatialUnit is the basic spatial entity of the QnD system. CSpatialUnits can be linked to one another and have a specific location. A CSpatialUnit can have either zero or any number of CSpatialUnits connected to it. One or more CHabitat objects exist within a CSpatialUnit and are not spatially defined, except via the relationship with the "homeSpatialU-nit." A CHabitat can hold any number of local instantiations of CComponent objects (CLocalComponents). These CLocalComponents have relationships with both "home" CHabitat and CSpatialUnit. With this basic QnD object architecture, both simple and complex designs are possible with both spatial elements and nonspatial elements.

PProcess objects provide all state changes and action within QnD. PProcess objects use DData objects as inputs, provide a calculation or series of calculations, and then write the resulting products into output DData objects, as illustrated in Fig. 6. PProcesses can be used individually as described in Fig. 6 or can be designed with constituent subprocesses within them to create a series of processes for more complex interactions. Table 1 shows the different types of processes that can be bound together in series within QnD.

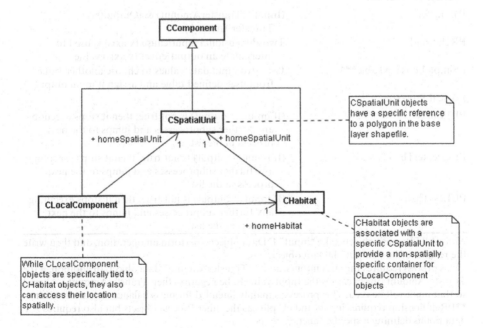

Fig. 5 Component objects are arranged in spatially explicit and nonexplicit configurations

Fig. 6 Each PProcess object
has specific DData input
and output relationships

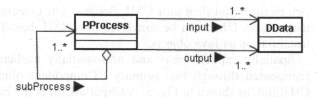

Table 1 PProcess objects for the QnD model

Process Type	Definition/Purpose
PProcesses for Calculation	
PAddValue	$Input1 + Input2 + Input3\ldots + Input_n = Output(s)$
PSubtractValue	$Input1 - Input2 - \ldots\ldots - Input_n = Output(s)$
PMultiplyValue	$Input1 \times Input2 \times Input3\ldots Input_n = Output(s)$
PDivideValue	$Input1/Input2/\ldots\ldots Input_n = Output(s)$
PExponentialValue	$Output = e^{(input)}$
PSetValue	$Output = Input$
PMeanValue	$Output = (Input1 + Input2 + Input3\ldots + Input_n)/n$
PTemporalMeanValue	$Output = \Sigma(Input1_t + Input1_{t+1} + Input1_{t+2}\ldots + Input1_{t+n})/t$
PTemporalRunningAverageValue	$Output = (Input1_t + Input1_{t-1} + Input1_{t-2}\ldots + Input1_{t-n})/Input2$
PCalculateCurrentValue	Within a given subprocess list, this object calculates all the subprocesses above it to get an updated and current DData object value.
Specialty PProcesses	
PTransfer*	(Input – TransferAmount) & (Output(s) + TransferAmount)
PRelationship**	Two-dimensional input/cause (x axis) is used to interpolate an output/effect (y axis) value.
PSimpleLookUpTable***	Uses two input data values to choose another value from user-defined table and assign it to an output.
Logical Processes	
PIfEquals	(If input = output) is not true, then it stops executing any further subprocesses and jumps to the next process in the list.
PIfGreaterThan	(If input > output) is not true, then it stops executing any further subprocesses and jumps to the next process in the list.
PIfLessThan	(If input < output) is not true, then it stops executing any further subprocesses and jumps to the next process in the list.

Note: All PProcess objects take "input" DData objects, perform an operation, and then write the results to the "output" DData object.

*This process requires Inputs, Outputs, and a "TransferAmount" data object. In addition, if the TransferAmount object causes the Input value to be negative, then TransferAmount is altered so that input will be zero. This process is mainly intended for mass balance style transactions.

**PRelationship requires inputs and outputs as the other Process objects but also requires XY data points defining a specific function shape.

***PSimple Look Up Table uses interpolation to determine precise output from set of discrete data points.

DData objects store all the relevant information for a specific QnD simulation. All DData objects are created from the input XML, GIS data files, or time-series files and represent a composite variable storing a set of double values. Each DData has several attribute variables that allow for various calculations. All available attributes are not always used for each DData as some data objects may use other attribute features whereas others do not. For example, a DData object that is linked with a time-series file (through its DriverLink attribute) may constantly change current values over time, whereas another may represent a static variable in the simulation and may not use any other attributes besides a single parameter value.

In addition to the primary SimulationEngine-related objects, several packages exist for various housekeeping and organization functionality. QnD *Control* objects (Fig. 7) are used mostly in the background and thus do not have the "C, P, D" typology of the SimulationEngine objects. The GameDriver object acts as a main simulator object to coordinate both the GameView and SimulationEngine. The PrimaryGameFrame object provides the main GameView frame. Both of these control objects utilize various factory-style objects (QnDModelCreator and qndMngReader) to read XML input and time-series files and to create the various constituent objects.

2.3 QnD Use-Case Designs: Three Actors, Many Roles

The basic actors within QnD simulations and software development fall into three general roles; *Players*, *Developers*, and *Coders*. A primary operating philosophy and basic interactions among the different actors have been

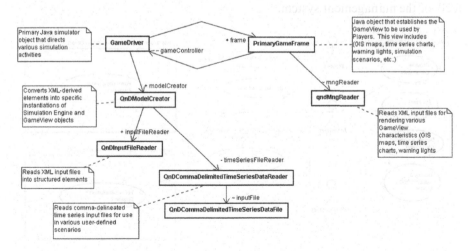

Fig. 7 Various control objects that implement the Simulation Engine and GameView elements

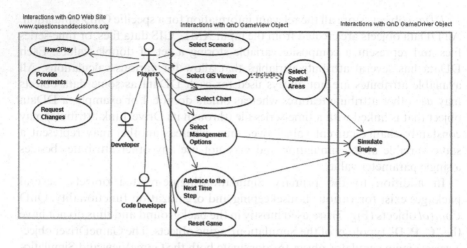

Fig. 8 Use-case diagram of QnD players

documented earlier in Ref. 17. Fig. 8–10 show the use-case diagrams for each of the primary QnD actors.

Players (Fig. 8) interact mostly with the GameView while playing and exploring the system, potential management responses, and trade-offs. Players can be stakeholders but can also be anyone who has an interest in the game. They see the simulated world as a larger, integrated ecosystem and have broad, varying interests. Although players may have some interest in technical simulation details, they mostly interact with the GameView elements via the map, charts, and management options. Players provide an important reality check to the overall design and function of the QnD system. Thus, they can provide feedback to other actors concerning the functionality as well as the "look and feel" of the management system.

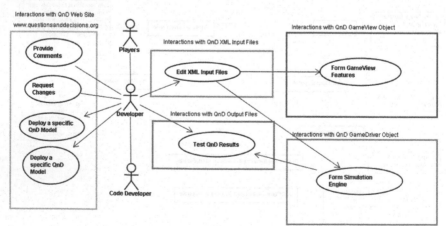

Fig. 9 Use-case diagram of QnD developers

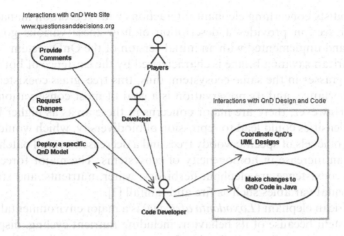

Fig. 10 Use-case diagram of QnD coders

Developers (Fig. 9) design and implement the game view and simulation engine objects using the XML input files. Although some developers might share a player role as well, their primary role is to translate the broader ideas of the players into functional object designs that are represented in the input files. Another fundamental role of developers is to provide any formalized calibration or validation of the simulation engine/game view. This confidence-building aspect is an important function in developing trust and interest into any simulation results that are seen by the overall group as critical. Developers decide how the game should be deployed to players, either through a Web site as a self-contained program or as stand-alone program on local computers for specific output file analysis.

Coders (Fig. 10) interact mostly with the Java source code and concern themselves with the overall applicability and expansion of the GameView and SimulationEngine parts as well as the functional deployment of the QnD models. Coders have control and responsibility of the overall design and evolution of the QnD system for all groups of players and developers. Coders may take specialized suggestions from players and developers and implement them at a broader, more abstract level within the source code to take advantage of new developments in the Java language, computer science concepts, or Internet technologies.

3 Questions and Decisions About Elephant–Vegetation Dynamics in the Kruger National Park, South Africa

The two following sections describe how overall the QnD concepts are translated into functional objects for simulation in a game-style format. The first section describes interactions with Kruger National Park, South Africa, and other scientists concerning elephant interactions with vegetation management. The latter section provides a description of how these various goals were designed and implemented with an initial version of the QnD system.

The African savanna biome is characterized by the existence of both woody trees and grasses in the same ecosystem; thus, this tree–grass coexistence is of much importance, and its preservation is a goal of most conservation authorities [3]. However, there are major concerns in these areas as higher levels of elephant densities could lead to repression of biodiversity, which would lead to reduction of levels of specific woody trees and a decreased overall patch density signaling an increase in homogeneity of ecosystems. The major forces acting upon this coexistence are elephant herbivory, water, nutrients, and stochastic environmental variables such as fire and rainfall [3].

The African elephant (*Loxodonta africana*) is a major environmental force in this ecosystem because of its behavior, including nutrient cycling, dispersal of plant offspring in seeds, and allowing for new plants to germinate in adequate space [22]. An important behavior that characterizes these populations is that during the summer (dry season), the elephant prefers to graze, or feed on grasses, whereas this behavior seemingly switches during the wet season, when the elephant turns to browsing leaves off of taller, woody trees [3, 5]. This browsing during the wet season has had a significant effect on woody trees, in that the habitat modification resulting from browsing has been ascribed to loss of canopy trees and an eventual transition to bushland dominated by shorter grasses and shrubs [9]. Thus, high elephant densities could irreversibly alter the structure and physiology of African savanna ecosystems [13]. Significant impacts from elephants upon tree–grass coexistence have been observed in the areas of South Africa, particularly in the northeast portion of the Republic of South Africa, within the Kruger National Park (KNP) [12].

3.1 KNP Elephant Model Development Strategies

The Kruger National Park Elephant Modeling Group (KNP-EMG) is a collection of international modeling researchers working to develop a collection of models to address issues concerning elephant and vegetation dynamics in the park. The KNP-EMG has provided the overall objectives for various modeling studies by issuing various objectives and goals for teams of cooperative elephant modelers including the following texts:

For the area or the Kruger National Park (~20,000 km²), at a spatial resolution of 10 × 10 km, can the model reconstruct the numbers and distribution of elephants over the period 1900–2000? Can the same model predict the density distribution for 2000–2007 (for which data will be withheld for testing purposes).

For the basaltic landscape of Central KNP, what is the shape of the long-term (50 year) trade-off curve between elephant biomass density and tree cover percentage, for tall trees (>6 m), short trees (2–6 m), and shrubs (0–2 m) of the following species: *Sclerocarya birrea, Acacia nigrescens,* and *Combretum imberbe*?

Thus, the focus of the KNP-EMG is to create a grid-based model of the KNP and also formulate a scenario-based simulation to track and predict the progression of these ecological systems. The study groups were directed to develop and test a new elephant–ecosystem model that incorporate important environmental variables and to effectively manage the future populations of the Kruger National Park.

Further informal conversations with various South African scientists and wildlife managers at the KNP Science Networking Meeting 2007 (Skukuza, South Africa) compiled the following tactical ideas for QnD modeling:

1. Build the initial simulation engine from the "Baxter" model [2, 3, 4].
2. Simplify the Baxter model where practical and possible.
3. Be clear and transparent as to what assumptions are being made or changed from the original Baxter model design and execution.

Thus, the purpose of this modeling group is to stimulate and support the development and testing of predictive models of elephant–ecosystem interactions in the KNP. The aim is to help improve the management and monitoring of savanna biodiversity levels, especially that of tree–grass coexistence, in reference to stochastic variables like fire and rainfall and, most importantly, elephant grazing and browsing. KNP environmental data will be incorporated into the KNP-EMG effort including (*inter alia*) rainfall, elephant densities, woody tree/ grass populations, rainfall totals, soil composition, and fire occurrence. Thus, the modeling effort will provide a visual analysis of the interrelationships between selected biotic and abiotic factors in the Kruger National Park.

At this initial model development phase, management options within the KNP are generally limited to those discussed in Ref. 17 and articulated within the KNP strategic planning [21]. The options include:

1. Fire management: via management fires or fire suppression.
2. Elephant population management: via capture/off-site removal or culling.
3. Surface water management: via opening or closing permanent water points (bore holes).

3.2 Design2Game: Translating Systems Designs and Previous Modeling Efforts into QnD SimulationEngine and GameView Implementations

The QnD model that has been developed for the initial version (QnD:EleSim) is a grid-based design based on simulation models developed by Baxter and Getz [2, 3, 4]. Collectively, these articles are further referenced simply as the *Baxter model*. According to the QnD modeling group objectives, the Baxter model was simplified in terms of algorithms to provide an initial simulation of elephant– vegetation dynamics over the entire 195,000 km^2 KNP area. This section provides an overview of the initial QnD:EleSim model design with occasional reference to the Technical Appendix at the end of this chapter. Thus, model elements can be viewed in conceptual and mathematical form in the main chapter with selected object representations available for further study in the Technical Appendix.

3.2.1 QnDEleSim SimulationEngine: Setting Spatial and Temporal Execution

The original Baxter model simulated $1 km^2$ of an African savanna ecosystem by dividing the area into 100 1-hectare grids and analyzed the effect of elephant herbivory on the tree and fire dynamics against one generic savanna tree species. The QnD:EleSim initial version will simulate the entire KNP with 195 10-km grid cells (CSpatialUnits).

Temporally, the original Baxter model simulated at 6-month intervals that correspond with wet $(t, t + 2...$etc.$)$ seasons and dry $(t + 1, t + 3...$etc.$)$ seasons. To allow for greater control of management activities at the beginning and ending of the wet/dry seasons, the simulation time step in QnD:EleSim was set to 1 month. All seasonal algorithms were subsequently disaggregated to monthly processes with cognizance to the general wet/dry seasonality.

3.2.2 QnDEleSim SimulationEngine: Setting Input Drivers and Scenarios

In terms of environmental factors, the Baxter model assumes that the moisture and nutrient availability are uniform throughout the entire grid system of $1 km^2$. However, in this study, GIS and temporal data sets from the Kruger National Park will be used to incorporate precipitation and the east/west divide between granite-based and basalt-based soils as well as the north/south rainfall gradient [21]. Table 2 provides a summary of spatial and temporal input data sets that are used by QnD for simulation.

3.2.3 QnDEleSim SimulationEngine: Setting CLocalComponents, DData, and PProcesses

This section describes the implementation of simplified Baxter model concepts into the XML-based, object-oriented structure of the QnD model. As described above, both temporal and spatial scales have been modified from the original concepts to allow for more active management interventions over time and space. The following paragraphs highlight selected elements for further detailed discussion. A more expansive and detailed description of object designs and calculation algorithms is currently being constructed for more extensive peer review, testing, and iteration.

Climatic Inputs

Two options are used to the control rainfall inputs throughout the QnD:EleSim program. One option can be used by configuring internal stochastic generator objects using Java objects developed by Dorai-Raj [11] and applying concepts

Table 2 Summary of input information

Input	Format
Rainfall (mm/month)	Time Series Files/Maps/Stochastic Relationships
Fire (intensity, % burnt)	Time Series Files/Maps/Stochastic Relationships
Soils (general class)	Spatial Input Map
Potential Grass Biomass (kg/ha)	Spatial Input Map
Initial Grass Biomass (kg/ha)	Spatial Input Map
Initial Grass Area (fraction)	Spatial Input Map
Initial Woody Seedling Population (no./ha)	Spatial Input Map
Initial Woody Sapling Population (no./ha)	Spatial Input Map
Initial Woody Shrub Population (no./ha)	Spatial Input Map
Initial Woody Tree Population (no./ha)	Spatial Input Map
Initial Woody Seedling Area	Spatial Input Map
Initial Woody Sapling Area	Spatial Input Map
Initial Woody Shrub Area	Spatial Input Map
Initial Woody Tree Area	Spatial Input Map
Initial Elephant Population (no./grid)	Spatial Input Map

Note: All maps are 10-km grid resolution, and all time series files include monthly data. All area maps are fractional coverage. ha, hectare.

by Matsumoto and Nishimura [23]. These objects can generate stochastic values according to various user-input statistical properties. The second option can utilize historical KNP climate data sets to directly input desired inputs. Given the modular structure of various CScenario objects, users can combine both time-series data sets and stochastic generator objects to construct rich and varied scenarios for further simulation.

In the Baxter model, 6-month rainfall values were normalized with the long-term mean value to create a relative rainfall for use in spatial simulations. Given the revised monthly time-step in QnD:EleSim, the monthly relative rainfall (RRfl) was calculated for each spatial area with the following equation:

$$RRfl(t) = \frac{Rfl(t)}{Rfl_{mean}}$$

where $Rfl(t)$ = rainfall (mm/month) and Rfl_{mean} = 30-year monthly average (mm/month).

Simulating Woody Plant Layer Growth

The original Baxter model incorporated one generic savanna tree species into nine different size classes ($i = 1...9$) ranging from Seedlings (<15 cm height) to

Fig. 11 Conceptual diagram of vegetation-related CLocalComponent objects present in each of the 195 CSpatialUnit/CHabitat objects

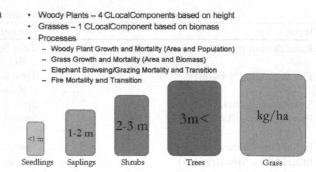

- Woody Plants – 4 CLocalComponents based on height
- Grasses – 1 CLocalComponent based on biomass
- Processes
  - Woody Plant Growth and Mortality (Area and Population)
  - Grass Growth and Mortality (Area and Biomass)
  - Elephant Browsing/Grazing Mortality and Transition
  - Fire Mortality and Transition

Woody trees (>5 m height) The tallest woody tree groups were considered fire-resistant. To simplify the Baxter model for initial testing with KNP data sets and woody species, QnD:EleSim uses four different woody height classes that were constructed from the original nine functional groups. As with the original Baxter model, the grass layer is represented with one class. This object layout is illustrated in Fig. 11.

This simplified structure was designed to allow linkage with limited KNP woody vegetation data sets as well as the inclusion of additional woody species in future model iterations. If additional subdivisions are required, QnD's modular structure can quickly be reexpanded to simulate all nine original Baxter size classes.

Wet and Dry Season Dynamics

Basic interaction diagrams for both wet and dry seasons are presented in Fig. 12 and Fig. 13. All plant growth occurs in the wet season (October–April) with fires

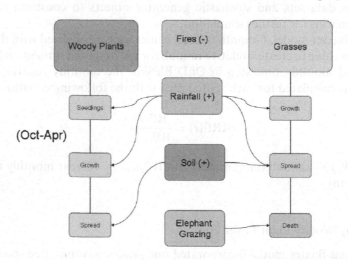

Fig. 12 Conceptual diagram of wet season dynamics as adapted from the Baxter model

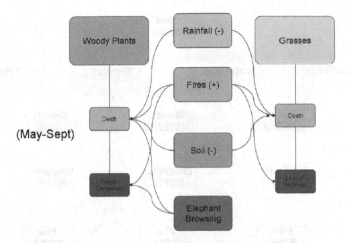

Fig. 13 Conceptual diagram of dry season dynamics as adapted from the Baxter model

less prevalent. A season's growth is closely related to rainfall and soil conditions. As the wet season draws to a close in April, overall senescence begins within grass layers, and woody layers tend to end active growth. Woody plant mortality in each size class is assumed to come through a base mortality level, fires, and elephants. As woody plants eventually progress into the dry season, the Baxter model proposes that lower levels of rainfall induce the death of some woody plants. Similarly, the prevalence of fires during the dry season leads to an average height decrease during these summer months ([10] p. 301). Also, elephant grass grazing occurs during the dry season, resulting in deaths and biomass decreases throughout.

A simplification of concepts from the Baxter model was instituted in the accounting of the relationships between the four woody tree metaclasses (Fig. 14). It is accepted that three primary forces—elephant effects, fire effects, and growth effects—can dramatically change the dynamics of the various populations of trees. Therefore, a system of nine *transition factors* has been set up to account for addition and loss of biomass to individual groups. For example, Growth Transition Factor 1 (GTF1) adds a certain amount of population and cover yearly to the sapling metaclass (because of upward growth) while subtracting that amount of population and cover from the seedling metaclass. Similarly, Elephant Transition Factor 1 (ETF1) subtracts a certain amount of population and cover from the sapling metaclass (because of browsing's effects on reducing average tree heights) and adds that amount of population and cover to the seedling class. Finally, for example, Fire Transition Factors (FTFs) behave the same as do ETFs in that they subtract population and cover from the "higher" metaclasses and add that amount of population and cover to the "lower" metaclasses. Figure 14 illustrates the conceptual design in which transitions between woody size classes are simulated. Transitions from smaller groups to larger groups are simulated through woody plant biomass

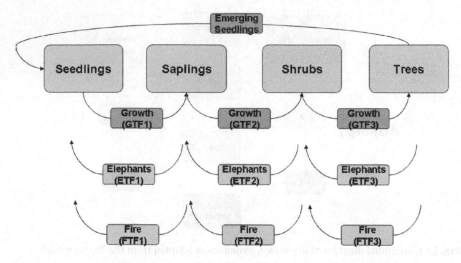

Fig. 14 A conceptual diagram of transition factors involved in woody plant size classes

and population transitions, whereas effects of fire and elephant browsing create transitions from larger to smaller size groups.

The following section provides a more detailed description of selected algorithms used in the QnD SimulationEngine. Woody seedling establishment was simulated with the following equation:

$$Rc_{seedl}(t) = RRfl(t) \bullet RcPot_{seedl}$$

where $Rc_{seedl}(t)$ = recruited or established seedlings (number/month), $RRfl(t)$ = monthly relative rainfall (unitless), and $RcPot_{seedl}$ = potential seedlings (mm/month).

Area-based competition, $Comp_i(t)$, from existing woody plants within the same size group or higher is included in the equation:

$$Comp_i(t) = 1 - \sum_i^{i=4} RCvr_i(t)$$

where $RCvr(t)$ = relative area coverage (fraction) of woody species (seedlings, saplings, and shrubs). (Note: In calculating the competition coefficient for the smallest woody size category (CSeedlings, $i = 1$), the relative area cover by grass is included in the summation.)

For saplings, shrubs, and trees, the adjusted growth rate (GTF) is described from

$$GTF_i(t) = RRfl(t) \bullet GTFbase_i(t) \bullet Comp_i(t) \bullet Pop_i(t)$$

where RRfl(t) = monthly relative rainfall (unitless), GTFbase$_i$ = base growth transition rate for each woody size class (i = 1. . . 4), Comp$_i$ = area-based competition from existing woody plants, and Pop$_i$(t) = the current population of woody plants in size class i.

Simulating Grass Layer Area and Biomass

The following section provides a more detailed description of selected grass growth algorithms used in SimulationEngine. An example configuration of Grass Layer PProcess and PSubProcess objects (along with concomitant input and output DData objects) is found in the Technical Appendix.

As before, grass biomass and cover growth are simulated based on the original Baxter model concepts with some alterations to match up with the KNP Vegetative Composition Assessment (VCA) database on the grass layer. During the rain season (October–April), higher rainfall levels result in growth of grass populations though limited by competition with woody trees resulting in higher death rates.

Monthly fractional area of grass cover (Area$_{gr}$) is calculated through the equation:

$$\text{Area}_{gr}(t+1) = \text{RRfl}(t)\left(1 - \sum_{i=1}^{3}\text{RCvr}(t)\right)(1 - \text{RGraz}(t))\left(1 - \sum_{i=2}^{3}\text{RCrowd}(t)\right)$$

where RRfl(t) = monthly relative rainfall (fraction), RCvr(t) = relative area coverage (fraction) of woody species (seedlings, saplings, and shrubs), RGraz(t) = proportion of grass in the grid cell grazed by elephants (fraction), and RCrowd(t) = proportion of crowding from saplings and shrubs (fraction).

For calculating monthly wet season grass biomass (kg/grid) (Bio$_{gr}$), the following equation is used:

$$\text{Bio}_{gr}(t+1) = S_{gr}^{w}\text{RRfl}(t)\left(1 - \sum_{i=1}^{4}\text{RCvr}_i(t)\right)(\text{Bio}_{gr}(t) + \text{Pr}_{gr}\text{Area}_{gr})$$

where S_{gr}^{w} = wet season survivability (fraction), RRfl(t) = monthly relative rainfall (fraction), RCvr$_i$(t) = relative area coverage (fraction) of woody species (seedlings, saplings, shrubs, and trees), Bio$_{gr}$(t) = grass biomass (kg), Pr$_{gr}$ = monthly wet season grass productivity (kg/grid) = (Total wet season potential productivity (kg/ha)/6 months), and Area$_{gr}$ = total grid area covered by grass (ha).

During the dry season (May–September), grass biomass or area growth does not occur while both area and biomass can be reduced by either fire or elephant grazing.

Simulating Elephant Populations

In the initial version of QnD:EleSim, elephants are conceptualized in two types of CLocalComponent objects: the Elephant Herd (a metapopulation class object that simulates a herd and its cell-to-cell movements) as well as the Tagged Elephant, in which one specific elephant will be "virtually tagged" and documented separately to simulate more individual reactions to ecological phenomena. Main processes undergone by elephants are grass grazing, tree leaf browsing, as well as birth, death, and a design of cell-to-cell movements.

As described above, elephant herds (CElephantHerd instances) are simulated as metapopulation classes existing temporarily in a 10 × 10 km grid cell for the entire month before having the option to move to another grid cell that has more desirable qualities in terms of vegetation or water. In reference to elephant movement between the spatial units, initial process designs include the calculation of habitat suitability indices (HSIs) [8, 30], for each of the 195 cells, based on factors such as woody tree biomass, grass biomass, and distance from watering holes. HSI values will provide a grid weighting for movement probabilities. Thus, elephant herd objects will be "set in motion" [11] as they navigate through various spatial units. This design allows algorithms for elephant movement to be tested ranging from pure random walks to probability weighted directional walks.

Simulating Fire

In reference to fire processes for burning vegetation, fire CLocalComponents were designed and implemented as if they were "ephemeral herbivores" consuming both area and population for woody size classes as well as area and biomass for the grass layer class. In some object designs showing a natural fires–only management scenario, fire, grass, and tree biomasses will be calculated and assigned to each spatial unit (0.0–1.0), and the fires will be randomly started within the grid. If the fire ignites in a specified cell, all linked cells will have the "opportunity" to catch fire; the adjacent cells with the higher biomass levels will have a higher probability to catch fire next. Other management scenarios call for spatially explicit fire scheduling and placement; these fires are simulated through the use of input maps, and thus fires in QnD:EleSim can be stochastic or user-defined.

3.2.4 QnDEleSim GameView: Setting the User Interface

As seen in Fig. 3, the initial version of the QnD:EleSim GameView includes a base map of the 10-km grid combined with selected additional GIS layers for added reference. QnD: EleSim version 0.0 models a simulation game-based environment: user-selected radio-button maps as well as tabbed display panes along the top of the map display various DData values for consideration. In the QnD GameView, a large map of the Kruger National Park (divided into 195 numbered 10-km^2 grids) will present itself, and various options can be selected from for viewing by selection of various radio buttons. For example, on the left-hand side, viewing of elephant densities, grass and tree biomasses, and rainfall totals will describe various phenomena using user-selected color gradients. In the lower-right-hand corner, a pie chart (specific to each spatial unit selection) will appear that contains a breakdown of the area cover for grasses, seedlings, saplings, shrubs, and trees in the cell. At the top-left of the screen, there will be a "play" button, which signifies running of either a 1-month or a 6-month time step, and thus, changes can be observed visually (in the map) or by selecting one of various X–Y charts that plots the abiotic and biotic factors of the KNP over time. As warning signs, red lights will appear next to "Rainfall," "Grasses," "Seedlings," "Saplings," "Shrubs," "Trees," and "Elephants" if any levels reach the KNP's defined levels of Thresholds of Potential Concern (TPC). Yellow lights will warn the user before TPCs, but as TPCs are attained, a red light will flash. Because the purpose of this QnD model is to create a management-based simulation of the KNP, to manage the ecosystem, toggle bars at the top allow the user to increase or decrease water point openings, decide whether to blot out or set off random fires, and, eventually, whether or not to capture or cull elephants.

3.3 Ongoing QnD:EleSim Calibration and Validation Activities

At this point in the design and testing process, QnD:EleSim v.0.0 has produced useful preliminary results, and more advanced calibration/validation/testing is under way. Figure 15 provides an example results map of the 195 Kruger National Park spatial units and documents "difference figures" between 1990 QnD simulation data of grass biomass and 1990 grass biomass data obtained from Kruger National Park's annual Veld Condition Assessment (VCA). Although the QnD:EleSim model tended to overpredict many cells' grass densities, the effect of fires and moving elephants are not included in this figure, thus a bias toward overprediction is expected until fire and elephant dynamics are finalized. Thus, the grass biomass local component of the QnD model has been initialized and gives plausible early results against historical KNP data. Continued testing, calibration, and validation with monitored grass, woody vegetation, and elephant population data is under way.

3.4 Serious Play: Playing Games for Systematic Analysis

Playing management games for heuristic development and exploration was the original goal of QnD development and use. Recent research efforts have expanded this role to include systematic analysis of sensitivity and uncertainty within these complex systems. Kiker et al. [19] presents a conceptual design for

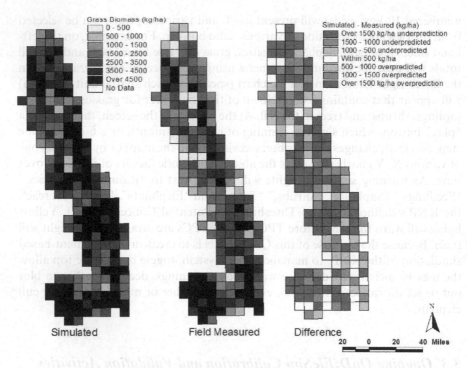

Fig. 15 Example map of KNP observed and initial QnD-simulated data (without fire and elephant effects) for 1990 grass biomass levels

linking simulation, sensitivity, and decision analysis tools together in service of adaptive management of transboundary water issues. The software package SIMLAB v2.2 [29] is used in the global sensitivity and uncertainty analysis of the Okavango QnD application. SIMLAB is designed for pseudorandom number generation–based uncertainty and sensitivity analysis. SIMLAB's Statistical Pre-Processor module provides potential parameter values using user-defined probability distribution functions (PDFs) to produce a matrix of sample inputs to run the QnD model. The QnD code has been altered to allow the incorporation of SIMLAB-derived matrices for automatic simulation. The program automatically substitutes the new parameter set into the input files, runs the model, and performs the necessary postprocessing tasks to obtain the selected model outputs for the analysis. The outputs from each simulation are stored in a matrix containing the same structure as the samples generated by SIMLAB. With the input and output matrices, SIMLAB is used to calculate the sensitivity indexes of the Morris [25] and extended FAST method [9, 20]. Finally the output probability distributions are constructed in SIMLAB based on the set of variance-based sensitivity run results to systematically quantify the uncertainty inherent within model parameters and its output. Given that models being applied to large complex environmental problems are often challenged by various groups to substantiate results and/or predictions, these model analysis tools are increasingly being used to scrutinize various assumptions of the model and the inherent limitations of the environmental data being used to drive them.

4 Conclusions

The QnD modeling software and its associated development methodology was created to quickly and efficiently construct a management/stakeholder-relevant model that integrates both explicit scientifically derived data and expert/anecdotal knowledge. Given QnD's object-oriented design and XML-based input files, systematic iteration with stakeholders is encouraged and promoted. New and novel ideas about the problem and potential solutions can be explored, adopted, or discarded to promote greater system learning.

Development of a QnD model is undertaken within a larger context of stakeholder engagement and public participation. When eliciting information to build QnD scenarios, many different perspectives are expressed, each with its own assumptions about cause–effect relationships and beliefs about what potential interventions would constitute ecosystem improvement. The development process, which involves actively working with stakeholders to build the model, play the game, and revise the model, is undertaken within a soft systems approach [7]. The soft systems approach distinguishes the QnD gaming and scenario-building process from the more traditional use of models simply as system predictors. The QnD development process can accommodate both hard data, such as field-measured experiments, and soft data, such as experiential learning, impressions, or general "rules of thumb." The model is used to facilitate dialogue and learning about the factors that influence the environmental system under consideration and to explore potential management actions.

This research has designed, implemented, and tested an elephant/vegetation/ fire model in an attempt to assess the KNP biodiversity challenges and how managers can begin to address ecosystem management. After the calibration period, a global sensitivity analysis using the program SIMLAB will be coupled with the QnD:EleSim model to test how sensitive the model is to changes in environmental phenomena and input parameters.

The applications for this project are widespread, in that after validation, it could potentially be used as a management tool for the Kruger National Park scientists or managers in reference to when potential Thresholds of Potential Concern will approach, and what types of management actions should take place. In the future, this QnD model can be improved upon by validating predictions of elephant and woody plant densities by comparison with the rich monitoring data sets available in the KNP. Thus, QnD:EleSim version

0.0 has been created as one of the many tools available to scientists and managers for addressing the complex savanna biodiversity challenges in the Kruger National Park, South Africa. Continued changes in the object structure and the strategic implementation of QnD are fully expected as KNP managers adaptively learn and test various strategies to answer questions and make decisions within this complex environmental system.

Acknowledgments Special thanks to the following people/organizations that allowed for the development of this model to be made possible: Dr. Judith Kruger, Dr. Rina Grant, and Dr Harry Biggs of the South Africa National Parks for supplying climate, soils, vegetation, animal census, fire, and other KNP data sets. In addition, their guidance and friendship are truly appreciated. The KNP Elephant Modeling Group (chaired by Prof. Robert Slotow and Dr. Robert Scholes) for coordination of elephant modeling research activities. The South African Weather Bureau for provision of climate data. Prof. David Saah of the University of San Francisco for his input and comments. Finally, the University of Florida Center for Precollegiate Education and Training for organizing and supporting this research opportunity for Mr. Thummalapalli.

Technical Appendix

This section contains a sample list of objects used in the Grass CLocalComponent (CGrass) object. The CGrass object exists within each of the 195 grid cells. In addition, the four woody vegetation objects (CSeedlings, CSaplings, CShrubs, and CTrees), potential fire objects (CFire), and elephant objects (CElephantHerd and CTaggedElephant) are also present in grid cells. They have a similar DData/PProcess object structure as CGrass, but they are not listed in this appendix. Overall in QnD:EleSim, approximately 7 CLocalComponent objects, 23 Process objects, 88 SubProcess objects, and hundreds of DData objects were programmed into each of the 195 CSpatialUnit/CHabitat combinations via the XML input files.

Grass Object (CGrass) DData and PProcesses

DData		
DBiomass	DProductivity	DNewAreaCovered
DAreaCover	DGrassBiomassSenesced	DBiomassPerUnitArea
DBaseProductivity	DWetSeasonSenescence	DNewGrassBiomass
DGrassValue1	DDrySeasonSenescence	DRelativeAreaCovered BySmallWoodyPlants
DGrassValue2	DGrassCrowdingCoefficient	
DGrassValue3	DMaxBiomass	
DGrassValue4	DRelativeGrazingIntensity	
DGrassValue5	DBiomassAdded	
DGrassValue6	DAreaCovered	

Process/SubProcesses/DData

PProcess: PWetSeasonProcessesCalculateGrassAreaCoveredAndBiomass
 PSubProcess: PIfWetSeason (Type = PIfEquals)
  If (Global.DWetSeason = = 1) **Then** continue to next PSubProcess – **Else** Go To
  Next Process
 PSubProcess: PCalcProductivity (Type = PMultiplyValue)
  CGrass.DProductivity = Global.GridArea x CGrass.DBaseProductivity
 PSubProcess: PCalcMonthlyProductivity (Type = PDivideValue)
  CGrass.DProductivity = CGrass.DBaseProductivity / Global.DSix
 PSubProcess: PCalcAvailableAreaCover (Type = PSubtractValue)
  CGrass.DGrassValue1 = Global.DOne – CSeedling.DRelativeAreaCover –
              Csapling.DRelativeAreaCover –
              CShrub.DRelativeAreaCover
 PSubProcess: PCalcInverseOfElephantGrazingIntensity (Type = PSubtractValue)
  CGrass.DGrassValue2 = Global.DOne –
              CElephantHerd..DRelativeGrazingRate
 PSubProcess: PCalcInverseOfCrowdingIntensity (Type = PSubtractValue)
  CGrass.DGrassValue3 = Global.DOne – CSapling.DCrowdingCoefficient –
              Cshrub.DCrowdingCoefficient
 PSubProcess: PCollectingTerms (Type = PMultiplyValue)
  CGrass.DAreaCovered =HomeSpatialUnit.DLocalRainfall x
              CGrass.DGrassValue1 x CGrass.DGrassValue2 x
              CGrass.DGrassValue3
 PSubProcess: PCalculateGrassSenesced (Type = PMultiplyValue)
  CGrass. DGrassBiomassSenesced = CGrass. DWetSeasonSenescence x
              CGrass.DGrassBiomass
PProcess: PCalculateDrySeasonGrassSenescence
 PSubProcess: PIfDrySeason (Type = PIfEquals)
  If (Global.DWetSeason = = 0) **Then** continue to next PSubProcess – **Else** Go To
  Next Process
 PSubProcess: PSetNewBiomassToZero (Type = PSetValue)
  CGrass. DNewGrassBiomass = Global.DZero
 PSubProcess: PCalcInverseOfElephantGrazingIntensity (Type = PSubtractValue)
  CGrass.DGrassValue2 = Global.DOne –
              CElephantHerd.DRelativeGrazingRate
 PSubProcess: PCalculateFireIgnition (Type = PSubtractValue)
  CGrass.DGrassValue6 = Global.DOne – CFire.DFireIgnition
 PSubProcess: PCalculateGrassFireLoss (Type = PMultiplyValue)
  CGrass. DNewGrassAreaCovered = Grass.DGrassValue6 x
              CGrass.DAreaCovered
 PSubProcess: PCalculateDrySeasonGrassSenescence (Type = PMultiplyValue)
  CGrass. DGrassBiomassSenesced = CGrass. DDrySeasonSenescence x
              CGrass. DGrassValue6 x CGrass.DGrassBiomass
PProcess: PCalculateFinalGrassBiomass
 PSubProcess: PCalculateFinalGrassGrowth (Type = PAddValue)
  CGrass. DGrassBiomass = CGrass. DGrassBiomass + CGrass.
              DGrassBiomassSenesced +
              CGrass.DNewGrassBiomass
 PSubProcess: PCalculateNewBiomassPerUnitArea (Type = PMultiplyValue)
  CGrass. DBiomassPerUnitArea = CGrass.DGrassBiomass / Global.DGridArea

References

1. Aldritch, C. (2005). Learning by Doing: A Comprehensive Guide to Simulations, Computer Games, and Pedagogy in e-Learning and Other Educational Experiences. John Wiley & Sons.
2. Baxter, P.W.J. (2003). Modeling the Impact of the African Elephant, Loxodonta africana, on Woody Vegetation in Semi-Arid Savannas. PhD Dissertation, University of California, Berkeley.
3. Baxter, P.W.J. & Getz, W.M. (2005). A model-framed evaluation of elephant effects on tree and fire dynamics in African savannas. *Ecological Applications* 15, 1331–1341.
4. Baxter, P.W.J. & Getz, W.M. (2006). Development and parameterization of a rain- and fire-driven model for exploring elephant effects in African savannas. *Springer Science* 9, 2–23.
5. Buss, I.O., & Smith, N.S. (1966). Observations on reproduction and feeding behavior of the African elephant. *The Journal of Wildlife Management* 14, 375–388.
6. Carpenter, S., Brock, W., & Hanson, P. (1999). Ecological and social dynamics in simple models of ecosystem management. *Conservation Ecology* 3(2), 4 [http://www.ecologyand society.org/vol3/iss2/art4].
7. Checkland, P.B. (1999) Soft Systems Methodology in Action. John Wiley & Sons.
8. Cook, J.G., & Irwin, L.L. (1985). Validation and modification of a habitat suitability model for pronghorns. *The Wildlife Society Bulletin* 13, 440–448.
9. Cukier, R.I., Levine, H.B., & Schuler, K.E. (1978). Nonlinear sensitivity analysis of multiparameter model systems. *Journal of Computational Physics* 26, 1–42.
10. Du Toit, J.T., Rogers, K.H., & Biggs, H.C. (2003). The Kruger Experience: Ecology and Management of Savanna Heterogeneity. Island Press.
11. Dorai-Raj, S. (2000). Java Random Number Generator. http://152.3.140.5/~dsv/comp-bio/ppstuff/ppinteraction/RandomNumberGenerator.java [Last Accessed Dec. 4, 2007].
12. Estes, R. (2004). The Safari Companion: A Guide to Watching African Mammals. Chelsea Green Publishing Company.
13. Holling, C.S., Gunderson, L.H., & Peterson, G. (2002). Sustainability and panarchies. In Gunderson, L.H. and Holling, C.S. (Eds), Panarchy: Understanding Transformations in Human and Natural Systems. Island Press. pp. 63–102.
14. Holdo, R.M. (2003). Woody plant damage by African elephants in relation to leaf nutrients in western Zimbabwe. *Journal of Tropical Ecology* 19, 189–196.
15. Jeppesen, E.I., & Iversen, T.M. (1987). Two simple models for estimating daily mean water temperatures and diel variations in a Danish low gradient stream. *Oikos* 49, 149–155.
16. Justice, L.C., Kiker, G.A., & Kiker, M.K. (2007, in press). Jamaican food security in a dynamic world: collaborative methods to envision food system models. *Caribbean Journal of Geography*.
17. Kiker, G.A., Rivers-Moore, N.A., Kiker, M.K., & Linkov, I. (2006). QnD: A scenario based gaming system for modeling environmental processes and management decisions. *Springer Science* 11, 151–186.
18. Kiker, G.A., & Linkov, I. (2006). The QnD model/game system: Integrating questions and decisions for multiple stressors. In Arapis, G., Goncharova, N., & Baveye, P. (Eds), Ecotoxicology, Ecological Risk Assessment and Multiple Stressors. Springer. pp. 203–225.
19. Kiker, G.A., Muñoz-Carpena, R.. Wolski, P., Cathey, A., Gaughn, A., & Kim, J. (2007, in press). Incorporating uncertainty into adaptive, transboundary water challenges: a conceptual design for the Okavango River Basin. *International Journal of Risk Assessment and Management* 10(3).
20. Koda, M., McRae, G.J., & Seinfeld. J.H. (1979). Automatic sensitivity analysis of kinetic mechanisms. *International Journal of Chemical Kinetics* 11, 427–444.

21. Kruger National Park (KNP). (2006). KNP Park Management Plan Version 1 (31 October 2006). [Online: www.sanparks.org/conservation/park_man/kruger.pdf Last accessed Dec 5, 2007].
22. Laursen, L., & Bekoff, M. (1978). Loxodonta Africana. *Mammalian Species* 92, 1–8.
23. Matsumoto, M., & Nishimura, T. (1998). Mersenne twister: A 623-dimensionally equi-distributed uniform pseudo-random number generator. *ACM Transactions on Modeling and Computer Simulation* 8(1), 3–30.
24. Morgan, M.G., Fischhoff, B., Bostrom, A., & Atman, C.J. (2002) Risk Communication: A Mental Models Approach. Cambridge University Press.
25. Morris, M.D. (1991). Factorial sampling plans for preliminary computational experiments. *Technometric* 33, 161–174.
26. NRC. (1991). Opportunities in the Hydrologic Sciences. National Academy Press, 348 pp.
27. Papajorgji , P.J., & Pardalos, M. (2005). Software Engineering Techniques Applied to Agricultural Systems: An Object-Oriented and UML Approach. Springer-Verlag. 247 pp.
28. Rittel, H., & Webber, M. (1973). Dilemmas in a general theory of planning. *Policy Sciences* 4, 155–169.
29. Saltelli, A., Tarantola, S., Campolongo, F., & Ratto, M. (2004). Sensitivity Analysis in Practice: A Guide to Assessing Scientific Models. John Wiley & Sons, Ltd.
30. Schamberger, M., Farmer, A.H., & Terrel, J.W. (1982). Habitat suitability index models: introduction. U.S.D.I. Fish and Wildlife Service. FWS/OBS-82/10. 2 pp.
31. Starfield, A.M., & Bleloch, A.L. (1991). Building Models for Conservation and Wildlife Management. Burgess International Group.
32. Van der Heijden, K. (1996) The Art of Strategic Conversation. John Wiley & Sons.
33. Walters, C., Korman, J., Stevens, L.E., & Gold, B. (2000). Ecosystem modeling for evaluation of adaptive management policies in the Grand Canyon. *Conservation Ecology* 4(2), 1 [http://www.ecologyandsociety.org/vol4/iss2/art1].
34. Wu, J., & David, J.L. (2002). A spatially explicit hierarchical approach to modeling complex ecological systems: theory and applications. *Ecological Modelling* 153, 7–26.
35. Wu, J., & Loucks, O.L. (1995). From balance of nature to hierarchical patch dynamics: a paradigm shift in ecology. *The Quarterly Review of Biology* 70(4), 439–466.
36. Yoe, C. (2002). Tradeoff Analysis Planning and Procedures Guidebook. U.S. Army Corps of Engineers, Institute of Water Resources Report (IWR 02-R-2). Prepared for U.S. Army Corps of Engineers, Institute of Water Resources by of Planning and Management Consultants, Ltd. Contract # DACW72-00-D-0001. http://www.iwr.usace.army.mil/iwr/pdf/tradeoff.pdf.

The Use of UML as a Tool for the Formalisation of Standards and the Design of Ontologies in Agriculture

François Pinet, Catherine Roussey, Thomas Brun, and Frédéric Vigier

Abstract For the past 20 years, ontologies have become more and more popular in various research fields such as Web technologies, databases, information retrieval methods, and so forth. The first goal of this chapter is to answer general questions about ontologies, such as: What exactly is an ontology? What is the purpose of ontology? Which types of systems use an ontology? The second goal of the chapter is to help readers understand how UML can be used to model ontologies in agricultural systems. UML and the Web Ontology Language (OWL) are compared, and an example inspired by the French project named Farm Information Management is presented.

1 What Is an Ontology?

This section will briefly describe the history of the word *ontology* starting from Aristotle's metaphysics to the more sophisticated Web technologies. Thus, readers will be able to gain a rapid overview of the use of ontologies in different domains. Then we will conclude by giving a general definition of ontologies.

Aristotle's definition of the word *Ontology* (with a capital *O*) is "the science of being *qua* being" (Aristotle's *Metaphysics*). This science is part of philosophy, dealing with the descriptions of existing entities. Its goal is to define the general categories or primitives used to classify all the entities in the world such as human beings, animals, plants, and so forth. In the early 1980 s, artificial intelligence researchers borrowed the term *Ontology* from the field of philosophy. Ontologies became the definition of domain knowledge. They provide the possibility to separate domain knowledge from operational knowledge [29].

F. Pinet (✉)
Cemagref, Clermont Ferrand, France
e-mail: francois.pinet@cemagref.fr

P.J. Papajorgji, P.M. Pardalos (eds.), *Advances in Modeling Agricultural Systems*,
DOI 10.1007/978-0-387-75181-8_7, © Springer Science+Business Media, LLC 2009

An interesting definition of ontologies in the field of artificial intelligence was proposed by Gruber in 1993: an ontology is "the specification of conceptualisations used to help programs and humans share knowledge" [13].

He developed a more precise definition: "...ontology is a formal explicit specification of a shared conceptualization." According to Gruber, *conceptualisation* refers to an abstract model of phenomena in the world after identifying the relevant concepts of these phenomena. *Explicit* means that the type of concepts used and the constraints on their use are explicitly defined. *Formal* refers to the fact that the ontology should be machine readable. *Shared* reflects the idea that ontology should capture consensual knowledge accepted by the communities [14].

In database and information system areas, ontologies are used to facilitate the interoperability of heterogeneous information sources. The ontology is the general schema organising all the entity properties described in a group of database schemas or information sources. Each information source is provided with a wrapper that maps its database schema to the ontology. Thus, users can query the ontology and obtain a result integrating all information sources needed to answer the query [34].

Information retrieval techniques use linguistic ontology such as indexing vocabulary in order to avoid word semantic ambiguity. Thus, document and query contents are represented by concepts (i.e., the meaning of terms) and not words (i.e., sets of characters). This technique means it is possible to improve the description of document (and query) contents and also system performance. One of the first operational systems was Ontoseek, which uses Wordnet taxonomy to describe yellow pages [15, 20].

The Semantic Web is an adaptation of previous technologies. Tim Berners-Lee, the inventor of the Web, defines the Semantic Web as "a web of data that can be processed directly and indirectly by machines." Indeed, the Semantic Web is about several things: the exchange and processing of data, documents, and services. The Semantic Web will integrate a community of agents capable of exchanging data and services from diverse sources in order to achieve a specific goal. More precisely, such Web services are reusable software components that implement a discrete functionality (like hotel reservation) accessible through the Web. Moreover, document description and content will be defined precisely thanks to series of data called *metadata* stored in ontologies. Thus, Web document retrieval is improved by this technique. The W3C [1] is in charge of developing a set of standards necessary for Semantic Web technology [30]. All these standards are based on the Extensible Markup Language (XML) (W3C Semantic Web activity).

In the agricultural domain, the well-known AGROVOC thesaurus is used to develop the Agricultural Ontology Service (AOS) project [1]. AGROVOC is a multilingual thesaurus concerning forestry, fisheries, food, environment, and

---

[1] W3C, World Wide Web Consortium; see www.w3c.org.

related domains. As presented in Ref. 1, "it consists of words or expressions (terms), in different languages and organized in relationships (e.g., 'broader,' 'narrower,' and 'related'), used to identify or search resources." AGROVOC was developed by the Food and Agriculture Organization (FAO) and the Commission of the European Communities in the early 1980s.

Using the knowledge contained in AGROVOC, "the AOS will be able to develop specialized domain-specific terminologies and concepts that will better support information management in the Web environment. A key objective is to add more semantics to the thesaurus, for example, by expanding and better specifying the relationships between concepts" [1]. For example, the term *pollution* is associated with the term *pollutants* by the relationship "Related Terms" in the AGROVOC thesaurus. This means that the term *pollution* is related to the term *pollutants*. It is possible to develop a more specific terminology. For instance, we can explicitly indicate that *pollution* is caused by *pollutants* in using a more meaningful relationship such as "caused by" [35].

To conclude, an ontology can take on various forms, and the use of the ontologies is also different, but we can state that an ontology should primarily contain:

- A vocabulary of terms;
- A set of term definitions that identify concepts and fix the term interpretation;
- A modelling of the domain of interest in order to represent relationships between concepts;
- An agreement of a community of ontology users about term definitions and the domain structure.

2 UML as an Ontology Language

The most common formalisms used to represent ontologies in the artificial intelligence community are the Knowledge Interchange Format (KIF) [22] and Description Logics (DL) languages [9, 21]. KIF is a computer-oriented language for the interoperability of knowledge among heterogeneous programs. It is based on the first-order logic and allows the formal definition of objects, functions, and relations. Description Logics are a family of knowledge representation languages. They are formal and can be used to model the knowledge of an application domain. The name *logic* refers to the logic-based semantics given to these languages. Until the 1980s, Description Logics were called terminological systems or concept languages. In the field of Semantic Web, the W3C proposes a language to define Ontology called OWL (Web Ontology Language) [31] developed as a follow-on from RDF (Resource Description Framework [4, 32]). OWL is composed of three increasingly expressive sublanguages: OWL Lite, OWL DL, and OWL Full. OWL Lite supports a user's basic needs. The part of the OWL language called OWL DL can be viewed as an XML representation of

Description Logics languages. OWL Full provides a very complete language; it is meant for users who want maximum expressiveness.

These types of languages provide interesting capabilities to develop a consistent set of concept classes; they support efficient reasoning and inference mechanisms to deduce new knowledge and compute the logical correctness of the conceptual structure. Unfortunately, these languages are little known outside artificial intelligence laboratories. Note that reasoning software is not able to support complete reasoning for every feature of OWL Full [31].

The Unified Modelling Language (UML) is a standard for modelling computational systems. It is used successfully in information system and object-oriented developments. As presented in Refs. 8, 11, 18, 19, and 27, some researchers propose to use UML as Ontology language. These studies have acknowledged the benefits of using a standard modelling tool such as UML in ontology construction.

The use of UML for ontology construction has only recently been present in specialised literature. There are several common features between UML and Ontology-based languages. One main advantage of UML is that it is widely used. Several tools are available for UML so designers can use them for describing their diagrams. UML is an open standard, and it is taught at many universities. Users and developers are likely to be more familiar with UML notations than with KIF, DL, or OWL. Moreover, whereas KIF, DL, and OWL have a textual representation, UML proposes several visual representations: class diagram, objects diagram, and so forth.

There are several common features between UML and Ontology-based languages [18] but the main drawback to UML is its lack of formal semantics; the traditional languages used to describe ontologies have formal semantics. The official document defining the semantics of UML contains an informal description in English [23]. Some researchers propose a mathematical model for UML (see, for instance, [5]). The work of Guizzardi [16, 17] concerns the development of different methodological tools (UML profiles, design patterns) in order to build an ontology using UML formally and correctly. Cranefiel and Purvis develop an ontology using UML and Object Constraint Language (OCL) [8, 24]; the concepts are described by UML classes, and constraints of concepts are described in OCL.

Because UML is widely used and supported by numerous tools, it could be viewed as a good candidate to develop ontologies. Nevertheless, UML is a semiformal formalism and will not be sufficient to represent all the details required by complex reasoning processes [7]. OCL is sometimes considered as "a variant of [first-order predicate logic] tuned for writing constraints on object structures" [6] and might overcome this limitation in the future; in the OMG specification of OCL [24], an annex presents a first version of a formal semantic. Currently, this annex does not describe how to deduce new knowledge or compute the logical correctness of OCL constraints.

3 Similarity and Differences Between UML and Traditional Languages Used to Describe Ontologies

Several studies compare UML and formal languages used to describe ontologies [3, 7, 18]. OWL is usually considered as the flagship language for ontologies [3]; thus this chapter focuses on a comparison between OWL (Full) and UML.

According to existing studies, there is a common part between UML and OWL (Fig. 1). The first subsection presents the main common concepts between the two languages, and the second subsection evokes several differences.

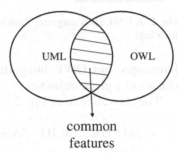

Fig. 1 UML and OWL

3.1 Mappings Between UML and Ontology Languages

In the OMG recommendation [18], mappings among the metamodels of OWL and UML are proposed. Several common features between UML and OWL are shown. Table 1 presents the main similarities between UML class diagrams and OWL.

3.1.1 Class and Subclass

UML and OWL are based on classes. In OWL, classes correspond with concepts defined in the ontology. Subclasses can also be modelled in both

Table 1 OWL and UML: The common features

UML	OWL
Class	Class
Subclass	Subclass
Instance	Individual
Attribute, association	Property
Multiplicity	minCardinality, maxCardinality, Cardinality

Note: See Ref. 18 for more detail.

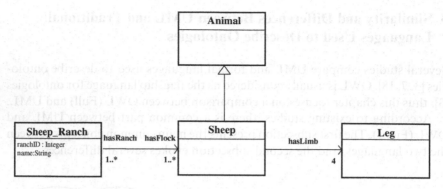

Fig. 2 A UML class diagram: Ranches have flocks of sheep, and a sheep is an animal having four legs

languages, but OWL integrates a universal class *Thing* that generalises all classes of a given ontology.

The class *Animal* of Fig. 2 is declared in OWL as follows:

<owl:Class   rdf:ID = "Animal"/ >

The syntax rdf:ID = "Animal" is used to declare a new name in the ontology (i.e., a new term). The Animal class can now be referred to as "#Animal" in the rest of the ontology.

The OWL taxonomic notation for classes is *subClassOf*. OWL *subClassOf* can be considered as similar to UML generalisation relationships [18]. According to the W3C OWL Guide [31]:

> It [i.e., subClassOf] relates a more specific class to a more general class. If X is a subclass of Y, then every instance of X is also an instance of Y. (. . .) If X is a subclass of Y and Y a subclass of Z, then X is a subclass of Z.

"A sheep is an animal" modeled in OWL:

<owl:Class rdf:ID = "Sheep" >
<rdfs:subClassOf   rdf:resource = "#Animal" / >
</owl:Class >

3.1.2 Object/Individual

In the object-oriented paradigm, instances of a class are referred to as *objects*. In OWL, an instance of class is called an *individual*. An individual is introduced by declaring it to be a member of a class. The next example declares the sheep named "Sheep_1" (i.e., the individual "Sheep_1" of the class Sheep):

<owl:Thing rdf:ID = "Sheep_1" / >
<owl:Thing rdf:about = "#Sheep1" >

```
<rdf:type rdf:resource = "#Sheep"/>
</owl:Thing>
```

The first line declares an individual "Sheep_1"; the other lines define that this individual is a member of the class *Sheep*.

3.1.3 Attribute, Association/Property

Class attributes (in UML) can be modeled with *properties* in OWL. *Properties* relate individuals to individuals (object properties), or individuals to data types, for example, integer, string, and so forth (datatype properties).

The next example illustrates datatype properties. The class Sheep_Ranch and its attributes (presented in Fig. 2) are declared in OWL as follows:

```
<owl:Class   rdf:ID = "Sheep_Ranch"/>
<owl:DatatypeProperty   rdf:ID = "ranchID">
<rdfs:domain   rdf:resource = "#Sheep_Ranch"/>
<rdfs:range   rdf:resource = "&xsd;integer"/>
</owl:DatatypeProperty>
<owl:DatatypeProperty   rdf:ID = "name">
<rdfs:domain rdf:resource = "#Sheep_Ranch"/>
<rdfs:range   rdf:resource = "&xsd;string"/>
</owl:DatatypeProperty>
```

The datatype properties are defined by a domain and a range. If we compare with UML, the datatype properties are attributes, the domains are their classes, and the ranges are their types. In the example, the presented ranges ("&xsd:integer" and "&xsd:string") belong to datatypes recommended by W3C. They correspond with XML Schema datatypes [33].

Links between individuals can be created in using object properties. For instance, it is possible to add an object property "hasFlock" to the class Sheep_Ranch:

```
<owl:ObjectProperty   rdf:ID = "hasFlock">
<rdfs:domain   rdf:resource = "#Sheep_Ranch"/>
<rdfs:range   rdf:resource = "#Sheep"/>
</owl: ObjectProperty>
```

The range of this property is the class Sheep (i.e., the values of the property *hasFlock* are instances of the class *Sheep*). Thus, a link is created between the instances of the class *Sheep_Ranch* and the instances of the class *Sheep*. In UML, this object property can correspond with:

- An attribute *hasFlock* in the class *Sheep_Ranch*; the type of the attribute is "*Sheep*," or
- An association end *hasFlock* between *Sheep_Ranch* and *Sheep* (as shown in Fig. 2).

The inverse of a property is modelled with *inverseOf* (e.g., the inverse of *hasFlock* is modelled as follows):

```
<owl:ObjectProperty rdf:ID = "hasRanch" >
<rdfs:domain rdf:resource = "#Sheep"/ >
<rdfs:range rdf:resource = "#Sheep_Ranch"/ >
<owl:inverseOf rdf:resource = "#hasFlock"/ >
</owl:ObjectProperty >
```

The inverse of *hasFlock* is *hasRanch*. The range of *hasRanch* is the class *Sheep_Ranch*, that is, the values of the property *hasRanch* are instances of the class *Sheep_Ranch*; *inverseOf* is used to declare that *hasRanch* is the inverse of *hasFlock*. Thus, both association ends ("*hasFlock ... hasRanch*") of the UML diagram are modeled in OWL.

3.1.4 Multiplicity/Cardinality

In UML, multiplicities are important to constrain the number of objects linked by associations. In OWL, any instance of a class may have an arbitrary number (zero or more) of values for a particular property [31]. To allow only a specific number of values for that property, cardinality constraints can be used. OWL provides the concepts *cardinality*, *minCardinality*, and *maxCardinality* for restricting the cardinality of properties. For example, we model that a sheep ranch has at least one sheep:

```
<owl:Restriction >
<owl:onProperty rdf:resource = "#hasFlock" / >
<owl:minCardinality rdf:datatype = "&xsd;nonNegativeInteger" > 1
</owl:minCardinality >
</owl:Restriction >
```

The cardinality constraint *minCardinality* belongs to the value space of the XML Schema datatype nonNegativeInteger. A *minCardinality* constraint describes a class of all individuals that have at least N semantically distinct values for the property concerned.

Let *Leg* be a class and *hasLimb* be a property of the class *Sheep* (i.e., a domain of *hasLimb* is *Sheep*). The following example constrains the number of values of the property *hasLimb*:

```
<owl:Restriction >
<owl:onProperty rdf:resource = "#hasLimb" / >
<owl:cardinality rdf:datatype = "&xsd;nonNegativeInteger" > 4
</owl:cardinality >
</owl:Restriction >
```

This cardinality restriction is similar to the multiplicity of the association *hasLimb* in the UML diagram shown in Fig. 2.

3.2 Differences Between UML and Ontology Languages

The author of Ref. 3 highlights the general differences between an ontology and a UML model. The main results of this study are presented in Table 2.

3.2.1 In OWL but Not in UML

OWL allows defining a class as the set of individuals that satisfies "restrictions" (e.g., a sheep is an animal belonging to the flock of a sheep ranch and having four legs). It is possible to infer from the properties of an individual that it is a member of a given class. OWL provides reasoning and inference mechanisms to deduce knowledge (e.g., determine that an individual belongs to a specific class) and compute the logical correctness of the conceptual structure (e.g., determine that the ontology is consistent).

Thus, it becomes possible to deduce that an animal is a sheep because it satisfies several properties: for example, it has four legs and it belongs to the flock of a sheep ranch. Currently, OCL can be used to model constraints (i.e., restrictions) on UML diagrams, but it does not integrate formal inference mechanisms.

3.2.2 In UML but Not in OWL

UML allows the specification of behavioural features. UML is able to model operations, parameters of methods, interface classes, and so forth. The inter-action diagrams, activity diagrams, and state diagrams can also be used to describe precisely dynamic aspects: sent messages, different states and life lines of objects, and so forth.

An ontology is mainly a model presenting relationships between concepts. Thus, OWL and the other ontology languages do not provide support for

Table 2 General differences between an ontology and a UML model [3]

Ontology	UML Model
Originated from the artificial intelligence world for the purpose of precisely capturing "knowledge"	Originated from the software engineering world for the purpose of simplifying the modelling of software
OWL is the main language	UML/OCL are the languages used to specify the models
Formal semantics (Description Logics)	Semiformal semantics (UML semantics are expressed by a metamodel)

expressing the behaviour of a system. This is the reason why only UML class diagrams and object diagrams are used to model ontologies.

Because UML class and object diagrams are widely used, it could be viewed as a good candidate to model visually the main features of ontologies; that is, classes, subclasses, properties, individuals, and so forth. (see Fig. 1 and Table 1). As presented in the next section, in the context of a French project, UML class diagrams have been used to specify ontologies in order to model consensual knowledge of the agricultural domain. The visual representations provided by UML had facilitated the design process of the ontology. The goal of the project was to define a standard to build a data interchange format for French agricultural information systems [12].

4 Farm Information Management Project

4.1 Exchange of Agricultural Data: A Need That Is Partially Met

In France, the main economic and institutional players have made commitments to developing diverse systems and standards, making it possible to control and manage the increasing flow of information linked to farm activities. Their main goal is to facilitate the interoperability between information systems.

As an example, the French Data Reference Centre for Water (SANDRE in French)[2] is in charge of developing a common language for water data exchange [26]. Data related to water in France are issued from thousands of organisations and public services. The SANDRE's priorities are to make compatible and homogeneous data definitions between producers, users, and databanks. For example, some themes considered by SANDRE are groundwater, hygrometry, fertiliser spreading, and so forth. SANDRE proposed "a common language concerning data involved in the French Water Information System. Specific terms relevant to water data are clearly defined and data exchange specifications are also produced to fulfill the communication needs between partners involved in the field of water" [26]. One of the SANDRE's goals is to define, at a national level, a common vocabulary concerning the field of water (SANDRE's common language). To fulfill this task, data models have been developed. They are associated with data dictionaries that gather all the definitions of data related to topic *Water*. XML-based exchange formats have also been proposed. The application range of SANDRE is larger than the farm activities.

---

[2] SANDRE is composed of a member from one of each of the signatory organisations of the French Information System for Water [26]: the French Ministry of Ecology, French Institute for Environment, Water Agencies, Fishing Council, French Research Institute for Exploitation of the Sea, "Electricité de France," Institute for Geological and Mining Researches, International Water Institute.

In another context, the association AgroEDI[3] Europe led several working groups to propose standards in agriculture for specific productions or activities (e.g., viticulture, phytosanitary treatments) [2].

These actions contribute to the development of professional information exchange and open the way to electronic management; they contribute to the progressive advent of a farming "network" by proposing several data interchange formats in some cases facilitating communication between certain types of agricultural information systems. However, the various actions that have been carried out have led to the definition of standards for each production chain or for each field of activity (e.g., water data management). A more general approach was necessary in order to propose a unified standard that covers all the main concepts used in the majority of agricultural information systems.

4.2 Ontology Definition in Agriculture: A Means of Communication

Thus the members of the new project named FIM[4] have studied a unified standardisation approach. The standard is not only limited to a production chain or a type of activities as in previous approaches; it can also be used in the majority of contexts in agriculture. The standard covers the common concepts used in the main production chains and is designed to be used in different fields of activities in agriculture in France. A final goal of the standard is to provide more complete data interchange formats in order to facilitate and to improve interoperability between agricultural information systems.

The first task of the project teams was to carry out an inventory of the various initiatives dealing with their themes. Then, different terms, concepts, and their relationships have been identified for each theme. An important part of the FIM project consists in integrating and enhancing the definition of concepts and work on standardisation already initiated by the various partners. The monitoring of these approaches and the participation in various work groups and their corresponding project committees are therefore fully integrated in the project.

An ontology has been chosen to formalise the standard. All the members of the project can propose new concepts to the developed ontology. Data interchange formats are also proposed on the basis of the vocabularies and the concepts of the ontology. The ontology is represented by UML class diagrams. UML has been chosen to model the ontology because the participants of the FIM project are familiar with UML. Figure 3 summarises the presented approach.

---

[3] EDI: Electronic Data Interchange.
[4] FIM: Farm Information Management (GIEA in French). The list of the different participants involved in the FIM project can be found at: http://www.projetgiea.fr/-Organisation.

Fig. 3 From standard to
data interchange format

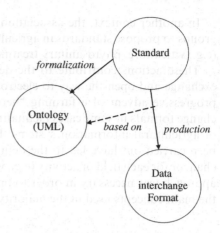

The developed ontology is composed of several parts. Each one is related to a specific theme [12]:

- Actors (modelling the main features of the actors involved in the agricultural production; farms, farmers, etc.)
- Foods (for animals)
- Analysis and monitoring of the production
- Soil analysis
- Buildings
- Animals
- Works related to the agricultural production (agricultural spreading, harvest, etc.)
- Identification and traceability of animals
- Agricultural inputs (fertiliser, water, etc.)
- Stockbreeding
- Regulations
- Land use
- Crops
- Livestock reproduction
- Animals and sanitary monitoring.

4.3 Use of UML

This section shows some examples taken from the FIM project. They illustrate certain concepts of the ontology represented with UML.

Figure 4 is based on the FIM project theme "Animals and Sanitary Monitoring." This theme deals with all concepts related to animal health: every curative and preventive measure taken for animals. This example presents the main concepts for sanitary event management (Animals, Interventions, Medical Treatments, Sanitary Products, Actors, etc.).

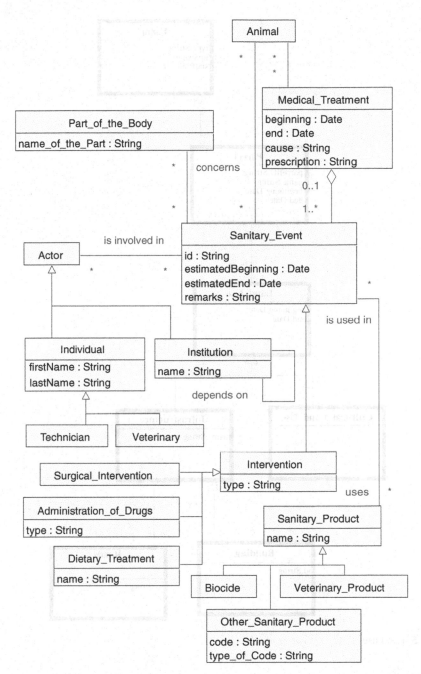

Fig. 4 Animals and sanitary monitoring

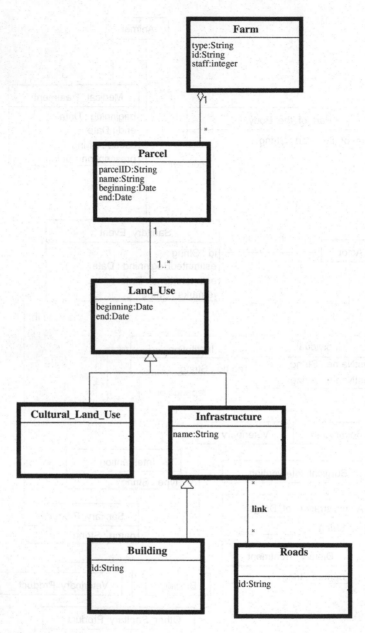

Fig. 5 Land use

The class Sanitary Event is the main concept of the diagram. It may correspond with surgical interventions, administration of drugs, dietary treatments.
Sanitary events concern animals and are associated with different categories of
products: biocide, veterinary products, and so forth. Actors can be involved in

the events (veterinaries, technicians, institutions). A medical treatment can be viewed as an aggregation of sanitary events.

Only the basic UML notations have been used (class, attribute, relationship). Each concept is represented by a class. Only the main attributes are on the diagram. Hierarchies are used to model the different types of interventions, sanitary products, and actors. These nomenclatures are represented in UML using the generalisation/specialisation relationship. This gives more flexibility to the model, providing the possibility of adding a new component for instance, and/or making existing concepts more specialised.

The second example is an extract from the theme "Land Use" (Fig. 5). It illustrates the use of hierarchies. It models that a farm is composed of agricultural parcels; the land use of their parcels is represented by a hierarchy (Cultural Land Use, Infrastructure, Building).

A documentation associated with classes and attributes is also described in natural language [12]. Thus, the ontology modelled in the FIM project presents:

- A vocabulary of terms;
- A set of term definitions that identify the concepts and fix the term interpretation;
- The relationships between concepts.

5 Conclusions

In the context of the FIM project, UML is used to formalise the different concepts common in agriculture (see [12]). The proposed ontology captures consensual knowledge accepted by the experts. UML enables the participants of the FIM project to discuss the concepts easily. Subsequently, these concepts are used to propose a standard data interchange format in order to facilitate and to improve the interoperability between agricultural information systems in France.

An interesting perspective will be to translate the ontology described in UML into a formal ontology (RDF, OWL, etc.) in order to offer new possibilities: produce reasoning, reach the requirement of the Semantic Web, integrate several database schemas, and so forth. In this case, one solution consists in translating the UML diagrams into a formal ontology language and then enriching the produced formal ontology. For example, it is possible to convert UML specifications into a formal ontology with Protégé and its UML Storage Backend Plug-In [28]. Then, additional specifications can be integrated into the ontology: creation of individuals, modelling of inferences, and so forth. For instance, the authors of Ref. 25 propose to start modelling an ontology with a UML class diagram. After that, this UML specification is translated into OWL with Protégé and its UML Storage Backend Plug-In. Then, inferences are defined with Description Logic in order to deduce new knowledge. A DIG reasoner is used to produce new information. A DIG reasoner supports the standard interface defined by the Description Logic Implementation Group (DIG; http://dl.kr.org/dig/). This standard provides a specification for connecting DL reasoners and different ontology modelling tools such as Protégé. A DIG reasoner provides an access interface that enables the reasoner to be accessed over HTTP.

References

1. FAO: AGROVOC – A Multilingual Agricultural Thesaurus. ftp://ftp.fao.org/gi/gil/gilws/aims/references/flyers/agrovoc_en.pdf.
2. Agro-EDI Europe: Agro-EDI Assocation Web Site. http://www.agro-edi.asso.fr.
3. Atkinson C.: Models versus Ontologies – What's the difference and where does it matter? VORTE2006, Regal Kowloon Hotel, Hong Kong, October 16th 2006.
4. Bekett, D.: RDF/XML Syntax Specification 2004. http://www.w3c.org/RDF.
5. Breu, R., Grosu, R., Huber, F., Rumpe, B., Schwerin, W.: Towards a precise semantics for object-oriented modeling techniques. In Object-Oriented Technology, ECOOP'97 Workshop Reader, 1997.
6. Clark T., Evans A., Kent S., Brodsky S., Cook S.: A feasibility study in rearchitecting UML as a family of languages using a precise OO meta-modeling approach. http://www.cs.york.ac.uk/puml/mmf/, 2000.
7. Cranefield, S.: UML and the Semantic Web. In: the International Semantic Web Working Symposium, Palo Alto, 2001.
8. Cranefield S., Purvis., M.: UML as an ontology modelling language. In Proceedings of the Workshop on Intelligent Information Integration, 16th International Joint Conference on Artificial Intelligence (IJCAI-99), 1999.
9. Donini, F., Lenzerini, M., Nardi, D., Schaerf, A.: Reasoning in description logics. In Gerhard Brewka, ed., Principles of Knowledge Representation, Studies in Logic, Language and Information, pages 193–238. CSLI Publications, 1996.
10. Drummond N., Shearer R.: The Open World Assumption or sometimes its nice to know what we don't Know. University of Manchester. http://www.cs.man.ac.uk/~drummond/presentations/OWA.pdf.
11. Gasevic D., Djuric D., Devedzic V.: Model Driven Architecture and Ontology Development. Springer, 328p, 2006.
12. GIEA: GIEA (FIM) Web site. http://www.projetgiea.fr.
13. Gruber, R.: Towards principles for the design of ontologies used for knowledge sharing. In International Workshop on Formal Ontology, Padova, Italy, 1993. Available as technical report KSL-93-04.
14. Gruber, R.: A translation approach to portable ontology specifications. Knowledge Acquisition vol. 5(2), pp. 199–220, 1993
15. Guarino, N., Masolo, C., Vetere, G.: OntoSeek: content-based access to the Web. IEEE Intelligent Systems vol. 14(3), 70–80, 1999.
16. Guizzardi, G., Falbo, R.A., Pereira Filho, J.G.: Using objects and patterns to implement domain ontologies. Journal of Brazilian Computer Society (JBCS), Special Issue on Software Engineering vol. 8(1), 2002.
17. Guizzardi, G., Wagner, G., Guarino, N., van Sinderen, M. An Ontologically Well-Founded Profile for UML Conceptual Models. Lecture Notes in Computer Science, vol. 3084. Springer, 2004.
18. IBM: Ontology Definition Metamodel, June 2006.

19. Philippe, M.: Translations between UML, OWL, KIF and the WebKB-2 Languages (For-Taxonomy, Frame-CG, Formalized English). Technical Report, May/June 2003.
20. Miller, G., Beckwith, R., Fellbaum, C., Gross D., Miller, K.: Introduction to WordNet: an on-line lexical database. International Journal of Lexicography vol. 3 (4), 1990, Revised August 1993. ftp://ftp.cogsci.princeton.edu/pub/wordnet/5papers.ps.
21. Nardi, D., Brachman, R.J.: An Introduction to Description Logics. In F. Baader, D. Calvanese, D.L. McGuinness, D. Nardi, P.F. Patel-Schneider, eds. Description Logic Handbook. Cambridge University Press, 2002.
22. National Committee for Information Technology Standards, Technical Committee T2 (Information Interchange and Interpretation): Knowledge Interchange Format, draft proposed American National Standard (dpANS). NCITS.T2/98-004. http://logic.stanford.edu/kif/dpans.html.
23. Object Management Group: Unified Modeling Language, version 1.5, March 2003. http://www.omg.org/docs/formal/03-03-01.pdf.
24. Object Management Group: UML 2.0, OCL specification, May 2006.
25. Pinet, F., Ventadour, P., Brun, T., Papajorgji, P., Roussey, C., Vigier, F.: Using UML for ontology construction: a case study in agriculture. In: the 7th AOS Workshop on Ontology-Based Knowledge Discovery: Using Metadata & Ontologies for Improving Access to Agricultural Information, Bangalore, India, November 2006.
26. SANDRE: SANDRE Web site. http://sandre.eaufrance.fr/rubrique.php3?id_rubrique=60&lang=en.
27. Schreiber, G.: A UML Presentation Syntax for OWL Lite. Technical Report, 2005.
28. Standford University: Protégé, 2005. http://protege.stanford.edu.
29. Studer, R., Benjamins, V., Fensel, D.: Knowledge engineering: principles and methods. IEEE Transactions on Data and Knowledge Engineering vol. 25, 161–197, 1998.
30. W3C: Semantic Web, 2001. http://www.w3.org/2001/sw/.
31. W3C: OWL Web Ontology Language:Overview. W3C Recommendation 2004-2-10. http://www.w3.org/TR/owl-features/.
32. W3C: Resource Description Framework (RDF): Concepts and Abstract Syntax W3C. W3C Recommendation February 2004. http://www.w3.org/TR/2004/REC-rdf-concepts-20040210/.
33. XML Schema Part 2: Datatypes Second Edition. W3C Recommendation, October 2004. http://www.w3.org/TR/xmlschema-2/.
34. Wiederhold G.: Mediators in the architecture of future information systems. IEEE Computer vol. 25(3), 38–49, 1992.
35. Wikipedia: AGROVOC - Web Site. http://en.wikipedia.org/wiki/AGROVOC.

Modeling External Information Needs of Food Business Networks

Melanie Fritz

Abstract Awareness of threats and opportunities in the business and competitive environment is crucial for sustainable economic success of every company. It becomes even more important in the food sector where companies are parts of interdependent business networks. Scanning and monitoring the business environment for competitive intelligence has received a substantial push by the emergence of the Internet and the information provided there. Efficiency considerations favor joint, industry-wide market and competition monitoring systems for companies in food networks. Therefore, a crucial prerequisite is modeling the network's external information needs. Modeling external information needs of agrifood networks is difficult because not all areas of the business environment are equally relevant to all companies, and every company has its own and distinct perspective on it. This chapter presents a guideline for modeling the differentiated external information needs in food networks and their transfer to a monitoring system infrastructure. The guideline consists of two phases: organizing the tasks and activities to perform and results to obtain. The first phase regards the analysis and differentiation of the external information needs in the business network; the second phase deals with the transfer of the differentiated information need to the processes and structure of a supporting software system and includes the design of a categorization scheme and appropriate personalization filters.

1 Introduction

In times of globalization and increasing competitiveness, awareness of threats and opportunities in the business and competitive environment is crucial for sustainable economic success of every company. It becomes even more important in the food sector where companies are parts of interdependent business networks. Networks

M. Fritz (✉)
International Research Center on Food Chain and Network Research and Department of Food and Resource Economics, Division for Business Management, Organization and Information Management, University of Bonn, Bonn, Germany
e-mail: m.fritz@uni-bonn.de

P.J. Papajorgji, P.M. Pardalos (eds.), *Advances in Modeling Agricultural Systems*,   149
DOI 10.1007/978-0-387-75181-8_8, © Springer Science+Business Media, LLC 2009

of companies are situations where companies are highly interdependent on one other [14, 21] and where awareness of developments of markets, products, business partners, and competitors becomes a crucial prerequisite for economic success for companies in networks in general (see [5]).

A prominent source for information about competition and markets is the vast number of distributed information sources provided through the Internet [1, 3, 4, 7, 15, 16]. However, a satisfactory and efficient use of these information sources for competition monitoring in food networks requires a focused, systematic, and automated scanning of their content and the linkage of scanning procedures and results with specific information needs of network companies. It is apparent that enterprises in food networks have, to some extent, comparable and analogous market-related information interests, equally relevant to all network companies. In this situation, feasibility and efficiency concerns ask for a monitoring system design where the scanning of the competitive environment builds on these analog information needs of the enterprises in the network. However, the needs of individual enterprises and their own and distinct perspectives on the respective external environment require a personalized, flexible, and specific information supply, which builds upon the individual perspective on the competition environment and easily adapts to changing situations.

Both the requirements for efficiency and individualization can be combined in dynamic and encompassing market and competition monitoring system applications for company networks where individualization and flexibility are provided through appropriate features for personalization and information filtering.

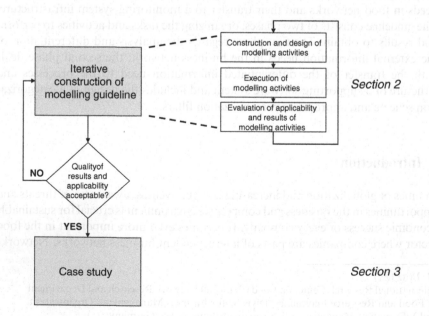

Fig. 1 Structure of the chapter

To facilitate personalization of the information for network firms, information requirements in the network need to be carefully analyzed to differentiate information needs common to the entire network from those specific to individual network firms. To successfully and systematically fulfill this task, a structured guideline for modeling the network's external information needs is required as a basis for the development of a monitoring system application.

However, existing literature does not provide a systematic guideline for this. It is the purpose of this chapter to present a guideline for modeling the external information needs of a business network to serve as a structured analysis and design guideline for the differentiation of external information needs in company networks and their transfer to an information system infrastructure. The chapter (Fig. 1) presents the modeling guideline resulting from an iterative construction process and its structure, activities, and methods to support the analysis and design (Section 2) and shows the results of its evaluation in a case study (Section 3).

2 Guideline for Modeling External Information Needs in Networks

An Internet-based market and competition monitoring system for food networks should automatically collect relevant information from Web sites offering high-quality information about trends and changes in the competitive and market environment. It should provide the information to network companies in a personalized way. In principle, Internet-based market and competition monitoring involves the process phases and process sequence shown in Fig. 2. This process is the operationalization of a conceptual framework integrating relevant knowledge bases and expertise from different scientific fields [8].

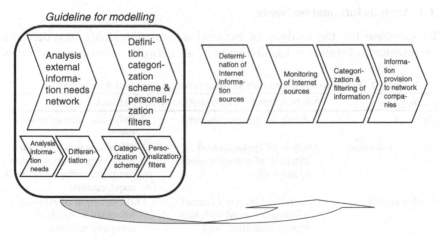

Fig. 2 Process phases for market and competition monitoring

The first two process phases regard the analysis of information needs in the network and the definition and design of a categorization scheme with personalization filters and represent the guideline presented in this paper. The outcome of the analysis and design activities of these process phases is a crucial input for the following phases; it decides on the quality of the information provision to network companies.

The guideline for modeling the external information need systematically structures analysis and design tasks in sequential process phases, links these tasks with suitable methods and tools, and defines results to acquire within each task [10, 18]. The construction of the guideline builds on the combination of available generic scientific expertise with complementary own developments for the particular situation in company networks. This development method is according to the principles of stepwise, cyclic system development [13]. In the following, the developed guideline for modeling the external information need and the underlying activities are presented with results of their application.

2.1 Analysis and Differentiation of the External Information Needs in Supply Networks

The first process phase of the guideline deals with the analysis and differentiation of external information needs in networks. The objective of this first phase is to analyze and map the external information needs of the network. The intended result is the representation of the differentiated information needs. An essential requirement in this phase is enabling the systematic differentiation of external information needs common to all network companies from external information needs specific to individual network firms. The first process phase consists of two subphases: organizing the tasks for analysis and differentiation of the external information needs in a network (Table 1).

2.1.1 Analysis Information Needs

The subphase for the analysis of external information needs considers the agglomerated information requirements of all network companies and takes

Table 1 First process phase with subphases for analysis and differentiation

Process Phase	Analysis External Information Needs Network	
Subphases	1. Analysis information needs	2. Differentiation information needs
Tasks and activities	Analysis of agglomerated external information needs in network.	Differentiation of map with agglomerated information needs according to network requirements.
Intended results	Hierarchical map of external information need with key topics, indicators, and required information.	Differentiation of external information needs in company network.

the perspective of the companies on their external environment. It can build on a well-established yet generic scientific basis. The critical success factors method defines an information need from companies' factors for success assessed in expert interviews and derives indicators for monitoring the current state of the success factors [22]; the five forces method assesses suppliers, buyers, rivalry among competitors, substitutes, and barriers to entry to analyze an industry [20]. This guideline for modeling the external information need integrates both approaches to facilitate an encompassing analysis of the external information need of networks, which complements specific information needs derived from the companies themselves with "ideal" forces for industry analysis. The intermittent analysis of companies' critical success factors allows for taking into account dynamically changing information needs of network companies.

Critical success factors analysis and the five forces indicate key topics in the company environment to be monitored such as, for example, competitors, distributors, suppliers, or product quality (see first column in Table 2). These key topics are the starting point for the construction of a hierarchical map of external

Table 2 Map of external information need for market monitoring in company networks

Key Topic	Indicators	Information Required
Access to distribution channels	Relation of used distribution channels to entire number	Market share of different distribution channels
	Contribution of own products to success of distribution partner	Turnover of own products
		Contribution of own products to earnings of distribution partner
		Image of own products
	Contribution of competitors' products to success of distribution partner	Turnover of competitors' products
		Contribution of competitors' products to earnings of distribution partner
		Image of competitors' products
	Personal relationships to distributors	Who in which position
Access to supply channels	Price developments input	Input prices
	Personal relationships to suppliers	Who in which position
	Economic situation of suppliers	Profit or losses of suppliers
		Expansion at suppliers
		Disinvestments of suppliers
Success of marketing measures	Change in consumers' product/ image perception as opposed to competitor	Survey results product group
		Consumer opinions to product group
	Relation demands to product characteristics	Survey results product group
		Consumer opinions to product group

Table 2 (continued)

Key Topic	Indicators	Information Required
Strategies of competitors	Market segments where competitors are active	Product launches of competitors
		Product changes of competitors
		Crises of competitors' products
	Marketing campaigns of competitors	New marketing campaigns of competitors
		Number of advertisements of competitors
		Marketing funds of competitors
	Economic situation of competitors	Turnover of competitors' products
		Profit or losses of competitors
		Expansion at competitors
		Disinvestments at competitors
	Personnel changes at competitors	Changes of executives at competitors
	Changes in company organization at competitors	Form of company organization at competitors
	Changes in competitors' production methods	Kind of production method at competitors
General and sector economic development	General economic development	Figures on economic situation
	Buying power	Available net household income
		Inflation rate
	Economic development in sector	Turnover products
		Selling prices products
	Development sector and consumer politics	News sector politics
		News consumer politics
Quality of own products	Quality of supplies	Reports regarding suppliers' products
	Customer quality perception	Results from market research
		Customer opinion on product quality
	Conformity products – regulation	Changes in laws and regulations
Success of substitutes	Market segments where substitutes are active	Product launches of substitutes
		Product changes of substitutes
		Crises of substitutes
	Marketing campaigns substitutes	New marketing campaigns for substitutes
		Number of advertisements for substitutes
		Marketing funds substitutes
	Changes in production method substitutes	Kind of production method
"Breaking news" in sector	News related to sector in the daily news	Content of the news

information needs of the supply network in a stepwise and systematic analysis and derivation process. To build the hierarchical map, appropriate information indicators have to be derived for the key topics; information required to monitor the indicators has to be delineated. Table 2 shows the stepwise delineation and mapping of the external information need of network companies from key topics and indicators.

2.1.2 Information Needs Differentiation

The second subphase deals with the differentiation of the agglomerated external information needs in the company network. It leaves the perspective of companies on their external environment and takes the view on the network "from the top" to systematically distinguish where information required for monitoring the indicators is shared by all network companies and where specific, different information is required to monitor indicators. It is essential to differentiate, for example, that the very same information is needed to inform all network companies about the general market development, but that, for example, different information content is needed to inform network companies about their respective competitors even if all companies need to be informed about their competitors; whether a company in the network is a competitor for another company or, for example, a supplier or buyer depends on the respective position of the company in the supply network.

This differentiation of the information need is the decisive key prerequisite for the realization of a market monitoring system for supply networks. It defines the configuration of the subsequent monitoring activities and adapts the monitoring system to the situation and external information needs in the company network. It therefore ensures the supply of network firms with personalized information about their specific competitive environment.

The development of activities for the differentiation of the information need of network companies could not build on an established knowledge base. In a combination of a heuristic approach and logical reasoning, activities were developed for the differentiation between

- parts of the hierarchical information needs map where the very same information content fits for the entire network and
- parts of the information needs map where specific, different information content is required for vertical and horizontal groups of companies in the network (Fig. 3) [12, 17].

The tool developed for differentiation is a matrix system with checklists serving as analysis instruments and supporting the extraction and acquisition of expertise regarding the situation in a specific network. Both matrix system and checklists represent the decisive key to the modeling of the differentiated information need and therefore the development of a market monitoring system for company networks.

The matrix system (Fig. 4) serves as instrument for the analysis and mapping of the structure of the regarded company network. The analysis and mapping of

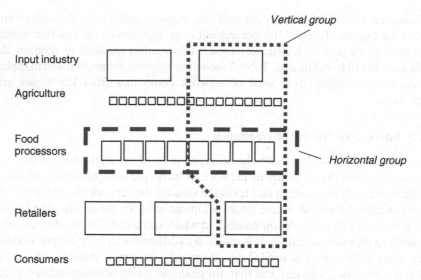

Fig. 3 Food production network with vertical and horizontal groups of companies

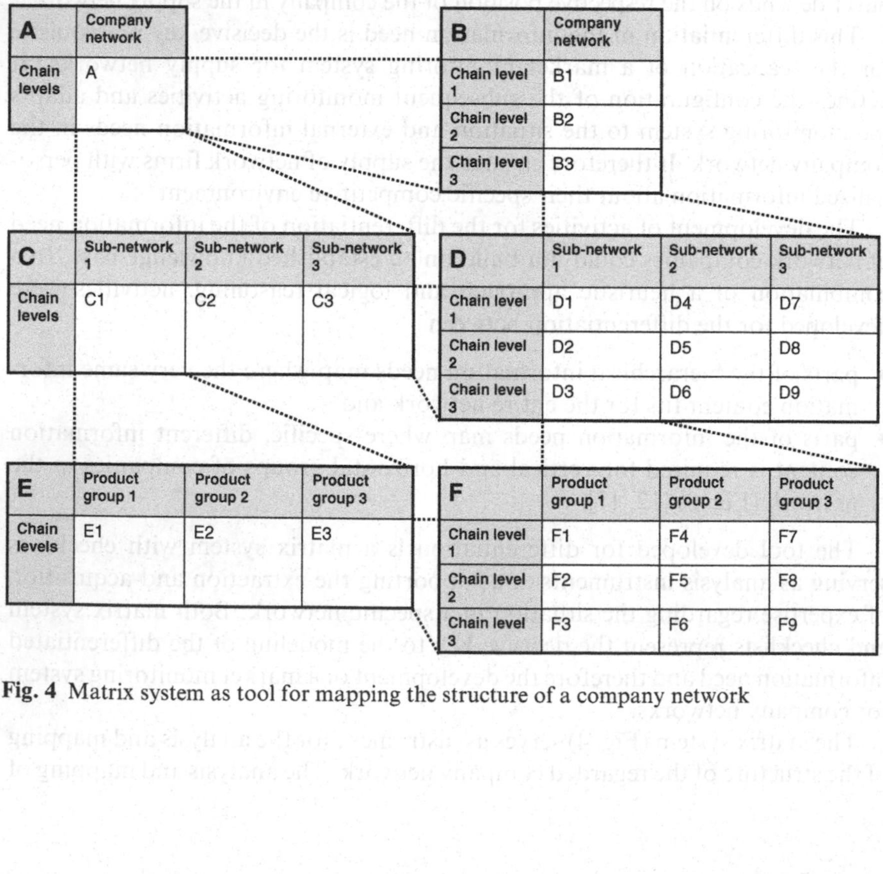

Fig. 4 Matrix system as tool for mapping the structure of a company network

the structure is necessary as a preparation for the modeling of the differentiated information need. To map the structure of the company network, the matrix system distinguishes

- The number and kind of horizontal chain levels;
- The number and kind of vertical subnetworks; and
- The number and kind of product groups as parts of the subnetworks.

Figure 4 shows the matrix system for a company network with three vertical and three horizontal segments and three product groups, respectively, with matrixes A, B, C, D, E, and F. Matrix A represents the entire network. The structure of this company network emerges from the subdivision of the network in different horizontal chain levels (B) and vertical subnetworks (C), and so forth. Matrix D shows the intersection of chain levels and subnetworks.

Once the structure of the network has been mapped in the matrix system, the hierarchical information needs map must be distributed across the appropriate matrixes in order to differentiate the external information need in the company network. This process is supported by checklists with guiding questions and selection criteria (Table 3). They are to be considered as "soft guidelines" supporting the extraction of expertise about the situation in a concrete supply network and show the way to modeling the differentiated external information need.

Table 3 Checklists for the differentiation of the external information need in company networks

Matrix	Objective of selection: all information need elements where the information content...	Guiding questions and selection criteria
A	...is equal and relevant for all companies in the network.	Information about economic situation, buying power, developments in politics relevant to the sector, "emergencies" in the sector
B	...is equal and relevant for all companies of a joint chain level.	Situation on the chain level: Different companies producing different product categories? Higher differentiation of the information needs regarding product-specific information required. Companies producing many product categories? Less differentiation of the information need required. Situation in the vertical chain: Different power allocation in the vertical chain and dependency on other chain level? Information about other chain level highly important! No dependencies? Information about this chain level less relevant. Otherwise: Information about competitors, customers/consumers.

Table 3 (continued)

C	...is equal and relevant for companies in a vertical subsector.	Information about developments specific to the subsector regarding the economic situation in the subsector, product quality in the subsector, legislation.
D	...is equal and relevant for companies in a joint intersection of subsector and chain level.	Information about development of input prices, suppliers.
E	...is equal and relevant for companies in a joint product group.	Information about product qualities.
F	...is equal and relevant for companies in a joint intersection of subsector and product group.	Situation on the chain level: Different companies producing different product categories? Higher differentiation of the information needs regarding product-specific information required. Companies producing many product categories? Less differentiation of the information need required. Situation in the vertical chain: Different power allocation in the vertical chain and dependency on other chain level? Information about other chain level highly important! No dependencies? Information about this chain level less relevant. Otherwise: Information about competitors, customers/consumers.

Figure 5 shows the application of matrix system and checklists for the differentiation of the external information need of companies in the food network across the matrixes mapping the structure of the network. The figure includes the indicators of the hierarchical information needs map (see Table 2). Their position in any one of the matrixes implies whether an indicator can be filled with the very same information contents for certain parts of the network or must be filled with different, specific information contents for other parts of the company network. For example, the results in matrix D show that the indicator "Price developments inputs" requires different information content for the subnetworks 1, 2, and 3 as inputs vary depending on the vertical production chain in the food network (milk, meat, or grain). The external information need of a particular company in the network is represented by the respective combination of matrixes and cells of matrixes according to the position of the company in the network.

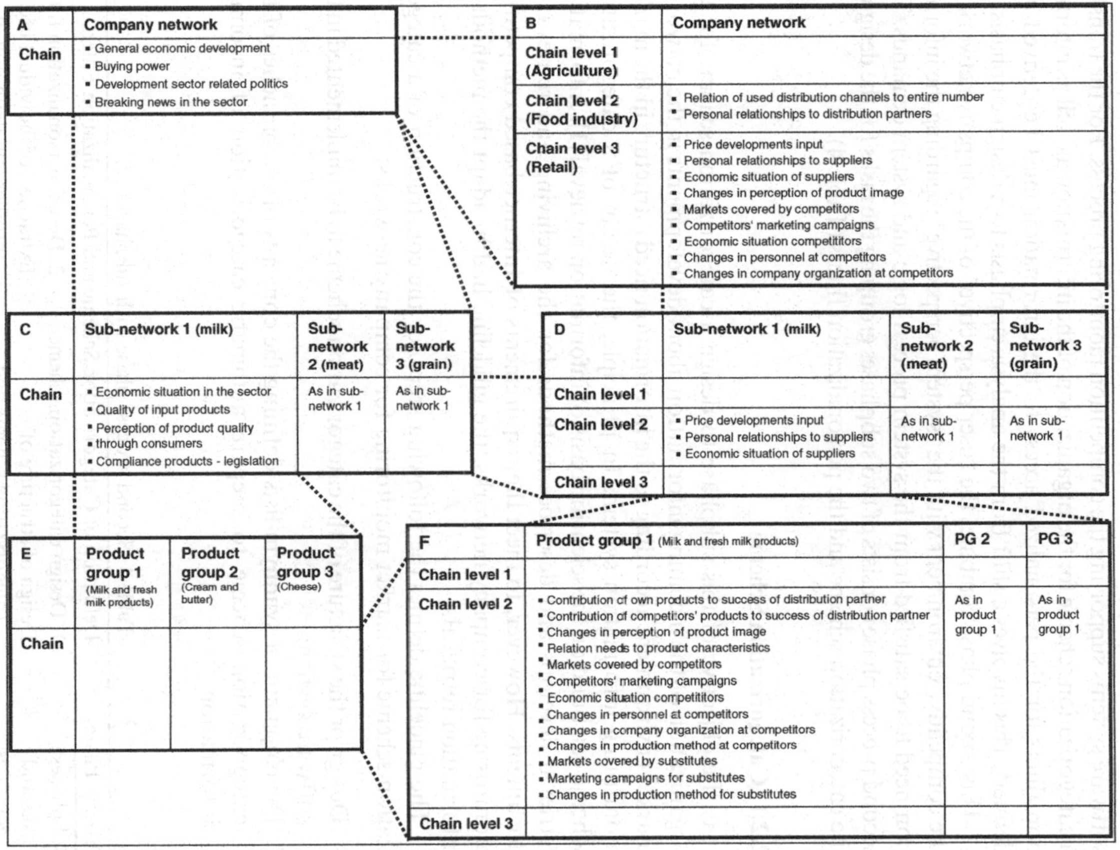
A
Company network
Chain
▪ General economic development
▪ Buying power
▪ Development sector related politics
▪ Breaking news in the sector
B
Company network
Chain level 1 (Agriculture)
Chain level 2 (Food industry)
▪ Relation of used distribution channels to entire number
▪ Personal relationships to distribution partners
Chain level 3 (Retail)
▪ Price developments input
▪ Personal relationships to suppliers
▪ Economic situation of suppliers
▪ Changes in perception of product image
▪ Markets covered by competitors
▪ Competitors' marketing campaigns
▪ Economic situation competititors
▪ Changes in personnel at competitors
▪ Changes in company organization at competitors
C
Sub-network 1 (milk)
Sub-network 2 (meat)
Sub-network 3 (grain)
Chain
▪ Economic situation in the sector
▪ Quality of input products
▪ Perception of product quality
▪ through consumers
▪ Compliance products - legislation
As in sub-network 1
As in sub-network 1
D
Sub-network 1 (milk)
Sub-network 2 (meat)
Sub-network 3 (grain)
Chain level 1
Chain level 2
▪ Price developments input
▪ Personal relationships to suppliers
▪ Economic situation of suppliers
As in sub-network 1
As in sub-network 1
Chain level 3
E
Product group 1 (Milk and fresh milk products)
Product group 2 (Cream and butter)
Product group 3 (Cheese)
Chain
F
Product group 1 (Milk and fresh milk products)
PG 2
PG 3
Chain level 1
Chain level 2
▪ Contribution of own products to success of distribution partner
▪ Contribution of competitors' products to success of distribution partner
▪ Changes in perception of product image
▪ Relation needs to product characteristics
▪ Markets covered by competitors
▪ Competitors' marketing campaigns
▪ Economic situation competititors
▪ Changes in personnel at competitors
▪ Changes in company organization at competitors
▪ Changes in production method at competitors
▪ Markets covered by substitutes
▪ Marketing campaigns for substitutes
▪ Changes in production method for substitutes
As in product group 1
As in product group 1
Chain level 3

Fig. 5 Differentiated external information needs model for food networks

2.2 Categorization Scheme and Personalization Filters

The objective of this second phase of the guideline for modeling the external
information need of a company network is to transfer the differentiated external
information needs model of the company network to the structure and processes of
software systems supporting the competition monitoring process. For the transfer,
a categorization scheme for the organization of the information as well as information filters for the personalized access to the information must be derived and
defined. This involves a shift from the analysis emphasis to a design emphasis.

This second phase combines the "user perspective" of the company network on
the competitive environment with the "system perspective" regarding the information need to be satisfied from the system process logic and system demands. The
second process phase consists of two subphases grouping the tasks for the design of
the categorization scheme and the personalization filters (Table 4).

2.2.1 Categorization Scheme

It is the objective of this subphase to design a categorization scheme for the
organization of the incoming information about developments in the network's
market environment according to the information needs structure in the market
monitoring information system. In principle, the design of a categorization
scheme can build on the scientific basis of information retrieval [1]. Information
retrieval develops and describes methods for the archiving and retrieval of
documents. However, to meet the requirements of Internet-based competition
monitoring for company networks, the guideline had to adapt the methods of
information retrieval.

The guideline defines the following steps for the construction of a categorization scheme for market monitoring for company networks:

1. Design of the structure of the categorization scheme to be implemented in the
   software system;
2. Development of synonym lists defining the concepts of the branches of the
   categorization scheme to prepare automatic categorization of incoming
   information.

Table 4 Second process phase with subphases

Process Phase	Definition of Categorization Scheme and Personalization Filters	
Subphases	1. Design categorization scheme	2. Design personalization filters
Tasks and activities	Design of structure of categorization scheme and development of synonym lists	Extraction of individual user profiles and definition of search queries
Intended results	Categorization scheme for differentiated information needs in network and synonym modules	Personalization filters for individualized information provision

To define the structure of the categorization scheme, the guideline integrates the generic map of the information needs of the company network (see Table 2) with the results of the differentiation analysis for the particular network (see Fig. 5). Result of this integration is an enriched hierarchical structure of the differentiated external information needs in the network. As the differentiation requirements vary for different food networks, the hierarchic map depends on the situation in a concrete network and may not be defined universally.

Table 5 shows an example of the integration of the external information need model with the differentiation results. The third column is the key to the particular structure of the categorization scheme for a specific supply network as it combines the generic information needs model with the differentiation results and therefore defines the structure of the categorization scheme. The letters and numbers in the third column link the information needs model to the matrix system mapping the differentiated information needs.

For the development of synonym lists for the branches of the categorization scheme, the guideline for modeling the external information need of the network combines available thesauri with expertise on the situation of a specific company network. A thesaurus is a vocabulary collection covering and organizing the terminology of a field. The guideline foresees a three-step route with generic economic thesauri, sector-specific thesauri, and scenario-specific expertise leading to three separate modules of generic synonyms to be used in market monitoring in general, synonyms specific to the sector, in which a network operates, and synonyms particular to a specific network scenario.

An example for a thesaurus containing general economic terms is "OECD Macro-thesaurus"; examples for a thesaurus containing words for, for example, the food sector are "Agrovoc" or the "CAB Thesaurus." In addition to the terms to extract from thesauri, scenario-specific expertise is required to include, for example, brand or company names and persons' names relevant to the network in the respective branches of the categorization scheme. Table 6 shows the three synonym modules for the example of the food network.

Table 5 Example for categorization scheme

Key Information Topics	Indicators	Information Content to Differentiate For
Access to distribution channels	Relation of used distribution channels to entire number	B2
	Contribution of own products to success of distribution partner	F2 F5 F8
	Contribution of competitors' products to success of distribution partner	F2 F5 F8
	Personal relationships to distribution partners	B2

Table 6 Example of synonym modules

Key Information Topic	Indicator and Information Need For B2	Concepts	Modules		
			Generic Synonyms	Sector-Specific Synonyms	Scenario-Specific Synonyms (for Germany)
Access to distribution channels	Relation of used distribution channels to entire number; market share of food distribution channels	Distribution channel	Distribution partner, distribution system, marketing channel, chain of distribution, sales channel, retail, discounter, discount shop, wholesale trade	Food retailer, supermarket, grocery shop, grocery store	Markant, Metro, Edeka, Rewe, Aldi, Tengelmann, Spar, Lidl, Norma, Dohle Handelsgruppe, Lekkerland-Tobaccoland, Schwarz-Gruppe, Penny
		Market share	Market volume, market leader, market power		

2.2.2 Personalization Filters

It is the objective of this subphase to superimpose the user perspective on the categorization scheme to realize the personalized information provision to network companies. This means that on the categorization scheme, which is the differentiated information need of the network "seen from above" and implemented in a software system structure, personalization filters realizing the "company perspective" for the different network companies on their competitive environment need to be imposed. The intended purpose of the personalization filters is to guide the incoming stream of information according to the information need of the individual network companies.

The design of the personalization filters could build on the knowledge from the research areas of information filtering and user modeling [6, 19, 11, 2, 23], which could be applied to the specific application scenario for the design of the personalization filters. The core of the fields of information filtering and user modeling is a user model representing a user's information need in the structures of an information system.

The categorization scheme containing the differentiated information need of the company network stands for the aggregation of the single user models of the companies in the network and contains their individual perspectives on their external environment. For the development of personalization filters for competition monitoring in networks, two tasks are defined:

- The extraction of the individual information needs of the network companies from the aggregated perspective; and
- The definition of search queries for the branches of the categorization scheme to represent the information need in the software system.

For the implementation of personalization filters, the single-user models need to be extracted from the aggregated categorization scheme according to the position of a company in the network and the respective relevant areas of the categorization scheme resulting from Fig. 5 and Table 5. The mode of functioning of the personalization filters is that for different companies only those areas of the categorization scheme relevant to them are visible and accessible resulting in an effective filtering of the information.

For the transfer of the users' information needs models of the food network to the system infrastructure of a market monitoring system, suitable search queries are required for every branch of the scheme. The queries can build on the synonym lists, which have been developed for the categorization scheme. Their definition depends on the specific rules and functionalities of the supporting technology tool working with, for example, free text search or search with Boolean operators such as AND or OR.

3 Evaluation of the Modeling Guideline

The evaluation of the guideline for modeling the external information needs of a business network has the objective to test its usefulness and applicability in the support of the analysis and differentiation of the external information needs in a company network and the design of an appropriate information infrastructure [9]. The guideline is a complex combination of activities and tasks and combines generic knowledge bases of different scientific fields with specific developments for the acquisition of specific expert knowledge regarding the scenario of market and competition monitoring for company networks. This complexity of the guideline for modeling inhibits the exact determination and analysis of embedded cause-and-effect relationships. Certain specificities of the guideline reduce the complexity of the evaluation needs: The different process steps build on distinct knowledge bases and generate separate, specific results. Therefore, the evaluation can focus separately on the single process phases and their results. As the guideline is linked to various scientific areas, the evaluation can bypass basic elements of the process rules where logical reasoning and scientific knowledge convince without further testing. For the guideline, this argument can be accepted for the approaches selected for the analysis of the external information need of companies.

The iterative development of the guideline (see Fig. 1) building on the cyclic inclusion of experiences already made represents a first testing of the applicability of the model. In a second step, a case study for the company network producing dairy products was performed with a convincing version of the guideline for modeling the external information need of a company network in the food sector to test the appropriateness of the rules and methods embedded for the activities and tasks for identifying and differentiating the information need of a company network.

In the case study, the modeling rules of the guideline were activated and the related activities and tasks performed to analyze and differentiate the external information needs of the network and to design and transfer and implement the categorization scheme and personalization filters to the technological system infrastructure of CleverPath Portal, a suitable market monitoring software from Computer Associates. The successful process activation for the case study represents in itself a testing of the applicability of the rules embedded in the modeling guideline. On the basis of the information provided by the market monitoring software system, the quality and purposefulness of the guideline in terms of personalization of the information for different network companies was evaluated together with experts regarding the information need of dairy network companies. The evaluation showed that the application of the guideline for modeling the external information needs had led to a market monitoring system that facilitated the personalization of the external market and competitive information according to the demands of the network companies [8].

4 Conclusions

In times of globalization and increasing competition, awareness of developments in the competitive environment has become a crucial element for economic success. Networks of companies are situations where companies are highly interdependent on one other and where awareness of developments of markets, products, business partners, and competitors is a crucial prerequisite for economic success for companies in networks in general. For market and competition monitoring systems that exploit economies of scale and build on joint monitoring activities for company networks, individualized information access needs to be supported through appropriate features for personalization and information filtering.

This chapter has presented a guideline for modeling the external information needs of a business network to serve as a structured analysis and design guideline for the analysis and differentiation of external information needs in company networks and their transfer to a monitoring system infrastructure. The guideline links suitable scientific research areas as generic knowledge bases with specific developments for the extraction of expertise on the situation in a company network. The modeling guideline was developed in an iterative process building on feedback loops and consists of two phases: organizing the tasks and activities to perform and results to obtain. The first phase regards the analysis and differentiation of the external information needs in the network; the second phase regards the transfer of the differentiated information need to the processes and structure of a supporting software system and includes the design of a categorization scheme and appropriate personalization filters. The guideline and the underlying activities were tested in a case study for the dairy network. The case study has shown that the modeling guideline supports the analysis and differentiation of the external information need in company networks in a purposeful way.

References

1. Baeza-Yates, R., Ribeiro-Neto, B. (1999). Modern Information Retrieval. ACM Press, New York.
2. Belkin, N, Croft, B. (1992). Information filtering and information retrieval: two sides of the same coin? Communications of the ACM 35 (12), 29–38.
3. Chen, H., Chau, M., Zeng, D. (2002). CI spider: A tool for competitive intelligence on the Web. Decision Support Systems 34: 1–17.
4. Choo, C.W., Detlor, B., Turnbull, D. (2000). Web Work. Information Seeking and Knowledge Work on the World Wide Web. Kluwer, Dordrecht.
5. Day, G.S., Schoemaker, P.J.H. (2005). Scanning the periphery. Harvard Business Review (November)11: 135–148.
6. Denning, P.J. (1982). Electronic junk. Communications of the ACM 25 (3): 163–165.
7. Desouza, K.C. (2001). Intelligent agents for competitive intelligence: survey of applications. Competitive Intelligence Review 12 (4): 57–63.
8. Fritz, M. (2005). Market and Competition Monitoring for Company Networks. New Potentials Through the Internet [in German]. DUV – Gabler, Wiesbaden.
9. Galliers, R.D. (1992). Choosing information system research approaches. Galliers, R.D. (ed.) Information System Research: Issues, Methods, and Practical Guidelines. Blackwell, Oxford: 144–162.
10. Glissmann, S., Smolnik, S., Schierholz, R., Kolbe, L., Brenner, W. (2005). Proposition of an m-business procedure model for the development of mobile user interfaces. Brookes,

W. Lawrence, E., Steele, R., Chang, E. (Eds.), Proceedings of the 4th International Conference on Mobile Business (mBusiness), Sydney, Australia, IEEE CS, pp. 308–314.

11. Hanani, U., Shapira, B., Shoval, P. (2001). Information filtering: overview of issues, research and systems. User Modeling and User-Adapted Interaction 11: 203–259.

12. Hausen, T., Fritz, M., Schiefer, G. (2006). Potential of electronic trading in complex supply chains: an experimental study. International Journal of Production Economics 104 (2): 580–597.

13. Heinrich, L.J. (2001). Wirtschaftsinformatik. Einführung und Grundlegung. 2nd edition. Oldenburg, München, Wein.

14. Lazzarini, S.G., Chaddad, F.R., Cook, M.L. (2001). Integrating supply chain and network analyses: the study of netchains. Journal on Chain and Network Science 1 (1): 7–22.

15. Liu, S. (1998). Business environment scanner for senior managers: Towards active executive support with intelligent agents. Expert Systems with Applications 15: 111–121.

16. McGoonagle, J.J., Vella, C.M. (1999). The Internet Age of Competitive Intelligence. Quorum, Westport.

17. Ménard, C., Klein, P.G. (2004). Organizational issues in the agrifood sector: toward a comparative approach. American Journal of Agricultural Economics 83 (3): 750–755.

18. Müller, G. (1981). Strategische Frühaufklärung. Ph.D. thesis, University of Munich, Munich.

19. Oard, D.W. (1997). The state of the art in text filtering. User Modeling and User-Adapted Interaction 7: 141–178.

20. Porter, M. (1980). Competitive strategy: Techniques for Analyzing Industries and Competitiors. The Free Press, New York.

21. Powell, W.W. (1990). Neither markets nor hierarchy. Network forms of organization. Research on Organisational Behaviour 12: 295–336.

22. Rockart, J.F. (1979). Chief executives define their own data needs. Harvard Business Review 58 (March-April): 81–93.

23. Stadnyk, I., Kass, R. (1992). Modeling users' interests in information filters. Communications of the ACM 35 (12): 49–50.

Enterprise Business Modelling Languages Applied to Farm Enterprise: A Case Study for IDEF0, GRAI Grid, and AMS Languages

Vincent Abt, Frédéric Vigier, and Michel Schneider

Abstract New requirements of society and demands imposed by the Common Agricultural Policy (CAP) reform have increased the complexity level required from farm management systems. New approaches for assisting farm management are needed to ensure a successful production system that is friendly to our environment and able to provide high-quality information management. Successful management of farm activities requires the collection, storage, and manipulation of a considerable amount of information. Farm enterprises are complex systems that we need to model to facilitate knowledge capitalization, business process reengineering and integration, change-management, and information system (IS) design and integration. Enterprise Modelling Languages allow business requirements models to be defined. Such business models help to align the business strategy, the organization infrastructure, and the IS infrastructure but are rarely used in the agricultural sector. We propose in this chapter to apply three Enterprise Modelling Languages from the industrial sector to farm enterprise. After a brief presentation of modelling language diversity, we will present the three Enterprise Modelling Languages (IDEF0, GRAI Grid, and AMS) and illustrate their application to farm enterprise through case studies. We will then discuss their usability and the interest of Enterprise Modelling Languages compared with other modelling languages (such as Unified Modelling Language) presented in other chapters in this book.

1 Introduction

The new requirements of society (e.g., traceability, multifunctionality, risk management, etc.) impose an important transformation of the European agricultural sector. Technical innovations, economic and environmental regulations, and Common Agricultural Policy reforms are also factors speeding up this transformation.

V. Abt (✉)
Cemagref, Clermont-Ferrand, France
e-mail: vincent.abt@cemagref.fr

P.J. Papajorgji, P.M. Pardalos (eds.), *Advances in Modeling Agricultural Systems,*
DOI 10.1007/978-0-387-75181-8_9, © Springer Science+Business Media, LLC 2009

Farm enterprises are particularly affected in terms of production, management, and organization. In today's highly competitive global economy, farm enterprises are confronted, as any manufacturing enterprise, with permanent changes in their internal and external environment. They must meet customer requirements and be innovative to produce quality products and services at low cost in a dynamic environment. Farm enterprises produce commodities but also raw materials, energy (fuel oils), and services (tourism, environmental services, research support, etc.). They need to ensure a successful production system and require a powerful management of farm activities through collection, storage, and manipulation of a considerable amount of information.

Farm enterprises are also complex systems we need to understand and formalize to facilitate knowledge capitalization, business process reengineering and integration, change-management, and information system (IS) design and integration [19]. Farm systems need to be modelled in terms of business processes, operations and decisions, organization and resources, and so forth. To identify business requirements, they need to be modelled regarding an enterprise system (ES) user viewpoint (e.g., farm operator, farm manager) without any consideration of supporting solutions or management tools.

In an IS design and integration perspective, business models are used to support IS specifications. Explicit business models should help to align the business strategy, the organization infrastructure, and the IS infrastructure. To provide successful and high-quality information management in farm enterprises, IS design must integrate business specifications to support appropriate farm business operations and decision making. Then IS designers and modellers establish IS models regarding an information system user viewpoint (e.g. IS operator).

Modelling languages are used to formalize enterprise and information systems regarding different aspects and viewpoints. Modelling languages, based on graphical syntax and defined semantics, facilitate communication between actors such as farm managers and farm operators, IS designers and IS users, economic advisers and technical advisers. Business modelling languages allow business requirements models to be defined at a "business level," which describes enterprise operations to be done in a business sense, in terms of enterprise operations, information, resource requirements, responsibilities, and authorities [19]. Such models and modelling languages are rarely used in the agricultural sector. They are either too informal (i.e., schematic) or too formal (i.e., computational) to make knowledge more explicit and to provide communication and support design.

We propose in this chapter to apply to farm enterprise three Enterprise Modelling Languages coming from the industrial sector. After a brief presentation of modelling language diversity, we will present the three modelling languages (IDEF0, GRAI Grid, and AMS) and illustrate their application to farm enterprise using case studies. We will then discuss their usability and the interest of business modelling languages compared with other modelling languages such as Unified Modelling Language (UML) presented in other chapters in this book.

2 Modelling Languages

2.1 Modelling Language Diversity

System designers need models to get an "as-is" and "to-be" representation of the system. A model is expressed in terms of a modelling language. A modelling language is a set of constructs for building models and can be described by its syntax and semantics. Syntax defines what are the legal constructs of the language. Semantics defines the meaning of the expressions written in the language [3].

According to their syntax and semantics, modelling languages are more or less understandable and easy-to-use and have sufficient expressive power or not. Depending on the goal of modelling, the selected modelling language should be adequate or competent for the purpose of the modelling task [3]. The most formal languages are mathematics, the less formal are natural languages. "In between, many forms of languages exist to model things or reality such as symbolic languages, graphical or diagramming languages, semi-formal languages, and formal description techniques" [19]. Semiformal languages are very good for common understanding and communications among people because of their graphical, easy-to-grasp syntax or formalism. Usually, they make use of diagrams comprising boxes, circles, and arrows and of established semantics. Examples of semiformal languages are Petri-nets, state-transition diagrams, and data flow diagrams (DFDs), but we will present below the UML set of languages and Enterprise Modelling Languages [2].

Modelling language features depend on life-cycle stages (e.g., specification, design, implementation) or systems aspects (e.g., data, function, resources) to consider. Modelling languages must above all be easy to interpret for the intended audience (e.g., IS or ES user, analyst or designer). They must make knowledge more explicit and facilitate knowledge transmission.

To achieve the goal of business requirements definition, it is essential to use an adequate formalism for representing farm enterprise according to business and management viewpoints (operations, resources, organization). Consequently, this representation is more easily understood by farm-related actors. Graphical formalisms, based on business semantics and a well-established syntax, are very useful to facilitate communication between these actors.

2.2 One or Several Modelling Languages?

Facing the diversity of existing languages, UML was created 10 years ago. UML is nowadays the reference for IS design and software engineering [13]. UML proposes a set of generic languages (use-case diagrams, class diagrams, activity diagrams, etc.). Syntax is mainly graphical, and semantics is based on object-oriented approaches.

In the model driven architecture (MDA) approach, UML is used to obtain system models for CIM (computation independent model) and PIM (platform independent model) levels [18]. UML is then used to represent IS requirements (CIM level) and IS design specifications (PIM and PSM levels). UML allows models transformation from one level to another. The communication between actors is facilitated by using the same modelling language (integration perspective).

The MDA approach introduces, however, difficulties to model business requirements with UML at the conceptual level (CIM) [4]. In the model driven interoperability (MDI) approach, the CIM level is then divided into a Top CIM level and a Bottom CIM level [4]:

- The Bottom CIM level aims at defining the requirements of the enterprise information system to develop. Models cover the part of the enterprise system that needs to be implemented in a computerized system, that is to say, to show the IS requirements but without linking them to any specific kind of technology. UML allows models to be defined at this level.
- The Top CIM level aims at representing a company from a "holistic" point of view. Models show enterprise domain, business, strategy, and so forth, at high level of abstraction and without considering any IS or software application features. Enterprise Modelling Languages (EMLs) allow models to be defined at this level. Different EMLs are suitable in this case to represent the diversity of business aspects and viewpoints. Through interoperability, conversion from one language to another is encouraged (interoperability perspective).

Two concepts exist of what business modelling languages should be. The first alternative promotes a unique and generic modelling set of languages (UML) in an integration perspective. The second alternative considers that UML is not sufficient to develop requirements models on all business aspects of the whole enterprise system [12]. It considers that it is suitable to get different languages mobilizing a whole range of syntactic and semantic constructs to represent the business of an enterprise. Interoperability between languages is then preferred to integration. In the next paragraph, we introduce EMLs.

2.3 Enterprise Modelling Languages

In the manufacturing sector, enterprise modelling is the "art of externalizing enterprise knowledge which adds value to the enterprise or needs to be shared" [20]. "An enterprise can be perceived as a set of concurrent processes executed on the enterprise means (resources or functional entities) according to the enterprise objectives and subject to business or external constraints" [19]. Enterprise modelling consists in making models of the structure, content, and behaviour of an enterprise using EMLs. EMLs are mostly semiformal languages and expressed in terms of a graphical language based on defined syntax and semantic. They facilitate in this way reusability, explicit knowledge making, and enterprise description. They allow a whole range of enterprise aspects (e.g., processes, operations, decisions, functions, resources, information, organization) to be modelled in a business sense (Table 1).

EMLs are used by one or several enterprise modelling methodologies according to modelling frameworks (e.g., CIMOSA, GIM, IDEF, ARIS) [20]. EMLs "define the generic modelling constructs for enterprise modelling adapted to the needs of people creating and using enterprise models. In particular EML will provide constructs to describe and model human roles, operational processes and their functional contents as well as the supporting information, office and production technologies" [11]. Constructs are business concepts such as business process, domain, enterprise activity, decision center, function, operation, resource, organization unit, and so forth. Each language uses a specific set of business concepts to focus on specific aspects. Coherence between constructs and languages in the same methodology is required to guarantee the coherence between models.

2.4 Case Study of Three Enterprise Modelling Languages

EMLs come from the industrial sector. They have been successfully used in this sector but rarely used in the agricultural sector. Are these languages not appropriate to this sector? We propose in this chapter to use three EMLs from the industrial sector to model farm enterprise systems: IDEF0, GRAI

Table 1 Examples of existing EMLs to model different aspects of the enterprise system

Enterprise Aspects	Enterprise Modelling Languages
Function	IDEF0 language
	CIMOSA Domain Decomposition
	CIMOSA Domain Definition
	IDEF3 language
	GRAI Net
Process	CIMOSA Process Definition
	GRAI Extend Actigrams
	ARIS Event Driven Process Chain
Information	IDEF1x language
	GRAI E/R Diagram
Decision	GRAI Grid
	GRAI Net
Organization	ARIS Organization Chart
	AMS language

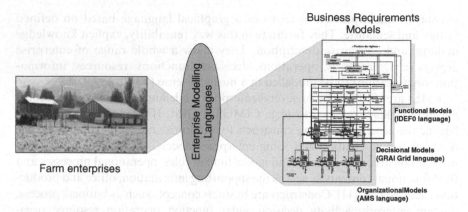

Fig. 1 EMLs to establish business requirements models of a farm enterprise

Grid, and AMS languages. Each of the three modelling languages covers a particular enterprise aspect to be modelled: function, decision, organization (Fig. 1). Some initiatives are using the IDEF0 language in the agricultural sector [6]. But as far as we know, GRAI Grid and AMS languages have not yet been used for farm enterprises. We will demonstrate their usability and interest through the elaboration of farm models, even if farm enterprise systems present particularities that must be taken into account.

The IDEF0 language (Integration DEFinition language 0) allows the functional aspect of the system to be described [16]. The GRAI Grid language (Graphes à Résultats et Activités Interreliés) enables production management and control decisions to be identified and organized into time periods and horizons [7]. The AMS language (Analyse Modulaire des Systèmes) enables production operations and controls to be identified and organized into workshop units [15]. Few modelling languages exist to model organizational aspects of an enterprise system, which is an argument to mobilize such an old language.

It is not the purpose of this chapter to serve as an exhaustive reference manual for the three languages. However, this chapter aims at serving as a clear introduction to the language properties (basic syntax and semantics). This chapter aims at presenting first farm models obtained with these languages. It is hoped that the chapter will quickly enable the reader to understand models and their interest and maybe allow him to construct simple models with these languages.

3 IDEF0 Language and Business Functional Models

3.1 IDEF0 Language Presentation

The IDEF0 language is based on Structured Analysis and Design Technique (SADT), developed by Douglas T. Ross and SofTech, Inc. [10]. The IDEF0 language belongs to the IDEF family of languages [14]. The language was developed with the goal of representing in a standard manner activities or functions that are typically carried out in an enterprise.

The IDEF0 definition of a function is "a set of activities that takes certain inputs and, by means of some mechanism, and subject to certain controls, transforms the inputs into outputs" (Fig. 2) [16].

The IDEF0 language semantic is then based on five major concepts:

- **Activities** are the functionalities of the system.
- **Inputs** are elements to be processed by the activity (e.g., files, documents, raw materials, products).
- **Controls** are elements like laws, policies, standards, and unchangeable facts of the environment. They control, direct, or force the execution of the activity but are not modified by it.
- **Outputs** are elements produced or modified by the activity (e.g., data, materials, products).
- **Mechanisms** are means to execute the activity. They are resources (human or material) that are used in bringing about the intended goals of the activity.

The IDEF0 language syntax is based on boxes and arrow segments. Activities are represented by boxes. Each box has a label: an active verb or verb phrase that describes the activity. Inputs, controls, outputs, and mechanisms are represented by arrows. Arrows may branch: branches may represent either the same thing or portions of the same thing.

Box and arrow segments are combined in various ways to form diagrams. The boxes in a diagram are connected by sequences of arrow segments. IDEF0 models are hierarchically arranged IDEF0 diagrams (Fig. 3). Unlike every other diagram in the model, the top-level diagram (context diagram, numbered A-0) contains only one box. This box represents, at the coarsest granularity, the single high-level activity that is being represented and decomposed in the IDEF0 diagrams. The parent–child relation holding between two diagrams signifies that the parent node is the decomposition of a box in a parent node. A decomposition of a box is a diagram that represents a finer-grained view of the function. Diagrams are numbered. This hierarchical decomposition results

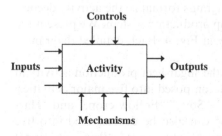

·**Input:** Elements to be processed by the activity (files, documents, raw materials, products)

·**Control:** Information which controls, directs, forces the execution of the activity but which is not modified by it (directives, rules, constraints)

·**Output:** Elements produced or modified by the activity (data, materials, products)

·**Mechanism:** Necessary means to execute the is activity (human or material resources)

Fig. 2 IDEF0 language and representation of an IDEF0 function

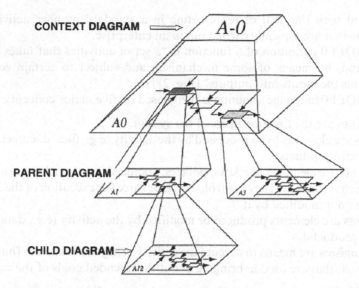

CONTEXT DIAGRAM

PARENT DIAGRAM

CHILD DIAGRAM

Fig. 3 IDEF0 hierarchical decomposition of boxes and diagrams

in both wide-scope and detailed representations of system activities. By convention, a detail diagram contains three to six boxes.

3.2 IDEF0 Business Functional Models

To illustrate the use of the IDEF0 language in the agricultural sector, we present IDEF0 models elaborated in a French crop–livestock farm. The farm considered in this example is located in the department of Puy de Dôme. The farm employs five full-time persons. It works 284 hectares (ha) of agricultural area (crops and grassland), owns a herd of 50 dairy cows, and fattens 200 bull-calves a year. For this case study, we will focus on crop production, and especially on sugar beet production.

IDEF0 language is used to elaborate models to understand and formalize the sugar beet production activity. IDEF0 diagrams formalize the activity decomposition and the main flows inside the crop production system. We present the hierarchical decomposition of the activity in Fig. 4. Each detailed diagram is then presented in Figs. 5, 6, and 7.

IDEF0 diagrams presented here allow the sugar beet production activity in the farm enterprise to be formalized and decomposed into five major activities: "Plan annual production," "Prepare soil," "Sow," "Follow crop," and "Harvest" (Fig. 6). The activity "Follow crop" can also be decomposed into five nonsequential subactivities: "Observe crop," "Fight weeds," "Bring additional fertilizing," "Fight plant diseases," and "Fight devastators" (Fig. 7).

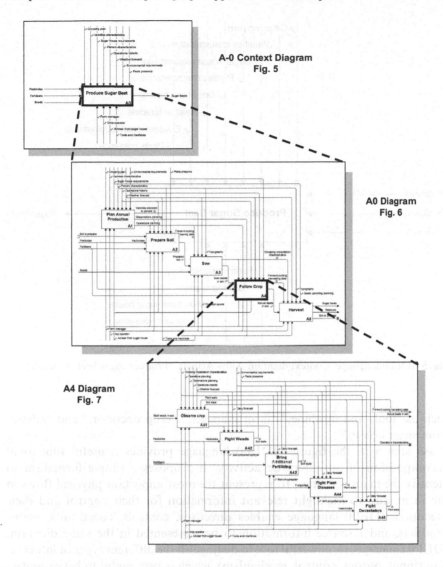

Fig. 4 IDEF0 language: The hierarchical decomposition of the activity "Produce sugar beet"

IDEF0 diagrams also present and characterize different physical flows (fertilizers, pesticides, seeds, sugar beets). They precise relevant information, controls, and resources needed to perform activities at different precision levels in the farm. For example, to "prepare soil" (Fig. 6), information is required on "observations and operations planning," "parcels characteristics," "operations historic," "weather forecast." These elements presented as controls show that they have to be integrated by the operator at this level of detail. Resources are also required

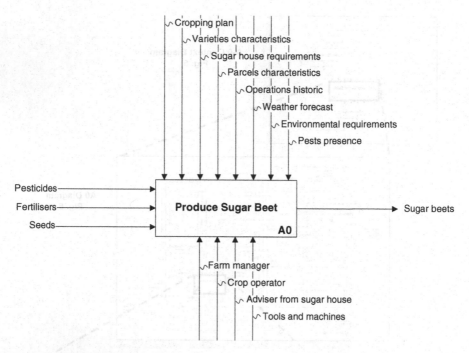

Fig. 5 IDEF0 language: Context diagram A-0 – Activity "Produce sugar beet" to model

such as "tools and machine," "farm manager," "crop operator," and "adviser from sugar house."

As shown in the figures, IDEF0 language provides a useful functional decomposition of an enterprise activity. It proposes a simple formalism to identify the main activities, to represent the most important physical flows in the farm, and to highlight relevant information for their control and their execution. IDEF0 language enables directives, controls, constraints, plans, products, and resource information to be represented in the same diagram. IDEF0 language is a powerful tool to distinguish the different types of information (input, output, control, mechanism), which is very useful to better understand the place of information in a system. IDEF0 language then allows information requirements to be defined at different precision levels using a business terminology. IDEF0 models can constitute in this way a semantically rich picture of requirements, understandable by different communities. They could potentially be complementary to other approaches and be reused by system designers. For example, they allow management tools requirements to be identified: Figure 8 presents the different management tools that could be developed or implemented to manage sugar beet production. IDEF0 models are in this way very useful to communicate between enterprise system analysts and information system designers.

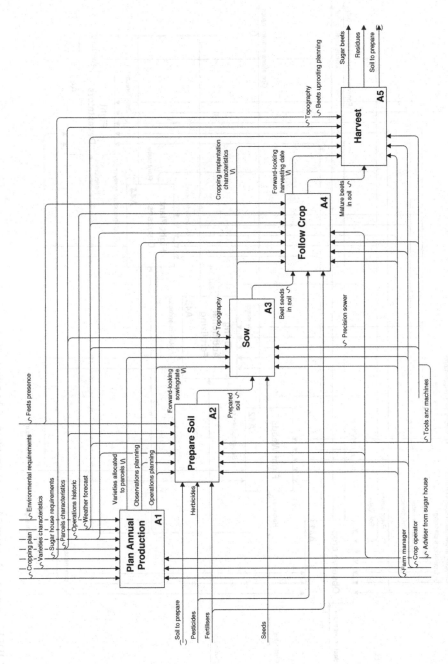

Fig. 6 IDEF0 language: Diagram A-0 – Decomposition of the activity "Produce sugar beet"

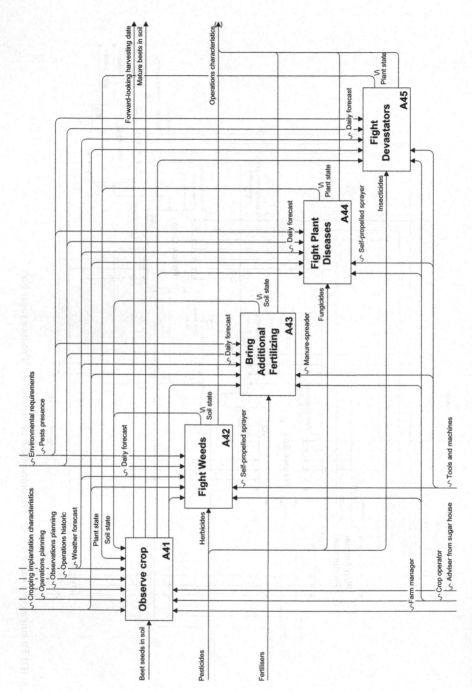

Fig. 7 IDEF0 language: Diagram A4 – Decomposition of the activity "Follow crop"

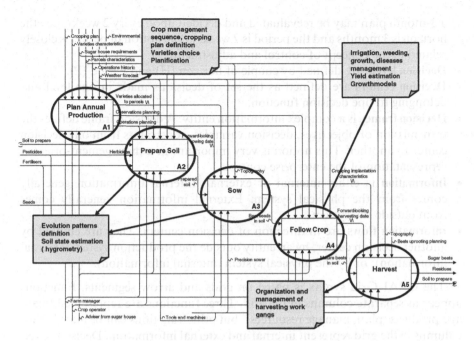

Fig. 8 IDEF0 language: IS and management tools requirements identification to manage sugar beet production (Diagram A-0)

4 GRAI Grid Language and Decisional Models

4.1 GRAI Grid Language Presentation

The GRAI Grid language was developed to analyze production management decisions and specify production management system [5]. The GRAI Grid language focuses on the decisional aspects of the management systems. "The GRAI Grid does not aim at the detailed modelling of information processes, but puts into a prominent position the identification of decision centres where decisions are made in order to manage a system" [7].

The GRAI Grid semantics is based on several concepts to model a decisional system [7]:

- **Functions** are the domains of decision. These functions (e.g., Plan production, Manage products, Manage resources) are fundamental for the management of any kind of system.
- **Horizons** (H) represent the expiration of the decision. A Horizon is the part of the future taken into account by a decision (e.g., when a plan is made for 3 months, the horizon is 3 months). The notion of horizon is closely related to the concept of planning.
- **Periods** (P) represent the frequency of questioning. A Period is the time that passes after a decision when this decision must be reevaluated (e.g.,

a 3-month plan may be reevaluated and decided upon every 2 weeks, i.e., the horizon is 3 months and the period is 2 weeks). The notion of period is closely related to the concept of control and adjustment.

- **Decision levels** are defined by couple of horizon (H) and Period (P).
- **Decision centers** are defined as the set of decisions made in one level and belonging to one decision function.
- **Decision frames** is a complex information entity. A decision frame defines the transmission of objectives, decision variables, constraints from one decision center to another. This notion is very important in a management-oriented representation of the enterprise control.
- **Information** is either internal or external. Internal information generally comes from the physical system. External information generally comes from outside the system.
- **Information flows** are in direction of decision centers. They are emitted by another decision center, by an entity outside the production system (external information), or by the physical system (internal information).

The GRAI Grid syntax is based on grids and arrow segments. Functions appear as separate columns in the grid. Three functions are represented (manage products, plan, manage resources), but other functions can be added. Two columns in the grid represent internal and external information. Decision levels are represented by the rows of the grid. A decision center is located at the intersection of a function (column) and a decision level (row). Two decision centers are connected using two types of links: information flows (simple arrow) or decision frame (double arrow) (Fig. 9).

The GRAI Grid is elaborated step by step through meetings gathering business managers and operators. Useful functions and decision levels are

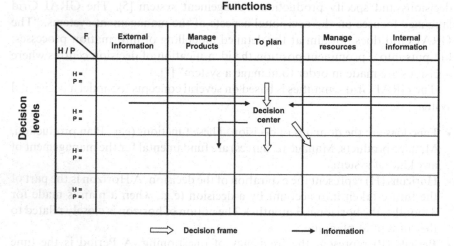

Fig. 9 The GRAI Grid description

defined. Progressively, decision centers are identified and discussed to elaborate the "as-is" GRAI decisional grid of the system. From this GRAI Grid, analysts are able to propose changes in the organization to provide a better management.

4.2 GRAI Grid Decisional Models

To illustrate the use of the GRAI Grid language in the agricultural sector, we present a GRAI Grid model elaborated in a French crop–livestock farm. The farm considered in this example is located in the department of Lot. The farm employs two full-time persons and one part-time person. It works 85 ha of agricultural area (crops and grassland) and owns a herd of 90 suckler cows. For this case study, we will focus on animal production, and especially the breeding unit.

The GRAI Grid focuses on the animal production breeding unit (Fig. 10). It allows production management to be understood and formalized. A detailed attention is given to animal production control (calves birth, fattening), resources management (suckler cows allocation), and "maintenance" (suckler herd replacement and feeding). Livestock feed supply and human/technical resource management are not represented in this grid.

The GRAI Grid presented in Fig. 10 structures the organization of the decisions into five management functions ("Manage products"; "Buy" and

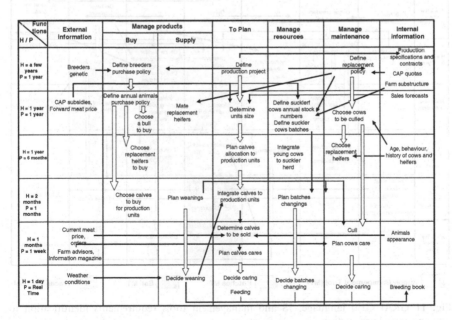

Fig. 10 GRAI Grid: Production management of a breeding unit

"Supply"; "Plan"; "Manage resources"; "Manage maintenance") and six decision levels from strategic to operational ones. The representation of decision centers inside the grid clarifies type and horizon/frequency of decisions that have to be taken. For example, focusing on the planning function, the grid informs that the definition of the production project is defined for several years (H = a few years) but can be modified every year (P = 1 year). Every year, the definition of this production project constrains (by a decision frame) the decision of determining the unit size, which constrains every 6 months the planning of calves allocation to production units, and so forth.

The GRAI Grid also presents and clarifies information flows that are very useful to make decisions. For example, "CAP subsidies" and "forward meat price" is mobilized to "determine unit size" and "define suckler cows annual stock number." Studying management decisions, the GRAI Grid gives also precision on information origin (internal/external) according to columns, and information mobilization frequency according to rows. It gives an idea how often the information is mobilized (and must be updated); for example, "Current meat price" and "orders" are mobilized weekly.

Furthermore, the GRAI Grid allows management tools to be better identified. Figure 11 presents the different management tools or decision support systems (DSSs) that could be developed and integrated to manage animal

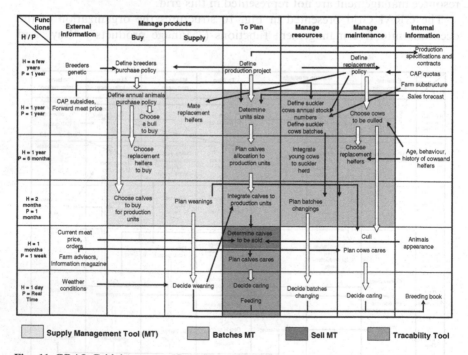

Fig. 11 GRAI Grid language: IS and management tools requirements identification to manage animal production

production according to functions and decision levels. Example: A "batches management tool" could be then designed to manage products at a tactical level corresponding with horizons between 1 year and 2 months.

5 AMS Language and Organizational Models

5.1 AMS Language Presentation

The AMS language was developed by Jacques Mélèse to analyze and design production management systems [15]. This language is 30 years old but proposes an interesting business modelling language to represent production and control subsystems as coordinated modules and different flows inside a system. It inspired particularly the GRAI Grid language. At present, the AMS language is not frequently used. Nevertheless, it is one of the rare existing graphical languages to represent organizational aspects. For this reason and because of the richness of its semantics and syntax, we are presenting this language in this chapter.

The AMS semantics is based on several concepts to model production and control subsystems:

- **Technical modules (TMs)** (or production modules) perform flows transformation according to enterprise objectives. A TM is the set of means and flows to provide the conversion of materials, products, and services. A TM (or production unit) can be decomposed into more detailed technical and control modules.
- **Control modules (CMs)** control and regulate the operations performed by the production module, mobilizing and creating different information types. A CM is the set of means and flows to provide the conversion of control and regulation directives.
- **Main flows** (input/output) are flows that justify the existence of TMs. They are transformed by the TM (e.g., materials, products, services).
- **Side flows** (input/output) are necessary flows to the transformation performed by TMs.
- **Operative and informative flows** (input/output) are necessary or additional information flows to support the activities of a TM (input) or to be produced by a CM (output) (e.g., production process range, assembly plan).
- **Control variables** are imposed directives. They set the objectives to attempt by the module, the rules, and constraints. A CM details, precises, and adapts control variables to the characteristics of a TM.
- **Regulation variables** are readjustment variables defined by the controller according to internal and external information.
- **Essential variables** are indicators to follow and evaluate the realization of the TM activities.
- **Internal and external information** is information coming from the TM (internal) or from outside the system (external).

The AMS syntax is mainly based on boxes and arrow segments. Modules are represented by boxes. Main flows are represented by thick arrows, and side flows by fine arrows. Operative and information flows as well as internal and external information are represented by dotted arrows. Control variables are represented by double arrows, regulation variables by wavelet arrows, and essential variables by arrows inside circles (Fig. 12).

AMS models are elaborated step by step. Useful technical modules are defined. Main and side flows are next identified. Control modules are then defined as well as all the other flows.

5.2 AMS Organizational Models

To present the use of the AMS language in the agricultural sector, we present an AMS model elaborated in a French crop–livestock farm. The farm considered in this example is the same as the one considered in the GRAI Grid example. For this case study, we will focus on the animal production unit. The farm sells three main beef productions: milk-fattened calves, fattened calves, and fattened

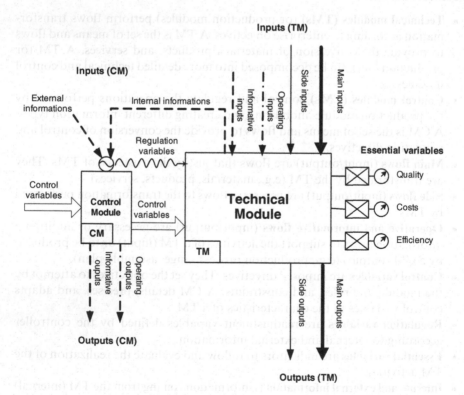

Fig. 12 AMS language: Modules and flows

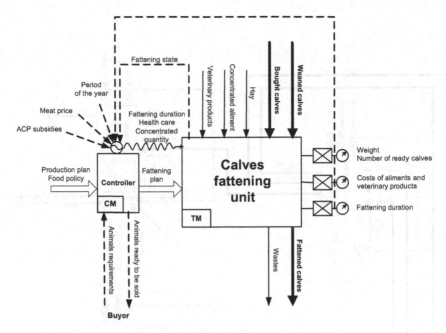

Fig. 13 AMS language: Zoom in the "Calves fattening unit and control"

cows. The AMS language allows production units (workshops) to be understood and formalized.

Figure 13 focuses on the "Calves fattening unit" and its control. It gives precision on flows processed by the workshop activity: weaned and bought calves, hay, concentrate aliments, and veterinary products are used to produce fattened calves and wastes. Figure 13 presents also main directives to control the fattening unit (production plan, food policy, fattening plan). It presents indicators (weight, number of ready calves, costs of aliments and veterinary products, fattening duration) and other information (period of the year, meat price, CAP subsidies) to regulate the fattening unit through regulation variables (fattening duration, health care, concentrate quantity).

Figure 14 presents the whole organization of animal production in the considered farm. The animal production is organized into five different units (workshops): "Breeding unit," "Suckling calves fattening unit," "Calves fattening unit," "Cows fattening unit," and "Replacement unit." Figure 14 presents the main "physical" flows (cows, heifers, calves) between units and gives an idea how the different units are linked to each other in order to sell three main productions: milk-fattened calves, fattened calves, and fattened cows. Every unit is described as well as the "Calves fattening unit" presented in Fig. 13. A precise description of how production units are controlled and what information is useful is then available.

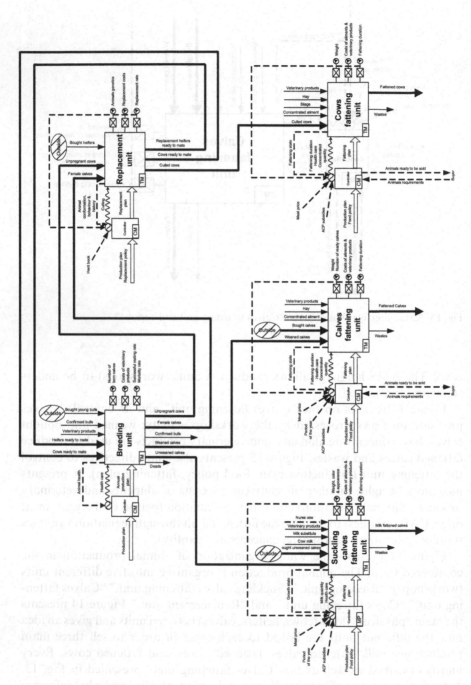

Fig. 14 AMS language: Farm organization of animal production units

6 Discussion

6.1 Interest of Enterprise Modelling Languages

The different presented models demonstrate the interest of EMLs to model farm enterprise. Each language enriches the traditional description of crop–livestock farms. It enables business requirements in terms of activities, organization units, functions, and decisions in the enterprise system to be defined. This business formalization is a powerful tool to first understand the enterprise system before designing IS. This study confirms in addition the success of the GRAI Grid language to a "new" sector: the agricultural one. This success is essentially due to the "comprehensive, one-page model generally sufficient to explain the structure of the management system" [7]. Despite the complexity of the language, this study confirms also the interest of the AMS language to analyze the consistency of the control system units. The decomposition into modules is quite similar to the farmer workshops representation, which helps the farmers to understand models and communicate with modellers.

Considering business aspects, this business formalization permits structured approaches and a quite exhaustive identification of any kind of information the enterprise system needs to manage. Information is then characterized in its context of production, mobilization, and transmission:

- The IDEF0 language allows the enterprise functional aspect to be modelled. Through the representation of enterprise activities, different natures of information can be identified: inputs, outputs, controls, and mechanisms. Management tools requirements can furthermore be pointed out (Fig. 8).
- The GRAI Grid language allows the enterprise decisional aspect to be modelled. Through the representation of enterprise decision centers, internal and external information can be identified and structured according to decision levels. Decision frames indicate also information transmission but do not permit, according to the formalism, to precise its content. Management tools and DSSs can nevertheless be pointed out (Fig. 11).
- The AMS language allows the enterprise organizational aspect to be modelled. Through the representation of organization units and controls, a wide range of different information and flows can be identified and characterized: main and slide flows, operative and informative flows, internal and external information, control, regulation, and essential variables. The language can also precise elements to design manufacturing execution systems (MESs), for example.

All these languages are complementary and focus on a particular enterprise aspect. Other modelling languages (e.g., Table 1) could be investigated to cover other aspects or better characterize farm enterprise requirements in terms of management tools and information.

6.2 Need of Complementary Modelling Languages

As discussed previously, EMLs are powerful tools for explicit enterprise system requirements according to an enterprise system user point of view. This is mainly due to their rich but structured semantics. Whereas UML and its business modelling extension [8] use few business constructs such as "process, event, resource, goal, information," EMLs suggest a wide range of structuring business constructs such as "activities, functions, processes, controls, mechanisms, horizons, periods, decision levels, decision centres, decision frames, technical and control modules, decision variables and indicators, etc." For example, UML and its extension do not propose adequate means to model GRAI grids or AMS control modules. As shown in Ref. 17, the business extension of UML can be used in place of IDEF, but IDEF, particularly at the level of IDEF0, offers more semantics. Thus, the models obtained with IDEF are more expressive and easier to understand for the enterprise system final user.

EMLs are then complementary to IS design modelling languages. This study confirms the need of EMLs in the agricultural sector to better specify and design farm information systems with UML. EML and UML are also complementary to establish models at respectively Top and Bottom CIM levels for a better response to ES and IS user requirements. Although "there is no direct one-to-one mapping of constructs," EML can constitute "a semantically rich picture of requirements that potentially can be reused by system designers, builders and developers on a piecemeal and incremental basis" [12]. Interoperability between EML and UML models is in this way an interesting research perspective to improve model driven engineering [4]. For example, "information captured in the form of IDEF0 models, such as information about input data sources and mechanisms, can be recoded in the form of UML use cases" for concept design and "UML class diagrams and activity diagrams for detailed design" [12]. Semantic information previously captured and coded in the form of IDEF0 models can then be mapped into UML. In this perspective, it would be interesting to promote tools for automating the mapping [9].

Modelling languages ergonomics is at last an important point to justify language diversity and complementarity. A language needs to be adapted not only to the modeller but also to the "as-is" system user (manager, ES and IS operator) and the model users (analyst, designer), as many actors do not feel comfortable with all kind of languages. This justifies the adaptation of concepts, but also formalisms, to the concerned actors in order to facilitate understanding and communication. As we observed in our field study for farm managers, UML syntax and formalisms are more difficult to understand than are those of AMS or GRAI because of primary origins of languages. "There is not one language which is equally suited for all purposes; each language has its individual strength to meet specific modelling requirements. Some languages might be applicable for a broad range of application while others are more specialised and purposed oriented" [3].

6.3 Necessary Adaptation to Farm Characteristics

EMLs were mainly designed for studying industrial manufacturing companies from primary and secondary sectors. Their use in the agricultural sector is rather promising, but difficulties can be encountered while using these languages with farm enterprises. This is linked to the richness of their semantics, which may be different from one sector to another.

Even if models presented here are in accordance with modelling languages, problems have to be solved to better fit to farm business characteristics such as those linked to natural resources management [1]. For example, the GRAI Grid language is mainly used for manufacturing productions, in a planned management framework with short or medium production cycles. The farm activities are nevertheless very seasonal and the production cycle is pretty long. This concept of seasonality is not represented in the GRAI Grid and imposes some approximations in the modeling. We decided for example to represent the decision center "Choose replacement heifers" at the decision level (H = 1 year, P = 6 months) (Fig. 10). In reality, these decisions are seasonal and are taken in June and in October. The Horizon is correct but the period should then be 5 or 7 months. We chose finally 6 months to get a coherent GRAI Grid model, but this represents a strong approximation, and decision seasonality is not represented.

It would also be useful to introduce a spatial representation of farm activities and resources to better model farm business management and organization. Our current research aims at identifying farm business characteristics and working with other EMLs to define EML adapted to farm characteristics and covering a whole range of farm enterprise aspects.

7 Conclusions

We presented in this chapter three EMLs successfully used in the industrial sector to represent enterprise business aspects and identify enterprise and information requirements.

We presented a first application of these modelling languages to the farm enterprises. Although farm enterprise presents some strong specificities, these languages allow farm business models to be defined. These models appear particularly useful to understand farm business and to represent in business semantics the farm enterprise aspects such as functions, decisions, and organization. Using graphical formalisms and business constructs, these languages suggest an innovative viewpoint regarding information. Facilitating identification of enterprise and information requirements, these languages propose additional approaches to better design information systems and management tools. We have shown also that these languages do not oppose languages such as

UML used for designing and implementing information systems. They are rather complementary.

We are now investigating other EMLs, covering the whole range of farm enterprise aspects to be modelled (such as function, decision, resources, environment, behaviour, organization, etc.), in order to propose an appropriate integrated farm business modelling methodology.

Acknowledgments A special acknowledgment to the farm managers, Guy Durand and Jean-Pierre Latron, for their contributions. A special thanks to Jean-Baptiste Bigeon, Cédric Durand, and Cédric Vittoz for their contribution to models elaboration.

References

1. Abt, V., Pierreval, H., Nakhla, M., 2007. Evolution du contexte et nouvelles perspectives pour l'exploitation agricole en génie industriel, 7e Congrès international de génie industriel (GI 2007), Trois-Rivières, Québec, Canada, 12 p.
2. Aguilar-Savén, R.S., 2004. Business process modelling: Review and framework. International Journal of Production Economics 90 (2), 129–149.
3. Bernus, P., Mertins, K., Schmidt, G. (Eds.), 2002. Handbook on Architectures of Information Systems. Springer, Berlin.
4. Bourey, J.-P., Grangel, R., Doumeigts, G., Berre, A.J., 2007. Deliverable DTG2.3 Report on Model Driven Interoperability, INTEROP-NOE Project.
5. Chen, D., Vallespir, B., Doumeingts, G., 1997. GRAI integrated methodology and its mapping onto generic enterprise reference architecture and methodology. Computers in Industry 33 (2–3), 387–394.
6. Cunha, G.J., Aguirra Massola, A.M., Saraiva, A.M., Lobão, V.L., 2006. Continental Malacoculture Chain Modeling and Traceability Requirements, 4th World Congress on Computers in Agriculture and Natural Resources (WCCA 2006), Orlando, FL, USA, 494–499.
7. Doumeingts, G., Vallespir, B., Chen, D., 2002. GRAI GridDecisional modelling. In: Bernus, P., Mertins, K., Schmidt, G. (Eds.), Handbook on Architectures of Information Systems. Springer, Berlin, pp. 322–346.
8. Eriksson, H.-E., Penker, M. (Eds.), 2000. Business Modeling with UML. Business Patterns at Work. John Wiley & Sons – OMG Press, New York.
9. Grangel, R., Ben Salem, R., Bigand, M., Bourey, J.-P., 2007. Interopérabilité guidée par les modèles: transformation de modèles GRAI en modèles UML, 7e Congrès international de génie industriel (GI 2007), Trois-Rivières, Québec, Canada, 10 p.
10. IGLTechnology (Eds.), 1989. SADT: un langage pour communiquer, Eyrolles, Paris.
11. IFIP-IFAC Task Force, 1999. GERAM: Generalised Enterprise Reference Architecture and Methodology, version 1.6.3 IFIP-IFAC Task Force, Rapport. http://www.cit.gu.edu.au/ubernus/taskforce/geram/versions/geram1-6-3/v1.6.3.html
12. Kim, C.-H., Weston, R.H., Hodgson, A., Lee, K.-H., 2003. The complementary use of IDEF and UML modelling approaches. Computers in Industry 50 (1), 35–56.
13. Letters, F., 2002. Modeling information-systems with UML Unified Modeling Language. In: Bernus, P., Mertins, K., Schmidt, G. (Eds.), Handbook on Architectures of Information Systems. Springer, Berlin, pp. 411–456.
14. Mayer, R.J., Painter, M.K., De Witte, P.S. (Eds.), 1992. IDEF Family of Methods for Concurrent Engineering and Business Re-Engineering Applications. Knowledge Based Systems Inc, College Station, TX, USA.

15. Mélèse, J. (Eds.), 1984. L'analyse modulaire des systèmes de gestion. Hommes et Techniques, Paris.
16. NIST, 1993. Integration Definition for Function Modeling (IDEF0). Federal Information Processing Standards – Publication 183, Springfield, VA, USA.
17. Noran, O.S., 2000. Business Modelling: UML vs. IDEF. Electronical report, Griffith University, School of Computing and Information Technology. www.cit.gu.edu.au/~noran.
18. OMG, 2003. MDA Guide version 1.0.1, Needham Heights, MA, USA.
19. Vernadat, F. (Eds.), 1996. Enterprise Modeling and Integration. Principles and Applications. Chapman & Hall, London.
20. Vernadat, F.B., 2002. Enterprise modeling and integration (EMI): Current status and research perspectives. Annual Reviews in Control 26 (1), 15–25.

A UML-Based Plug&Play Version of RothC

Petraq Papajorgji, Osvaldo Gargiulo, James W. Jones, and Sibiri Traore

Abstract This chapter presents the stepwise conversion of the FORTRAN-based RothC Soil Organic Carbon model into a plug&play component amenable to use as part of larger modeling frameworks. As a first step, RothC was converted into a stand-alone Java modular application to ensure consistency with the parent model. The plug&play component was then developed based on the Unified Modeling Language (UML). The plug&play component provides services that other system/components can easily use. The behavior of RothC is presented through interfaces that other system/components can implement. The use of interfaces to express behavior of components facilitates the collaboration of teams located in different geographic regions. Various UML diagrams present static and dynamic aspects of the system. UML facilitates model documentation, and the plug&play architecture facilitates implementation by other researchers, who can integrate the RothC component into their studies or systems without making extensive structural changes or recompilation of their entire modeling frameworks.

1 Introduction

RothC is a model developed to study the dynamics of organic matter in soils [7]. The RothC model has been widely used to simulate changes in SOC (soil organic carbon) content on a variety of soil types [8] and for a variety of land uses including arable land, grassland, and forestry [2]. It has also been used at the regional scale [14] and at the global scale [6, 15]. RothC is been tested in 18 different experimental treatments on 6 long-term experimental sites in Germany, England, the United States, the Czech Republic, and Australia [2]. RothC simulates the soil organic carbon trend over time by dividing it into compartments [1, 5]. Each compartment decomposes by a first-order process with its own characteristic rate. The five RothC organic carbon pools are characterized by different decomposition times. Incoming organic carbon is

P. Papajorgji (✉)
Center for Applied Optimization, University of Florida, Gainesville, FL, USA
e-mail: petraq@ufl.edu

P.J. Papajorgji, P.M. Pardalos (eds.), *Advances in Modeling Agricultural Systems*,   193
DOI 10.1007/978-0-387-75181-8_10, © Springer Science+Business Media, LLC 2009

accumulated in one of two pools of plant material organic carbon: decomposable plant material (DPM) and recalcitrant plant material (RPM). Incoming organic carbon passes through these compartments once only, and all incoming carbon is assumed to belong to one or the other compartment. DPM and RPM both decompose to the same products: CO_2 (lost from the system), microbial biomass (BIO), and humified organic matter carbon (HUM) [7].

Our objectives were to:

1. Redesign RothC, originally developed in a traditional programming language (FORTRAN), using Unified Modeling Language (UML) to create a stand-alone version implemented in Java programming language [13]. Furthermore, our goal is to create an object-oriented version that can be implemented in more comprehensive models that need to simulate the SOC.
2. Design a plug&play component so that other researchers can integrate it in their models without making extensive structural changes or recompilation of their entire modeling frameworks. The Java RothC version was redesigned using the UML approach [12]. We believe that a detailed description of this approach is useful to other researchers as it provides architectural solutions that can be built with today's requirements and be flexible enough to meet future needs. On the other hand a component that works in harmony with other parts of an existing system is very promising. The plug&play architecture allows functionality encapsulated in a class/component to be replaced by an alternative solution when appropriate. Moreover with the plug&play technology, it is possible to add new functionalities without changing the existing system [10].

The plug&play version of RothC could be linked to software such as DSSAT (Decision Support System for Agrotechnology Transfer) [4] and SEAMLESS [17]. First, a stand-alone system was developed to ensure the accuracy of the calculations. UML diagrams were constructed to study the relationships between concepts involved in this system. The results obtained by the stand-alone system were compared with those obtained by the RothC implemented by Traore et al. [18].

2 The RothC Model

This section presents a short description of the RothC model. RothC approach simulates the soil organic matter trend by dividing it into five compartments as shown in Fig. 1.

The first two of the five compartments considered in RothC (DPM and RPM) represent plant carbon added monthly from crop residues. Incoming organic carbon passes through these compartments once only, and all incoming carbon is assumed to belong to one or the other. DPM and RPM both

Fig. 1 The structure of the RothC model. DPM, decomposable plant material; RPM, recalcitrant plant material; BIO, microbial biomass; HUM, humified organic matter carbon; IOM, inert organic matter; CO_2, carbon dioxide (lost from the system); FYM, farmyard manure; FOM, fresh organic matter

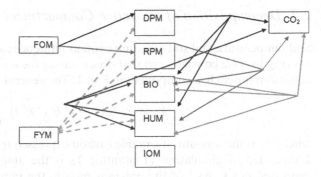

decompose to the same products: CO_2 (lost from the system), BIO, and HUM. When HUM decomposes, more CO_2, more BIO, and fresher HUM are formed. The additional amounts of products created are calculated using the typical decomposition rates of each fraction. The soil is also assumed to contain a small organic compartment that is inert to biological attack. This compartment is referred to as inert organic matter (IOM) [7]. The incoming plant material is partitioned between the DPM and RPM pools, according to the vegetation type of the soil, by a predefined DPM/RPM ratio. The reader is referred to the original RothC model papers [2, 3] for details of this soil carbon model. In this chapter, we focus on the implementation of the model and describe the inputs and interface necessary to run the model.

2.1 RothC Data Requirements

The RothC model requires the following input data [3]:

1. Monthly rainfall (mm).
2. Monthly evapotranspiration (mm).
3. Average monthly mean air temperature (°C).
4. Clay content of the soil (%).
5. An estimation of the decomposability of the incoming plant material; the DPM\RPM ratio.
6. Soil cover: necessary to indicate whether or not the soil is vegetated. Carbon decomposition has been found to be faster in fallow soils than in crop soils, even when the crop soil is not allow to dry out [8].
7. Monthly input of plant residues (ton Carbon per hectare (t C/ha)): the plant residue input is the amount of carbon that is put in the soil per month (t C/ha), including carbon released from roots during crop growth.
8. Monthly input of farmyard manure (FYM) (t C/ha). If any amount is input, then it is treated quite differently from fresh plant residues inputs, because of different decomposition rates.
9. Depth of soil layer sampled (cm).

2.2 Decomposition of an Active Compartment

Soil temperature and soil moisture content influence carbon decomposition by altering the rate constants, so that decay during the month follows exponential equations for each C component in Fig. 1. The general form of the equations is

$$Y_1 = Y_0 \, \mathrm{EXP}(-a^* \, b^* \, c^* \, k^* \, t)$$

where Y_1 is the amount of organic carbon expressed in t C ha^{-1} obtained after 1 time-step of simulation (1 month); Y_0 is the amount of organic carbon expressed in t C ha^{-1} of the previous month, the initial condition value; a is the rate-modifying factor for temperature; b is the modifying factor for moisture; c is the cover rate-modifying factor; k is the decomposition rate constant for that compartment; and t is the time step; the time step used in RothC is the month.

Decomposition is assumed to proceed at the maximum rate until there is a water tension of -100 kPa in the topsoil (roughly corresponding with a 20-mm moisture deficit in the top 23 cm of soil at Rothamsted). Water deficit is calculated as the balance between mean monthly evapotranspiration and mean monthly rainfall, assuming that the soil emerges from winter at field capacity, with a zero soil moisture deficit. Temperature and moisture factors are calculated on a monthly basis, from the specific weather data of the research area using the mean monthly air temperature, rainfall, and open pan evaporation (see [2, 3, 16]).

2.3 State Variables and Outputs

The goal of the simulation at each step is to calculate the values for the rates and corresponding state variables. The list of state variables and rates used in RothC is shown in Table 1.

3 RothC Stand-alone Model

To depict the concepts participating in RothC model, a detailed examination of the data required by the model was carried out. Based on the model requirements, the list of depicted concepts is presented as follows:

1. Weather (temperature, rainfall, evapo transpiration (ETP).
2. Soil (soil depth, soil type, soil cover, %clay).
3. Plant (total biomass, root/shoot ratio, root distribution factor [RDF], % organic carbon in plant).
4. Management (crop residues, FYM amount).
5. RothC (FYM DPM, FYM RPM, FYM HUM).

Table 1 State variables in RothC and their definitions

State Variables	Definition
DPM TOTAL	The organic carbon included in the decomposable plant material fraction. It is made up by the DPM Carryover, $DPM_{ROOT+CROP\ RESIDUES}$, and finally by the $DPM_{MANURE+COMPOST}$.
$DPM_{ROOT+CROP\ RESIDUES}$	The decomposable fraction of organic carbon included in the roots and in the crop residues.
$DPM_{MANURE+COMPOST}$	The decomposable fraction of organic carbon included in the manure and in the compost
RPM TOTAL	The organic carbon included in the recalcitrant plant material fraction. It is made up by the RPM Carryover, $RPM_{ROOT+CROP\ RESIDUES}$, and finally by the $RPM_{MANURE+COMPOST}$.
$RPM_{ROOT+CROP\ RESIDUES}$	The recalcitrant fraction of organic carbon included in the roots and in the crop residues.
$RPM_{MANURE+COMPOST}$	The recalcitrant fraction of organic carbon included in the manure and in the compost.
BIO TOTAL	The microbial biomass organic carbon fraction. It is composed of the microbial biomass fraction of the decomposable pool, the resistant pool, the microbial pool, and also by the humified pool.
BIO DPM	The microbial fraction of the decomposable pool.
BIO RPM	The microbial fraction of the recalcitrant pool.
BIO BIO	The microbial fraction of the microbial pool.
BIO HUM	The microbial fraction of the humified pool.
HUM TOTAL	The humified organic carbon fraction. It is composed of the humified fraction of the decomposable pool, the resistant pool, the microbial pool, and also by the humified pool.
HUM DPM	The humified fraction of the decomposable pool.
HUM RPM	The humified fraction of the recalcitrant pool.
HUM BIO	The humified fraction of the microbial pool.
HUM HUM	The humified fraction of the humified pool.
CO_2 TOTAL	The total CO_2 released during the organic carbon transformation processes. It considers the CO_2 emission from the decomposable pool, the resistant pool, the microbial pool, and also by the humified pool.
CO_2 DPM	The CO_2 emission from the decomposable pool.
CO_2 RPM	The CO_2 emission from the recalcitrant pool.
CO_2 BIO	The CO_2 emission from the microbial pool.
CO_2 HUM	The CO_2 emission from the humified pool.

3.1 Weather

The Weather class is modeled as a container to hold daily Weather Data objects. This particular implementation of the weather data is valid for a RothC component developed to test the results. SEAMLESS framework would use its own weather data system.

The parameters included in this class are

(a) Monthly rainfall (mm).
(b) Monthly evapotranspiration (mm).
(c) Average monthly mean air temperature (°C).

These parameters are used by the RothC class to evaluate the soil moisture decomposition and the temperature decomposition factor.

3.2 Soil

This class is modeled to present soil data and behavior.
The parameters included in this class are

(a) Clay content in the soil as a percentage (clay content is used to calculate how much plant-available water the topsoil can hold; it also affects the way organic matter decomposes).
(b) Soil cover (a parameter that express if the soil is bare or vegetated) (0–1).
(c) Soil depth of the layer analyzed (cm).
(d) Soil type is used by the software to decide the DPM/RPM ratio; as an example, for agricultural zone and improved grassland DPM/RPM is 1.44, for unimproved grassland it is 0.67, and for woodland it is 0.25.

The DPM/RPM ratio of each particular incoming plant material is used to split the incoming organic carbon in the DPM or in the RPM pool.

3.3 Management

In the management class, the organic carbon from the crop residues and from the farmyard manure is to be considered.
The attributes included in this class are

(a) Monthly input of FYM amount in t C ha^{-1}. In the model, the FYM is assumed to be more decomposable than is the plant material. It is split between the different pools in this way: 49% goes to the DPM, 49% goes to the RPM, and 2% goes to the HUM compartment.

   1. FYM DPM represents the fraction of the FYM going to the decomposable plant material pool.
   2. FYM RPM represents the fraction of the FYM going to the resistant plant material pool.
   3. FYM HUM represents the fraction of the FYM going to the humified plant material pool.

(b) Monthly input of crop residues amount expressed in t C ha^{-1}. This parameter represents the plant material that goes to the soil monthly and the carbon released by the roots during the growth.

3.4 Plant

This class is modeled to present plant data and behavior.

These are the parameters included in this class: Root carbon (t C ha^{-1}). To obtain this value of carbon in the soil, we have to consider the following parameters:

1. Total biomass (t C ha^{-1}).
2. Percentage of carbon in the plant (%).
3. Root distribution factor (RDF), which is the fraction of root mass in the top layer of the soil (20 cm).
4. Root/shoot ratio.

3.5 Decomposable Plant Material

In this class are considered all the parameters related to the DPM fraction.

In this class are imported parameters values from the "RothC" class, such as:

DPM Carryover: this parameter considers the modification of the initial organic carbon value.

DPM Decay: the difference between the total carbon of the previous month and the DPM Carryover of the following month.

DPM Initial: the initial organic carbon value.

DPM Manure: the organic carbon that comes from the applied manure.

DPM Root: the organic carbon that comes from the roots.

Total DPM: the total value of the decomposable plant material fraction.

4 RothC Plug&Play Component

The concepts presented above (Plant, Soil, Weather, DPM, etc) represent distinct and homogeneous entities with well-defined behavior and are modeled as separate objects. Figure 2 represents these objects and their relationships. As presented in Fig. 2, at the center of the designed system is the *ROTHC* object. The data stored in this object are different state variables and rates of carbon exchange among different pools. This object plays the role of the simulator controller [11], in the sense that it will send the right message at the right time to all other objects needed to carry out the simulation. Therefore, object *ROTHC* needs to have access to all other objects involved in the simulation to use data and behavior from those objects.

From the RothC model [8], the total organic carbon is divided into active and inert compartments. The active compartment comprises the decomposable plant material (DPM), the recalcitrant plant material (RPM), the microbial biomass (BIO), and the humified organic matter (HUM). These compartments have numerical characteristics (data) and behavior used to calculate different rates. As shown in Fig. 3, decomposable plant material, recalcitrant plant

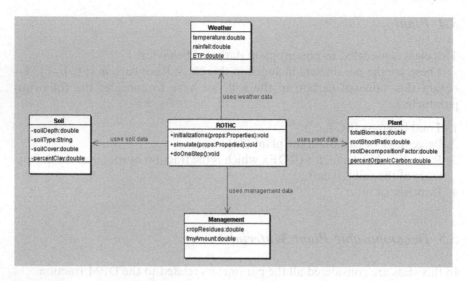

Fig. 2 Conceptual model of *ROTHC*

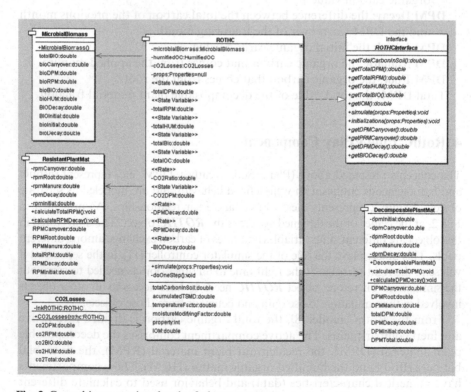

Fig. 3 Core objects creating the plug&play component

material, microbial biomass, and humified organic matter are modeled as separate objects as they represent specific data and behavior. These objects are, at the same time, attributes of the **ROTHC** object. With their data and behavior, they define the complex behavior of object **ROTHC**. **ROTHC** serves as an "orchestra director" that controls the execution of operations from different objects according to the simulation logic [11]. As an example, object **DecomposablePlantMat** provides behavior for calculating total decomposable plant material, decomposable plant material manure, decomposable plant material carryover, and decomposable plant material decay. As it has access to object **DecomposablePlantMat**, **ROTHC** can then send messages to activate methods for calculating total decomposable plant material, decomposable plant material decay, or other calculations defined in **DecomposablePlantMat**. **ROTHC** makes available the calculated values of state variables and rates during one step of the simulation to other components in the surrounding environment. The same can be said about the relationships among **ROTHC** and **HumifiedOC**, **MicrobialBiomass**, and **ResistantPlantMat** objects.

Object **CO2Losses** models the data and behavior needed to calculate carbon losses. These data and behavior could have been attributed to object **ROTHC** but we decided to model them as a separate object to avoid having an overloaded **ROTHC** object. As these data and behavior pertain to CO_2 losses, we choose to attribute them to object **CO2Losses**. We believe this modeling choice facilitates understanding of our model and will help other researchers to use and integrate it into their systems more easily. The relationship between **ROTHC** and **CO2Losses** is a little different from the rest of the objects in Fig. 3. **ROTHC** has access to data and behavior of **CO2Losses** and at the same time **CO2Losses** has access to data and behavior of **ROTHC**. **ROTHC** should have access to **CO2Losses** for the same reasons that it has access to all other objects of the diagram shown in Fig. 3, as it is the main class of component **ROTHC**. **CO2Losses** must have access to **ROTHC** to obtain data about CO2Ratio that is stored in **ROTHC** calculated using the following equation:

$$CO2Ratio = 1.67 * [1.85 + 1.60 * e(-0.0786 * percentageClayInSoil)]$$

Although the calculation of CO2Ratio requires soil data, the percentageClayInSoil, **ROTHC** can perform the calculation as it has access to object **Soil** as shown in Fig. 2.

It is important to note that the data and the behavior of objects **DecomposablePlantMat**, **HumifiedOC**, **MicrobialBiomass**, **ResistantPlantMat**, and **CO2Losses** are invisible to the outside user. The user or outside software components that communicate with **ROTHC** do not have to know implementation details about these objects, their data and their behavior. All the communications with the outside environment are carried out through the unique object **ROTHC**. This architecture facilitates the maintenance of the system and its reusability.

In the compartment of behavior (the second compartment of object *ROTHC* in Fig. 2), there are two methods that provide *ROTHC* with simulator controller behavior. The methods are *initialize (Properties props)* and *simulate (Properties props)*. The method *initialize (Properties props)* is used to create instances of objects *Soil*, *Plant*, *Management*, and *Weather* and establish relationships among them as shown in Fig. 2. At the same time, this method initializes objects *DecomposablePlantMat*, *HumifiedOC*, *MicrobialBiomass*, *ResistantPlantMat*, and *CO2Losses* with their initial values. The method *simulate (Properties props)* starts the simulation process with initial values passed as parameters. These initial values are then passed to method *initialize(Properties props)* to start the initialization process.

ROTHC attributes representing state variables and rates have respectively the stereotype <<State Variable>> and <<Rate>>, as shown in Fig. 3. *ROTHC* provides the necessary behavior to make the values of state variables and rates to user/components in the surrounding environment.

4.1 Dependency Injection Design Pattern

Software systems are often organized as a set of collaborating components that depend upon other components to successfully complete their intended purpose. In this intensive dialogue among components, some of them play the role of clients and some others play the role of service providers. Usually, a component "Client" needs to identify or obtain certain information about components providing these services, referred to as "Services." First, a client needs to know the components that can provide the needed services. Second, a client needs to know where to locate the needed services and, finally, the communication protocol used to dialogue with service providers. Only when these three issues are addressed can the communication between "Client" and "Services" be fruitful. In the case that the identity of the service provider, its location, and/ or the communication protocol with the service provider is modified, then potentially, a substantial amount of code needs to be changed on the client side to reestablish connection with the service provider.

This problem can be approached in different ways. One solution could be to have the client implement the code for locating and instantiating the required services. Another solution could be to have the clients declare their dependency on services and pass some additional information responsible for locating and instantiating the services. Service reference is then provided to the client when needed. In the second case, the client is not obliged to change the code when references to the services change. This is an implementation of dependency injection (DI) design pattern [9], and the "external" piece of code referred to earlier is likely to be either hand coded or implemented using one of a variety of DI frameworks. Interfaces are central to the implementation of DI design pattern.

4.2 Communication with Other Components

The *ROTHC* component is designed according to the design pattern DI [9]. According to this design pattern, components dialogue with each other knowing only their interfaces. Therefore, the communication with other outside components is established through an interface that class *ROTHC* should implement. This interface is presented in Fig. 3 as *ROTHCInterface*. This interface serves as a communication bridge between our component and SEAMLESS or other modeling frameworks. It will receive all messages sent to our component and delegate them to the appropriate objects contained in the component. It is important to note that the development of SEAMLESS and RothC are totally independent and carried out by groups of researchers working in different geographical regions.

An interface only defines the kind of services an object should provide for use by other objects [11]. It represents the set of messages that can be sent to an object created by a class that implements this interface. Therefore, an object created by the class *ROTHC* will respond to any of the messages defined in *ROTHCInterface*. Objects in the SEAMLESS framework that would like to use the functionalities defined in object *ROTHC* are forced to pass through this interface. Thus, the functionalities defined in *ROTHCInterface* should be defined through an intensive dialogue between the future users of *ROTHC* and its developers. In the case that there is some functionality defined in *ROTHC* but not in *ROTHCInterface*, then the users will not be able to use it unless the appropriate definition of this functionality is defined in *ROTHCInterface*. It is important to note that *ROTHCInterface* contains only the definition of the functionality, and the implementation of this functionality is provided by the class that implements the interface, in our case *ROTHC*. Figure 4 shows the communication bridge between *ROTHC* and *ROTHCInterface*. In order for SEAMLESS framework to use *ROTHC* component, it only has to declare an object of type *ROTHCInterface* and send to this object messages defined by this interface.

Another important modeling issue that needs to be addressed is how the SEAMLESS framework will pass parameters such as soil, plant, and management data to *ROTHC*. One potential solution is to pass to *ROTHC* a long list of all (plant, soil, weather, and management) parameters. *ROTHC* needs to re-create the appropriate objects (plant/soil/weather/management) and populate them with parameters obtained from the SEAMLESS framework. This solution may work well in cases when no behavior from plant/soil/management is needed in *ROTHC*. The inconvenience is that this solution requires the re-creation of the same objects that already exist in the SEAMLESS framework and, therefore, it is not very efficient.

In the case that *ROTHC* needs to use behavior defined in plant/soil/management in SEAMLESS framework, objects created from *Plant/Soil/Weather/Management* need to be passed to *ROTHC*. Developers of both teams

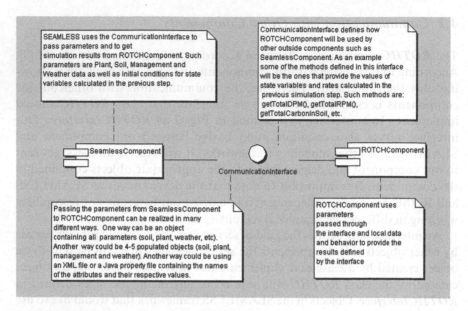

Fig. 4 Communication bridge between components

(*ROTHC* and SEAMLESS) need to have a high level of interaction; every detail of design of any object should be agreed upon by both teams. Considering that both teams are geographically far from each other, the implementation of this solution would be rather difficult.

Here again, the interfaces can provide an elegant solution to this problem. Both teams need only to agree on defining interfaces *InterfacePlant*, *Interface-Soil*, *InterfaceWeather*, and *InterfaceManagement*. The SEAMLESS team will create the classes that implement these interfaces (*Plant/Soil/Weather/Management*). *ROTHC* developers will only use the agreed upon interfaces and use the functionalities they define to perform the necessary calculations required by the simulation process. Again, details of implementing classes *Plant/Soil/Weather/Management* are of no concern to the developers of *ROTHC*.

An efficient way to pass parameters to *ROTHC* component is to store them in a *Properties* file. A *Properties* file is a Java artifact that allows for storing data using a key/value pair as shown in Table 2. As an example, the instance of *Soil* can be stored and retrieved using the key "soil," and an instance of *Weather* can be stored and retrieved using the key "weather." This file will be created and populated with the required parameters in the SEAMLESS environment. *ROTHC* is provided with the behavior to retrieve objects (soil, plant, weather, etc.) and to use their data and behavior for the simulation process. Thus, SEAMLESS framework can perform a simulation using the operation *simulate (Properties props)* defined in *ROTHC* as shown in Fig. 3.

Table 2 Data structure used to pass parameters to *ROTHC* component

Key	Value
"soil"	Soil instance
"plant"	Plant instance
"weather"	Weather instance
"dpmInitial"	Value of dpmInitial
"rpmInitial"	Value of rpmInitial
"humInitial"	Value of humInitial
....	....

ROTHC is provided with another operation referred to as *doOneStep( )* that represents actions that are carried out during one step of the simulation (i.e., using weather data for a particular month). During this time, several rates (referred to as carryover variables) and state variables are calculated, and the SEAMLESS framework could obtain their values by sending appropriate messages to *ROTHC* component. These messages are defined in interface *ROTHCInterface* as shown in Fig. 4. As an example, SEAMLESS framework can ask *ROTHC* to calculate decomposable plant material carryover, decomposable plant material decay, resistant plant material carryover, resistant plant material decay, and so forth. The SEAMLESS framework will use these values to calculate the initial values for the next simulation.

The following code in Java presents the mechanism of passing simulation parameters to *ROTHC* from the SEAMLESS environment.

Line 1 creates an instance of the *Properties* object that will hold the parameters. This instance operates as a dictionary as shown in Table 2. Line 2 creates an instance of component *ROTHC*. Lines 3 to 7 represent the initialization of parameters. Lines 8 to 15 create an instance of *Plant*, populate it with initial values, and store it onto the *Properties* file. Lines 16 to 23 create instances of *Soil* and *Management*, populate them, and store them onto the *Properties* file.

```
 1 Properties props = new Properties();
 2 rothInterface = new ROTHC();
 3 dpmInitial = 0.1655;
 4 rpmInitial = 3.8397;
 5 bioInitial = 0.5296;
 6 humInitial = 11.4859;
 7 iomInitial = 0;
 8 //initialize plant
 9 plant = new Plant();
10 plant.setTotalBiomass(0.0);
11 plant.setPercentageCarbon(0.4);
12 plant.setRootDecompFactor(0.58);
```

```
13  plant.setRootShootRatio(0.13);
14  //add plant to properties
15  props.put("plant",plant);
16  //initialize soil
17  soil = new Soil();
18  soil.setSoilType("cultivated area");
19  //add soil to properties
20  props.put("soil",soil);
21  //initialize management add management to properties
22  management = new Management();
23  props.put("management",management);
24  weatherInterface = new Weather("weather.dat");
25  while (weatherInterface.hasNext()) {
```

```
    1.  weatherInterface.next();
    2.  //add initial values DPM,RPM,IOM, BIO,HUM
    3.  props.put("dpmInitial",new Double (dpmInitial));
    4.  props.put("rpmInitial",new Double (rpmInitial));
    5.  props.put("bioInitial",new Double(bioInitial));
    6.  props.put("humInitial",new Double (humInitial));
    7.  props.put("iomInitial",new Double (iomInitial));
    8.  //send message to ROTHC Component
    9.  rothInterface.initializations(props);
   10.  rothInterface.simulate(props);
   11.  dpmInitial = rothInterface.getTotalDPM();
   12.  rpmInitial = rothInterface.getTotalRPM();
   13.  bioInitial = rothInterface.getTotalBIO();
   14.  humInitial = rothInterface.getTotalHUM();
```

```
26  }
```

Lines 24 to 26 (including the subsequence of lines 1 to 14) represent the loop over the weather data to calculate different carbon rates. Line 25.1 obtains the next daily weather data to be used in the simulation process. Lines 25.2 to 25.7 store the initial data for the current session onto the *Properties* file to be used in the initialization step shown in line 25.9. Line 25.10 shows the command sent to *ROTHC* component to perform the simulation for 1 month with the parameters stored onto the *Properties* file.

This is a simulation of how other frameworks, such as SEAMLESS, will use *ROTHC* component. The iteration starts with obtaining the initial values for different carbon layers that are passed to *ROTHC*. The values of rates calculated for the current step are used as initial values for the next simulation step. Each step of the simulation is independent, as *ROTHC* receives as parameters all data required for the simulation. Therefore, it is possible in *ROTHC* to go back in the simulation and redo calculations with the same or different values.

5 Conclusions

Today there is a greater need than ever for reusing existing components and assembling together disparate components to form a cohesive architecture. This software modeling approach allows for better re-use of existing functionalities and facilitates software construction. The use of design patterns increases the quality of software as it makes it flexible, reusable, and easy to maintain.

ROTHC component is developed to be a plug&play component to be used in different agricultural and environmental modeling frameworks. Functionalities are defined through interfaces that other frameworks need to use. Appropriate design patterns allow separating declaration of behavior from its implementation. Therefore, the implementation of functionalities can be provided as services by different teams that may have no contact at all with the team in charge of developing a modeling framework.

ROTHC is developed in Java and is tested and provides the same results as other of its implementations in FORTRAN or other programming environments.

References

1. Bostick, Mc N., Bado, B.V., Bationo, A., Soler, C.T., Hoogenboom, G., Jones, J.W., 2007. Soil carbon dynamics and crop residue yields of cropping systems in the Northern Guinea Savana of Burkina Faso. Soil & Tillage Research 93, 138–151.
2. Coleman, K., Jenkinson, D.S., Crocker, G.J., Grace, P. R., Klir, J., Korschens, M., Poulton, P.R. Ricther, D.D., 1997. Simulating trends in soil organic carbon in long term experiment using RothC-26.3. In: Evaluation and Comparison of Soil Organic Matter Models using Data from Seven Long-Term Experiments, eds. P. Smith, D. S. Powlson, J. U. Smith and E.T. Elliot. Geoderma, 81, 29–44.
3. Coleman, K., Jenkinson, D.S., 1995. RothC-26.3 a model for the turnover of carbon in soil - Model description and users guide. IACR, Rothamsted Harpenden Herts.
4. DSSAT. http://www.icasa.net/dssat/.
5. Falloon, P., Smith, P., Coleman, K., Marshall, S., 1997. Estimating the size of the inert organic carbon matter pool from total soil organic carbon content for use in the Rothasmed carbon model. Soil Bio Biochem 30(8/9) 1207–1211.
6. Jenkinson, D.S., Adams, D.E., Wild, A., 1991. Model estimates of CO2 emission from soil in response to global warming. Nature 351, 304–306.
7. Jenkinson, D.S., 1990. The turnover of organic-carbon and nitrogen in the soil. Philos Trans R Soc Lond B Biol Sci 329, 361–368.
8. Jenkinson D.S., Hart P.B.S., Rayner J.H., Parry L.C., 1987. Modelling the turnover of organic matter in long-term experiments at Rothasmed. INTECOL Bulletin 15, 1–8.
9. http://martinfowler.com/articles/injection.html.
10. Papajorgji, P. 2005. A plug and play approach for developing environmental models. Environmental Modelling & Software 20, 1353–1357.
11. Papajorgji ,P., Pardalos, P. 2005. Software Engineering Techniques Applied to Agricultural Systems an Object-Oriented and UML Approach. Springer, New York.
12. Papajorgji, P., Beck, W.H, Braga, J.L., 2004. An architecture for developing service-oriented and component-based environmental models. Ecological Modelling 179 (1), 61–76.

13. Papajorgji, P., Shatar, M.T., 2004. Using the Unified Modeling Language to develop soil water balance and irrigation-scheduling models. Environmental Modelling & Software 19, 451–459.

14. Parshotam, A., Tate, K.R., Giltrap, D.J. 1995. Potential effects of climate and land-use change on soil carbon and CO2 emissions from New Zelands indigenous forests and unimproved grasslands. Weather and Climate 15, 3–12.

15. Post, W.M., Emanuel, W.R., Zinke, P.J., Staggenberger, A.G. 1982. Soil carbon pools and world life zones. Nature 298, 156–159.

16. Robertson, W.K., Hammond, L.C., Johnson, J.T., Boote, K.J. 1980. Effects of plant-water stress on root distribution of corn, soybeans, and peanuts in sandy soil. Agron J 72, 548–550.

17. SEAMLESS. http://www.seamless-ip.org/.

18. Traoré, P.C.S., Bostick, W.M., Jones, J.W., Koo, J., Goïta, K., Bado, V., 2008. A simple soil organic matter model for biomass data assimilation in community-level carbon contracs. Ecol Appl 18(3): 624–636.

Ontology-Based Simulation Applied to Soil, Water, and Nutrient Management

Howard Beck, Kelly Morgan, Yunchul Jung, Jin Wu, Sabine Grunwald, and Ho-young Kwon

Abstract Ontology-based simulation is an approach to modeling in which an ontology is used to represent all elements of a model. In this approach, modeling is viewed as a knowledge representation problem rather than a software engineering problem. Ontology-based techniques can be applied to describe system structure, represent equations and symbols, establish connections to external databases, manage model bases, and integrate models with additional information resources. Ontology reasoners have the potential to automatically compare, organize, search for, and discover models and model elements. We present an environment for building simulations based on the Lyra ontology management system, which includes Web-based visual design tools used for constructing models. An example application based on a model of soil, water, and nutrient management in citrus that uses the approach is also presented.

1 Introduction

There is a need to better communicate model structure and elements to the worldwide community of model builders. There is also a need for models to communicate and interact automatically on a machine-to-machine basis over the Internet. Modeling methodology can be improved by using ontologies for all aspects of model building including the design, documentation, development, and deployment of models. Ontologies are formal representations of the concepts and their interrelationships within a particular domain. Application of ontologies to modeling and simulation results in a new approach called ontology-based simulation. This chapter explores several ways in which ontology-based simulation can be applied, as illustrated through a specific application in the area of soil, water, and nutrient management.

H. Beck (✉)
Agricultural and Biological Engineering Department, Institute of Food and
Agricultural Sciences, University of Florida, Gainesville, FL, USA
e-mail: hwb@ufl.edu

P.J. Papajorgji, P.M. Pardalos (eds.), *Advances in Modeling Agricultural Systems*,
DOI 10.1007/978-0-387-75181-8_11, © Springer Science+Business Media, LLC 2009

An ontology can be viewed as being a formal representation of concepts and their relationship within a particular domain. In our case, the domain is an agricultural or natural system including hydrological, biological, physical, and chemical transformation and transport processes. The ontology in this domain includes concepts such as plant, soil profile, soil layer, water content, nitrogen concentration, and many others. Ontologies attempt to precisely define each concept (what water concentration is, how it is measured), and much of this is expressed through relationships among concepts (e.g., how is water concentration related to water content and soil volume). Ontologies are based on formal languages, meaning the concepts and relationships are expressed in a language that is well defined. A leading World Wide Web Consortium (W3) standard for such languages is the Web Ontology Language [26]. Ontologies can also be thought of as machine-interpretable dictionaries, as they provide formal definitions for domain concepts. Machine-interpretable means that the computer can make inferences about the relationships among concepts. Ontology reasoners have the potential to provide some very useful functions including automatic classification. They can be used as a basis for model discovery, that is by locating existing model elements to be used for some new purpose and in the future assisting in constructing new models by assembling model elements automatically.

In ontology-based simulation, model building is considered to be a knowledge representation problem and not a software engineering problem. Software engineering is a formal technique for designing and building software systems. Traditionally, building a model involves writing computer code in a particular programming language. Although advances in programming languages, including object-oriented programming and even the most recent Unified Modeling Language (UML) methodologies such as model driven architecture (MDA), have been used to improve the process, most modeling is still viewed from the standpoint of how to best implement software to realize the model. In ontology-based simulation, the problem of modeling is raised an additional level, to where the software implementation and associated software engineering concerns are irrelevant. Modeling becomes an abstract design problem in how best to represent knowledge about the model structure and behavior. Knowledge representation consist of two parts: (1) creating data structures that represent what we know about a problem or domain, and (2) defining inference operations that can automatically analyze these data structures to draw new conclusions. Ontologies are one approach to knowledge representation in which data structures are created that define concepts and relationships among concepts in a domain and in which ontology reasoners can automatically process these data structures to draw additional conclusions about concept relationships. Although ontologies continue to advance the tradition of object-oriented design, ontology languages are not programming languages. Ontology objects contain no variables, methods, or other program code. They are purely declarative descriptions of concepts.

As with all object-oriented design methodologies, ontologies adopt a systems analysis approach through which complex systems are decomposed into smaller, interacting elements. Modularizing agricultural and natural systems into smaller units is essential to model processes that occur within them. In ontologies, dynamic behavior of physical and biological systems can be described not by program methods or constraint languages but through mathematical equations. Ontologies decompose systems to their smallest elements, resulting in equations and the symbols appearing in those equations. But simple objects recompose to form complex objects at all levels from the fine-grain symbols to the whole-systems view. We use the term *model element* rather than *component* because the term *component* has an established usage in software engineering and is defined as a modular piece of program code with well-defined input and output characteristics. In general, there is no attempt to represent knowledge about the internals of software components, rather they are treated as reusable parts. Specification of libraries of components including metadata and interface standards for components is currently an active area within the modeling community. In contrast, model elements in an ontology are explicitly represented at all levels, from coarse to fine, and the ontology naturally forms a library of reusable model elements based on behaviors represented by mathematical equations. Note that we neither claim nor deny that *all* behaviors can be described mathematically, but the behaviors exhibited by physical and biological systems, including the behaviors exhibited by the vast majority of models built in agriculture and natural resources, can all be described by a set of mathematical equations. This is a fundamental principle behind engineering system analysis.

Ontology-based simulation addresses several problems related to modeling and simulation. These include management of model libraries called model bases. They also include representing models at fine levels of detail, down to the symbol and equation level. System structure at a coarse level can be described by ontology objects as well. The problem of attaching models to data sources is addressed by building an ontological description of available data sources. Ways in which ontology reasoners can be applied to automatically classify, compare, and locate model elements can address problems related to model management. Solutions to these problems, some of which we have implemented and others which are proposed directions, are described in greater detail below.

Our ontology-based simulation environment is implemented within an ontology management system (OMS) that provides a platform and tools for creating and managing ontologies, including those used for simulation. We have constructed an OMS, which we call Lyra, for addressing a wide range of information management requirements in the area of agriculture and natural resources. Lyra includes support for modeling and simulation. The ontology management system is a database management system built entirely around an ontology language rather than traditional relational or object database languages. Ontologies represent models at a high level of abstraction that explicitly exposes knowledge contained in models. The ontology database supports management of large collections of ontology objects, including reasoning facilities that help in organizing and searching for model elements. Web-based visual authoring tools enable modelers to design and run simulations by using ontology objects. The resulting environment integrates ontology-based simulation within an even larger knowledge management system that supports research, extension, and education applications.

We illustrate many of the concepts introduced above through a citrus water and nutrient management system (CWMS) for modeling soil, water, and nutrients with respect to soil physics and chemistry and demand for water and nutrients by the citrus tree. This was designed and implemented using the Lyra ontology-based simulation environment. In the sections that follow, an overview of ontology-based simulations and the problems addressed is provided. After that is a description of CWMS and the way it was implemented in the Lyra OMS. We conclude with recommendations for future directions in the rapidly developing field of ontology-based simulation.

2 Ways in Which Ontologies Can Be Applied to Modeling Agricultural and Natural Resource Systems

2.1 What Is an Ontology?

An ontology is a formal specification of the concepts and relationships among these concepts within a particular domain. Concepts and relationships are defined using an ontology language. The language is formal in that it is well defined, but typically ontology languages are relatively simple in that they contain a limited number of language constructions. If the ontology language is too complex, the reasoners become computationally too expensive. Keeping the language small also reduces the risk of creating illegal, inconsistent, or otherwise incoherent concept descriptions.

We will introduce some basic ontology language constructions using Web Ontology Language (OWL). All ontology languages include the notion of a concept. A concept can be generic, in which case it is represented as a class, or it can be a specific occurrence of a class, in which case it is represented by an individual (in OWL, an individual plays the role that is typically defined as an instance in other object-based languages). Individuals are the actual things in the world, and classes are categories that group together similar individuals and describe how they are similar. The set of all classes in an ontology comprise what is referred to as the "T-BOX," or terminology box, because the classes define the terms used in the domain. The individuals belong to an "A-BOX," or assertion box, which are statements about the actual things in the world, expressed using terms defined in the T-BOX. The word *object* is not part of the OWL language but is often used to refer to individuals, or more loosely to both individuals and classes.

Classes and individuals are further defined through properties and other relationships. Generalization is described using superclass and subclass relationships with superclasses being more general forms of a class and subclasses being more specific. This gives rise to the well-known generalization taxonomy common in most object-oriented languages. Individuals can have properties that describe the qualities of the individual (qualities are primitive properties such as strings and integers) and relationships to other individuals. A property has a domain (source) and range (target). The domain is the individual (actually a set of possible individuals) for which the property is being asserted, and the range is the set of possible values being asserted for the property.

Classes have property restrictions rather than properties. Classes are generalizations about sets of individuals. Property restrictions are generalizations of the properties of these individuals. Property restrictions define the possible values that a property can have. Property restrictions (when possible) provide a set of necessary and sufficient conditions that an individual must satisfy to belong to the class. Ontologies do not use inheritance, by which individuals obtain properties from their classes, on the contrary, classes arise by discovery of common properties among sets of individuals. Often it is not possible to define a class using a definitive set of necessary and sufficient conditions, and such classes are called primitive classes.

In contrast with object-oriented programming languages, ontologies do not contain variables or methods and do not use inheritance. Individuals and subclasses can belong to more than one class. Whereas there are some notions in ontologies that are common to UML, mainly because they are both object-design languages, it is a mistake to claim that UML can be used to model ontologies, or that ontologies can be used to design software in a way that is similar to UML. UML and ontologies are aimed at different purposes. UML does not support reasoning about concepts, and ontologies do not support software design. Although ontologies can be used to organize source-code libraries as in LaSSIE [6], the objectives are quite different.

Typical reasoning facilities provided by ontologies include subsumption and classification. Subsumption is used to determine whether one class is a superclass of another. A subsumes B if B is a subclass of A and logically implies that every instance of B is also an instance of A. This determination is automatic; rather than being told that A subsumes B, SUBSUME(A,B) is a test that returns true if A subsumes B and false otherwise. The test is conducted by determining if the property restrictions in B satisfy the property restrictions in A, that is, the property restrictions of B are the same or more restrictive than the property restrictions in A. Using subsumption tests, it is possible to automatically determine where in the taxonomy a new class should be placed (below the most specific classes that subsume the new class, and above the classes subsumed by the new class), a process known as classification. Subsumption can also be used to test the consistency of an ontology by determining if a manually created taxonomy logically violates any property restriction.

There are other possible reasoning facilities. Subsumption and classification can be extended to individuals. Realization determines automatically whether individual C belongs to class A (true if the individual satisfies the property restrictions for A). Conceptual clustering can be used to automatically induce a new class (whereas classification is a deductive reasoning process based on inference over existing classes). Given a set of individuals, conceptual clustering automatically creates classes that are generalizations over the properties of the individuals. Conceptual clustering can be used to cluster together similar individuals by identifying how individuals are similar or different. Classification can be used for query processing. A query is represented as a new class, the "query" class, and is classified to identify individuals belonging to the new class. These individuals are the results of the query. However, much work remains in applying these techniques, and some potential applications are described below.

2.2 Literature Review

Recently, ontologies have received much attention for implementing mathematical models and building simulation systems. The aim of adapting ontologies to simulation systems is similar across various related projects, but the design and implementation of an ontology is different depending on the problem domain [3].

Miller et al. [22] noted that for modeling and simulation, an ontology provides standard terminology, which increases the potential for application interoperability and reuse of simulation artifacts. Furthermore, semantics represented in an ontology can be used for discovery of simulation components, composition of simulation components, implementation assistance, verification, and automated testing. They proposed a Web-accessible ontology for discrete-event modeling (DEMO), which defines a taxonomy of models by describing structural characterization (state-oriented, event-oriented, activity-oriented, and process-oriented models) and a model mechanism explaining how to run the model.

Although Miller et al. focused on the creation of an ontology for general stochastic models such as Markov processes or Petri nets, Fishwick and Miller [11] placed emphasis on capturing mostly object or instance-based knowledge. They presented a software framework, RUBE, which provides an integration method for the phenomenon of model and model object and multiple visual modes of display to provide interfaces for developing dynamic models. Three-dimensional visualization was used by Park and Fishwick [27] to animate the models. An ontology is used to define a schema of model types and models, and a sample air reconnaissance scene is represented with the Web Ontology Language, OWL.

Some researches [4, 18, 29] address the use of an ontology that describes aspect of the world focusing on entities and in a simulation and on the data and the rules governing the simulation. They understood that data used by a model is a key characteristic of semantics, which an information system ontology should define, rather than building an ontology that is independent from simulation form or contents. For example, ontology-based task simulation

[29] uses an ontology for evaluating the usability and utility of a task or data for the decision-making process. JOntoRisk [4], which is an ontology-based simulation platform in risk management domain, developed the meta risk ontology for validating or reviewing the meta structure.

SEAMLESS [16] is a component-based framework for agricultural systems that is used to assess agricultural and environmental policies and technologies from the field-farm level to the regional level in the European Union. For SEAMLESS, an ontology is designed to relate different concepts from models, indicators, and source data at different levels and to structure domain knowledge and semantic meta-information about components for retrieving and linking knowledge in components. It also is used to check the linkage between components through input and output variables in the system. The Model Interface Ontology encapsulates knowledge of biophysical agricultural models. Static and dynamic models are included, and the system dynamics approach, which describes a system with stocks and flows, is applied to conceptualize models. This approach to model ontologies provides advantages that include the simplicity of model representation by using states, inputs, and outputs, but it has limits on representing mathematical expressions of models and manipulating models to build a complex system. SEAMLESS does not attempt to represent models based on their mathematical equation form in the ontology.

A Web-based simulation using an ontology in the hydrodynamic domain [15] shows how an ontology can be used in simulation. To solve the governing equations for a two-dimensional hydrodynamic model, an ontology was created to describe a numerical model and define a specific metadata set that describes hydrodynamic model data, which is used to search and retrieve metadata information. Instances of a simulation ontology created during the simulation process are stored and retrieved by a relational database.

The Modeling Support Tool (MoST) [32], a software framework for supporting the full modeling process, offers an ontological knowledge base (KB). The KB is a collection of knowledge on modeling for various domains of water management, which is developed by domain experts. They adopt ontological approaches to develop a knowledge structure, store the knowledge to the KB following an ontological structure, and build software applications to use the KB.

A model base is a large collection of models and model components. As the number and scale of models grows, the conception and role of models within a problem domain widen and become complex. Some models may be considered as an integration of related process models, while previously a single-process model itself was enough to make a simulation. As various concepts are applied to develop an ontology to build a model, it becomes a challenge to develop an ontology that contains different views and to manage models [10].

As there are diverse aspects to understanding and describing models in a specific domain, it is not easy to reuse an existing model with other models or to replace a model with other models that satisfy the same requirements of input data and parameters. In large-scale problem domains, the need increases for comparing and evaluating models in order to locate an adequate model for a given environment. Lu et al. [19] compared different models for estimating leaf area, and Eitzinger et al. [9] performed an evaluation and comparison of water balance components in different models. To provide a model base, there is an effort to develop a set of crop models for various crops and integrate models with a farm decision support system [30]. A modular approach to model development [17] contributes to categorizing and organizing models as software components, an executable unit of independent production [7, 8] in the agro-ecological domain. Although there are useful ideas on categorizing and reusing the existing components, they cannot fully address the difficulties of model management because they are developed for a specific programming environment.

These predicaments make it important to organize a model base that can compare similar yet different models and components. It will be useful to categorize and organize models into a well-designed structure for the purpose of locating and reusing models. There have been many efforts to construct model bases, and recently ontologies are being applied to this purpose because of their strength in categorizing and organizing knowledge. Watershed modeling is considered as an aggregation system of unit hydrology and chemical processes, which includes precipitation, infiltration, evapotranspiration, and erosion. Haan et al. [14] presented a collection of generic processes and practical models that have been used to study the hydrologic cycle in watersheds. The MoST model ontology was developed following the structure of components in the system to manage models, and it made it possible to switch one model with other models in the same process level for seeking appropriate model composition resulting in an adaptable conclusion [32]. But, the complexity of the representation is not enough to describe processes in detail, and the large scale of the system makes it difficult to manage models. Although it enables model switching, it is limited to simple models.

Research is under way to examine the decision-making process over a farm region or water management area [1, 32]. This has resulted in a library of models that allows a user to build up a simulation system easily with unit process models. The library contains an ontology for storing the model knowledge that is gathered from references or experts. Usually, in those cases, models can be repeatedly used for building up a system, but there are limitations in modifying or creating another model from known models, even models that the system provides. A simple case is that an ontology is not designed originally to allow any manipulation, and this problem is usually found at the multiscale simulation model.

To solve the difficulties of managing model in ontologies, SEAMLESS built a model ontology [1, 31] that contains multiscaled categories over an agricultural domain. An interface was provided for managing model knowledge, which is an authoring tool that supports the capability to create and categorize a model and to modify model knowledge. Model knowledge appearing in the interface includes a model description, creator, a components list using selected model, and model elements. Model elements describe model input, output, and state variables that can be used to select models. Although input, output, and state variables can be dictated in the interface, it does not represent the detailed and complicated mathematical relations between them. A model ontology just contains knowledge of concepts related with a mode as input/output or state variables, and their mathematical relationship is coded or internally described in the system. To resolve these limitations, it is required to focus on realizing a model ontology based on their mathematical representation and meaning explicitly.

2.3 System Structure

System structure can take many forms. One form is the logical system decomposition into subsystems. A plant may have subsystems including photosynthesis and carbohydrate maintenance, water uptake and transpiration, phenological development stages, damage caused by diseases and pest, and others. There is also geometric structure. A plant is composed of leaves, stems, roots, and flowers. Soil geometry comes in many forms and can be decomposed into soil profiles and individual soil layers within a profile. These can be modeled in one, two, or three dimensions.

The subsystems comprising the system structure can all be described as concepts in an ontology. Relationships between subsystems can be modeled as relationships between ontology objects. Geometric structure can be modeled using "part-of" relationship. For example, a soil layer is part of a soil profile. Categories of subsystems, such as the familiar source, sink, storage, and flow elements of Forrester diagrams [12], can be modeled as ontology classes. Further decomposition of elements within a subsystem (including sub-sub systems) can be modeled to any desired level of resolution using additional objects.

The ability to decompose a system into smaller, simpler subelements is not unique to ontologies; it is an approach that has been used extensively in object design and by modular approaches to modeling. However, ontologies enhance the approach by bringing formal definitions and more precisely representing these subsystems. For example, the terminology can be made more precise, as in representing the difference between a soil layer and a soil horizon, and the terminology can change as needed by local convention. Ontologies enable concept sharing across modeling communities in spite of terminology differences.

2.4 Representing Symbols and Equations

Unlike object-oriented programming languages, ontology languages do not include methods for performing computations. Ontology languages are not programming languages, they are declarative knowledge representation languages used for building descriptions of concepts. Including methods or even variables in such languages destroys the formality of the representation language and eliminates any possibility of applying inference techniques to manipulate objects. In short, there is no way to know how a particular segment of programming code behaves (it is undecidable). This also implies that ontologies cannot be used to perform computations. However, this disadvantage is outweighed by the benefits of precise representation of equations and symbols.

Fortunately, it is not necessary to rely on the use of methods to represent model behavior. In fact there is a better way; utilizing classic mathematical notation. All models of dynamic systems in agriculture and natural resources can be defined by a set of mathematical equations. By taking one additional step, using the ontology to represent the equations, we can describe model behavior within the ontology. Although the ontology cannot perform calculations necessary to implement a simulation of the model, the process of building and running a simulation can be fully automated once the model has been defined. Thus modelers to not need to be concerned with building the simulation (other than needing to specify parameters that control the performance of the simulation) and are free to focus on abstract modeling activities.

There are many existing systems that enable modelers to design and build models at the mathematical level such as Mathematica [35] and Simile [33]. Several of these include specialized languages used to define the mathematics. In our approach, it is possible to express models using classic and standard mathematical notation. The advantage of this approach is that no special programming skills are required to develop the models. Another advantage is the ability to model mathematical equations using ontology objects. Once again, the ontology provides a way to better define the symbols used in equations and has the ability to categorize these symbols into a taxonomy.

Equations are composed of symbols, some of which are operators. Symbols can be expressed in many ways. One would be the mathematical form, for example,"t" can by a symbol representing time. But the actual symbol used is just a term, the concept is what counts. Thus, an ontology object for "time" could be termed using "t," "time," or "zeit" to use a multilingual example. Time can be in different units (calendar date or Julian date, hours, minutes, etc.), and it can be discrete or continuous. Whatever the symbol, all associated knowledge about the symbol can be represented in the ontology object. Note that traditionally associated knowledge must be manually written into documentation ("t represents time"), but this documentation is only human readable, not machine readable.

Symbols that are operators ($+$, $-$, $\times$, $/$, $=$, and many others) can be represented by ontology objects having associations to the arguments needed by the operator. For example, divide requires a dividend and divisor. Equal requires two arguments for the two things being equated (a left side and a right

Fig. 1 An equation (NH_4^+ = $N_t - NH_3$) represented in the ontology as a set of objects, one for each symbol. Relationships among operators results in a tree structure

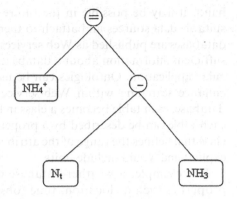

side). An equation represented by a set of ontology objects takes the form of a tree as shown in Fig. 1.

Once equations are represented in this form, they can be part of the ontology model base, and the ontology can hold many equations needed for a particular model or for many different models. Different models can share the same equations.

There are many possibilities for reasoning about equations represented in this form. One would be the comparison of equation structure. By comparing different trees (as in Fig. 1) for different equations, similarities and differences between equations used in different models could be determined. Also, equations in this form can be automatically converted to different formats. One of these is Extensible Markup Language (XML), thus equations can be converted to MathML [20] or OpenMath [25] for exchange with other systems. Another conversion is from equation to program code that can solve the equation. This technique is used to build programs needed to execute the simulation and is one solution of the problem of how to perform the calculations. In the future, ontology standards might expand to incorporate such calculations as part of ontology reasoner. For now, it is possible to automatically generate program code in any desired target language (although our experiments have been limited to Java, the same technique could generate code in C + +, C#, or FORTRAN).

2.5 Connecting to External Databases

Models typically need to attach to external sources of data such as weather observations, data on soil characteristics, or information about production practices such as details of irrigation or fertilizer application. Identifying existing sources of data, understanding their exact format, and adopting them to work within a particular model can be a tedious manual process. On the other hand, it may be possible in the future for models to search the Internet for suitable data sources and attach to them automatically. This could be done if databases are published as Web services, and the Web service registry provides sufficient information about a database to determine its suitability for a particular application. Ontologies can be used to represent database schemas and enhance searching within Web-service registries. For an existing relational database, each table becomes a class in the ontology, and each attribute within each table can be described by a property between the table class and another class that defines the range of the attribute. The class-defining range can be very explicit and would include units.

For example, a weather database could be described by a class having properties such as location, time (observation time and observation interval), temperature, humidity, rainfall, and descriptions of other attributes. The domain for temperature could be a class including units (C or F) and value for the temperature. There can be different subclasses of temperature (e.g., temperature at ground level, temperature at 3 feet, soil temperature). Each record in the database table corresponds with an instance of the weather class.

Describing the database in this way would enable a model to automatically search for suitable databases. This would be done through a query that attempts to match up the class and instances for the database with symbols in the model (parameters and inputs) required by the model. Details such as observation interval and units could also be resolved. When a Web service is located that provides a suitable database, the Web service then makes the data available to the model in XML format through standard interfaces.

2.6 Integration with Other Information

Much additional information can be associated with a model. Additional documentation in the form of graphs and figures or additional text descriptions can be incorporated within appropriate objects in the model ontology. Data gathered through experiments for use in model estimation and validation can be included through a database interface such as described in the previous section. Research publications associated with the model can be integrated within this framework as well. When simulations are used in eLearning, the instructional design considerations, training scenarios, and assessment items can all be included. The advantage of this approach is that information is better organized. Different objects can be used for different purposes. For example, the objects representing equations and symbols in the model can be transformed directly into HTML, PDF, or other formats for automatic generation of documentation. This eliminates the need to write documentation as a separate, independent document and creates a strong connection between the model and associated documentation.

2.7 Ontology Reasoning

Applications of ontology reasoning are based on comparison of object structure. This identifies how objects are alike or different. If one class is a special case of another, it can automatically be classified through a subsumption relationship. Similar objects can be grouped together to make new classes through conceptual clustering. These techniques can also be used for search and query processing by automatically identifying objects that satisfy a query class. Imprecise queries that locate objects that may be similar to but not identical with a target description are also possible.

A technique for locating data sources was described in a previous section. Similar search techniques could be applied to locate existing model elements, equations, or symbols satisfying some desired requirement. Clustering techniques could be used to compare the structure of two models to identify differences and similarities. Generalizations over different models can be abstracted in the formation of high-level classes. For example, there have been many models written on irrigation scheduling that take into account soil–water balance. With these models written in traditional format (coded in programming languages), it is impossible to do comparisons of these models except through tedious manual procedures. If these models are represented as sets of ontology objects, then automatic techniques could be applied for this purpose.

2.8 Model Base

A model base is a database of many models, model elements, equations, and symbols. There is a need for model bases as a way to organize the collection of models developed by many modelers over many projects. There is also a need to share and re-use models, subsystems, and elements. In general, many similar but different models are developed by different modeling teams to address a similar problem (such as soil–water balance). When these models share the same elements, it is desirable to identify and re-use those elements. It is also important to know where these models differ so that the appropriate model can be selected for use in a particular situation. During model development, it is convenient to be able to easily switch among different elements for modeling the same process in order to evaluate the behavior under each element. At the highest levels of abstraction, it is possible to show the most fundamental processes of physical systems and how they appear in different, lower-level situations (e.g., a high-level abstraction such as capacity occurs in specific low-level situations such as water capacity for soil moisture, the concentration of nutrients in leaves of a plant, and in the capacitor of an electrical circuit).

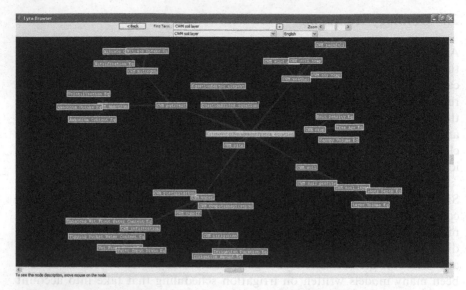

Fig. 2 A model base of equations used in the Citrus Water Management System

The ontology facilitates construction of model bases. The taxonomic organization allows for classification of different models and model elements at different levels of abstraction with the top of the taxonomy being most abstract and with more specific instances of models at the lower levels. Classification can facilitate organizing and locating model elements. Elements modeling the same subsystem or providing the same value for a symbol can be clustered within the same class. Figure 2 shows a taxonomy of equations for the soil, water, and nutrient modeling domain. A particular model can select subsets of these equations as needed for a particular situation.

3 Example: A Soil, Water, and Nutrient Management Model

This section describes an environment that we have created for building ontology-based simulations. It is based on the Lyra ontology management system, which uses Web-based authoring tools for creating models. We provide a full example of ontology-based simulation using a model of soil, water, and nutrient management in citrus.

3.1 Lyra Ontology Management System

We have created an environment for developing ontology-based simulations that is part of a larger environment for managing ontologies in general. Lyra is an ontology management system that was created for the purpose of applying ontologies to a wide range of knowledge and information management problems in agriculture and natural resources. Lyra is based on the principle that ontologies offer a way to organize knowledge, information, and data within and across disciplines. An ontology management system is a database management system in which the data modeling language is an ontology language. This contrasts with traditional database systems in which a relational data modeling language is used where information is viewed as a set of tables. While relational databases can be absorbed within an ontology management system, relational databases are not capable of storing complex objects that are needed for modeling and many other applications. Lyra provides a solution to creating and managing ontology-based simulations and other applications where large number of objects must be created, manipulated, and efficiently stored, retrieved, and distributed.

3.1.1 Lyra Database Management Facilities

Lyra includes features commonly associated with database management facilities. Central to Lyra is an ontology language based on OWL. Lyra contains a set of language constructs for creating classes, individuals, property restrictions, properties, and data types. An efficient physical storage system optimized for ontology objects enables objects to be rapidly accessed and brought into main memory for processing. We are currently implementing ontology reasoners based on classification and subsumption, and we have implemented a SPARQL [34] query facility for filtering and searching. In order to publish ontology objects to make them available on the Internet, Lyra provides several Web service and object request broker technologies. Lyra databases are wrapped inside Java Remote Method Invocation (RMI) servers that allow remote Java applications to attach to the server to send and retrieve objects. As a more standard language-independent solution, Lyra also supports Web services that publish methods for sending and retrieving objects in XML format. Java servlets also provide a simple URL-based technique for retrieving objects. Lyra supports full XML import and export so that the contents of the database can be shared. This allows application programs to access Lyra objects from anywhere on the Internet.

3.1.2 Authoring Tools

Lyra supports a variety of authoring tools to enable modelers to directly create and manipulate objects. Lyra includes some general-purpose object editors (LyraBrowser and ObjectEditor), as well as domain-specific editors (RuleEditor, LanguageEditor, SimulationEditor, and EquationEditor). Modelers should be able to interact with objects in ways they are most familiar with, and domain-specific editors have proved to be the easiest for modelers to use, whereas general-purpose object editors provide a generic view of the object database.

The authoring tools have several important common features. First is that they are graphic-based, enabling modelers to manipulate objects visually. They are also Web-based meaning that these tools are accessible through any Web browser (that has the Java plug-in). The tools are cross-platform so that they run in many different browsers on different hardware platforms and operating systems. They all communicate to remote databases using Web services, the Java RMI interface, or URL-based techniques. This results in a wiki-style collaborative development environment in which modelers at geographic locations around the world can work together to develop models.

LyraBrowser is a general-purpose editor for visualizing and creating objects. An animated graphic interface allows authors to inspect objects and their relationships. Objects are displayed as nodes in a graph, and related objects are displayed using link. Authors can navigate the graph by clicking on nodes that then expand to show additional related objects. Editors for specifying object properties pop up when authors click on object nodes.

ObjectEditor is similar to the LyraBrowser except that the displays are static. Sets of related objects are prearranged in a map that is created manually by the author. The database is segmented into modules of related objects.

RuleEditor is a domain-specific editor for creating expert systems. A rule editor allows authors to create IF-THEN rules using complex Boolean expressions. A fact editor contains a list of facts and possible value. The RuleEditor automatically generates rule files that can be processed using the Jess inference engine [13]. Rules, facts, and associated symbols are stored in the Lyra OMS.

LanguageEditor is a domain-specific editor used for natural language processing. It enables creation of linguistic databases containing dialogues, phrase patterns (grammars), phrases, words, and morphemes.

SimulationEditor is an editor used to create system diagrams based on source, sink, storage, and flow components. This editor, along with the EquationEditor, is used to create ontology-based simulations.

EquationEditor is a familiar template-based equation editor for creating equations and fully specifying associated symbols. It differs significantly from other equation editors in that the underlying representation of equations and symbols is defined by the ontology.

Our approach to ontology-based simulation focuses on model authoring facilities and system validation tools. In the following section, background technologies that enable modelers to develop ontology-based facilities are presented, and the SimulationEditor and the EquationEditor are described in greater detail. Additionally, system validation tools, the symbol referencing flow diagram, and the sensitivity analysis tool, which provide facilities for model management, are described.

The EquationEditor

The EquationEditor is a tool for defining equations associated with a model and properly defining symbols appearing in these equations. It provides a facility for creating, browsing, and inspecting all equations, symbols, and units appearing in the model. It uses an interface that resembles other equation editors such as Microsoft Office Equation Editor [21] and MathType [5] but differs significantly because all the equations and symbols are represented by using ontology objects. This provides a way to represent the meaning of equations and symbols that are not possible with other equation editors. The EquationEditor is composed of three subeditors, Symbol Editor, Mathematical Expression Editor, and Unit Editor.

The Symbol Editor

Symbol Editor (Fig. 3) is an editor for individual symbols appearing in equations and includes a symbolic expression of a symbol, a quantity of measurement, and a description of the linguistic and programmatic properties of the symbol. A symbol is implemented as a class in the ontology that has a unique meaning within a specific domain. Often the same term (string of characters) is

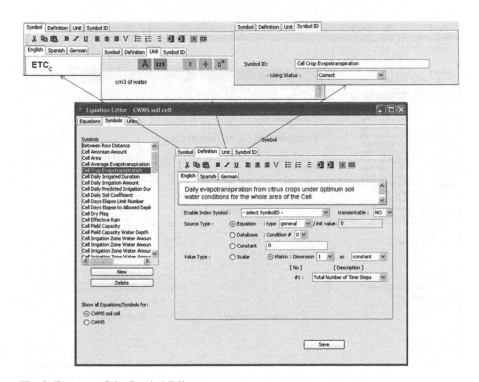

Fig. 3 Features of the Symbol Editor

used over different domains and refers to different symbols and thus has different meanings. Because a symbol has a unique identifier and is associated with a specific concept in the ontology, use of the same term for different symbols is permitted, and the domain ontology can be used to resolve their ambiguous meaning.

The value of a symbol is determined by one of three methods: from an equation, from a database, or from a constant that is directly assigned to the symbol. In the case where the symbol value is determined by an equation, there must be an equation in the database in which this symbol appears alone on the left side. To obtain the value from the database, some constraints may be required in order to locate and query a database to obtain the value (e.g., a current time and a soil layer number for querying a soil temperature at a specific date), and these constraints can be specified as a part of the symbol's properties (Fig. 4). If the symbol value is a constant, value of the constant is stored directly with the symbol.

Symbols can also be arrays when a symbol is used in different discrete intervals in space and time. For example, soil water content can be expressed in different soil layers that occur in different soil profiles, characterized by the depth from the soil surface, the soil profile number, and time.

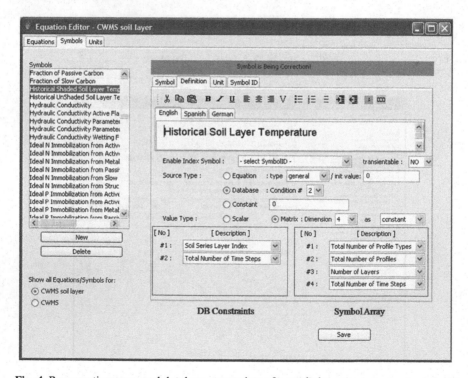

Fig. 4 Representing array and database constraints of a symbol

The Mathematical Expression Editor

The mathematical expression editor is designed to graphically create an equation with mathematical operator templates and symbols (Fig. 5). An equation is an expression that has a hierarchical tree data structure composed of operators and symbols. The equal operator is the root node of the tree. One constraint in the EquationEditor is that all equations have a single symbol on the left side. The value of the left-side symbol is defined by the calculation of the right-side expression. Thus the equation is assumed to be a function that has symbols as arguments. The editor provides many templates that describe specific operators. There are eight operator groups used to compose an equation (Table 1).

The Unit Editor

The Unit Editor is used to create and maintain the unit for a symbol (Fig. 6). A unit includes not only the generic collection of global standard units such as the metric unit system (SI) and the English unit system but also domain-specific units such as "cm^3 of soil." It is very important to carefully track the units

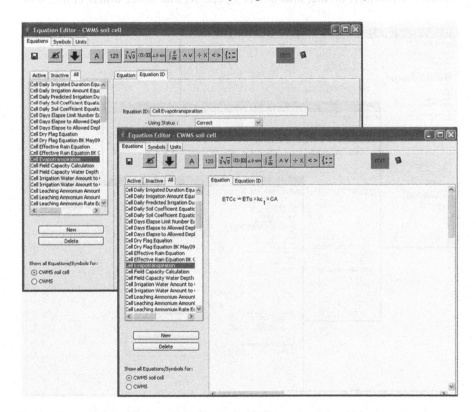

Fig. 5 Features of the Mathematical Expression Editor

Table 1 List of operators in the EquationEditor

Exponential operators	Subscript, double subscript, superscript, exponent, subscript and superscript, function, square root, root, log
Fence operators	Parenthesis, bracket, brace, absolute, ceiling, floor
Trigonometry operators	Sine, cosine, tangent, arcsine, arccosine, arctangent
Calculus operators	Limit, differential, indefinite, definite, summation, product, maximum, minimum
Logic operators	And, or, not
Arithmetic operators	Add, subtract, multiply, divide, negation
Relation operators	Less than, greater than, less and equal, greater and equal, equal, equivalent, not equal, not equivalent, less than and less than equal to, less than equal to and less than
Case operators	n-case, matrix

associated with symbols because different models may use the same symbol but with different units. A unit is not represented by a simple string but by a composition of symbols (like an equation) that is associated with the class defining the symbol. The unit can be expressed using a composition of limited operators (multiply, divide, and power operator) and other units. Thus, basic

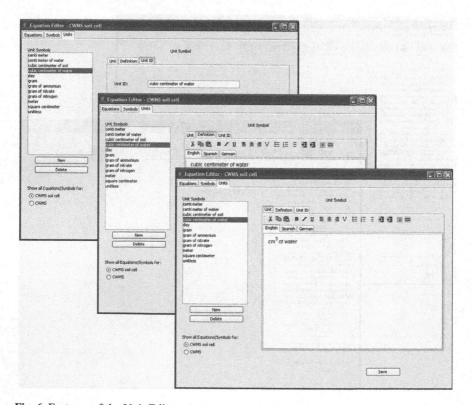

Fig. 6 Features of the Unit Editor

units such as length and weight can be reused for creating a composite unit. This makes it possible to automatically convert one unit to another (e.g., convert an English unit to a metric unit).

The SimulationEditor

The SimulationEditor is designed to represent models of dynamic systems using graphic elements such as source, sink, storage, and flow. It adopts concepts from the compartmental modeling technique [28] and Forrester notation [12], which are widely used in agriculture and natural resource models. However, like the EquationEditor, these concepts are represented internally using the ontology. The SimulationEditor is used for specifying the overall model structure in the form of elements and incorporates the EquationEditor described in the previous section in order to build equations associated with each element. The SimulationEditor also contains facilities for automatically generating and running simulations and generating reports. The SimulationEditor provides a graphic user interface to create and maintain a simulation system that includes a structure design interface, a simulation control interface, and a simulation result reporting interface.

The Structure Editor

The structure editor is the main interface of the SimulationEditor and provides functionalities that enable a modeler to create and maintain a simulation project by designing the structure of a system and to interact with the EquationEditor and the simulation controller. Structural design of a simulation system is a procedure by which a modeler creates physical or environmental elements and relationships in the system by using graphic elements. The concept of graphic elements in the structure editor is based on the compartment elements of Forrester: source, sink, storage, and flow. For example, a three-dimensional soil geometry may be described (Fig. 7) as a composition of soil cell (production unit), soil profile (horizontal), and soil layer module (vertical). These three elements may be defined as an instance of storage element, and relationships between these elements are represented by "part of" properties. Irrigation is realized with the flow element representing the flow of water into the cell. Equations and symbols associated with each element in the structure diagram are created using the EquationEditor. Thus the model is completely specified using these two editors.

The Simulation Controller

The simulation controller is a simulation engine used to generate and run a simulation based on the mathematical model, under specified conditions, with results presented in tabular reports or graphic displays. To generate a simulation, the simulation engine automatically converts ontology objects to program source code. It then compiles and runs the generated program to execute the

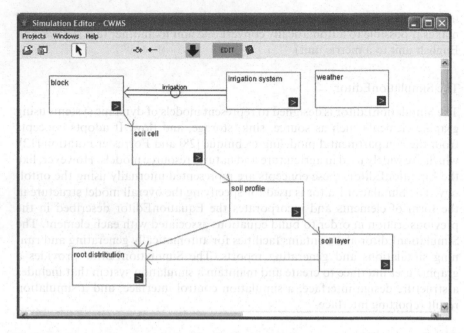

Fig. 7 System structure created using the SimulationEditor

simulation. Currently, Java is the target language, although in theory code from other languages could also be generated. It is not necessary for the modeler to examine, work directly with, or otherwise be concerned about the generated program source code. This process is completely internal to the operation of the software and transparent to the modeler. The compiled source code can be used as a component that can be accessed by other software environments after the model is developed.

The data object conversion and generation of program source code follows these steps:

- A class representing an element in the SimulationEditor forms a single class in Java. The class contains member variables and methods for all the symbols and equations in the element.
- Each symbol in the element is declared as a member variable named after the name of the symbol. If the symbol is a matrix, the member variable is declared as an array with the same dimensions as the symbol. A method is created that contains code for obtaining the value of the symbol. The name of this method is based on the name of the symbol.
  - If a symbol is a constant, the return value of the method is a constant for the symbol's value.
  - If a symbol obtains its value from a database, the method returns a value obtained by querying a database for the value of the symbol, subject to constraints specified in the symbol's properties.

- If a symbol obtains its value from an equation, the method contains code for solving the equation to obtain the value. Because the equation contains other symbols, the values of these symbols (on the right-hand side of the equation) are obtained by calling methods for determining their values.

The generated source code can be used independently as part of a component library and inserted into other application. For example, the resulting simulation application can be integrated into a desktop application used by growers, it can be part of a larger decision support system such as Decision Information System for Citrus (DISC) [2], and it can be part of a Web-based simulation environment (users can run the simulation through a Web page), or the simulation can be part of a Web service that is part of a distributed simulation environment.

Running a simulation involves compiling and executing the automatically generated source code. The simulation is controlled by recursively evaluating the value of a target symbol. Within the SimulationEditor, there is an interface to communicate with the model code library, which contains a method for calling the target symbol's method, which results in execution of the simulation. The generated source code contains variables for storing all values of variables that are retrieved by a report generator to display model results when the simulation has finished executing.

The report generates displays simulation results by showing the values of specific symbols in the form of a table or a graph as a function of time and proper dimension described in a symbol. A list of available symbols is provided to create reports. A designed report form can be categorized and maintained in the ontology and is created by selecting and adding to the target variables list and the dependent variable.

Additional Model Publishing Tools

The XML generator is a tool to generate an XML representation of the model. XML enables the model to be shared outside of the Lyra OMS environment. Two forms of markup language, MathML [20] and OpenMath [25], can be generated. MathML is an application of XML for describing mathematical notation and capturing its structure. It aims at integrating mathematical formulas into Web documents and is a recommendation of the World Wide Web Consortium (W3C) math working group. OpenMath is a document markup language for mathematical formulas. Among other things, it can be used to complement MathML, which mainly focuses on the presentation of formulas, with information about meaning. To generate these XML formats from equations in the ontology, each operator template class that is declared in the EquationEditor has a method transforming operator and arguments to a string containing the XML tag expression. An operator template can have other operator templates as arguments. The XML generator traverses this tree from

```
[ Source Equation]
A = B + C

[ generated documents]

<!-- MathML -->
<math display="block" xmlns=' http://www.w3.org/1998/Math/MathML'>
<mrow>
  <mi> A </mi>
  <mo>=</mo>
  <mrow>
    <mi> B </mi>          <!-- OpenMath -->
    <mo> + </mO>          <OMOBJ xmlns= "http://www.openmath.org/OpenMath" version ="2.0" cbase= "http://www.openmath.org/cd" >
    <mi> C </mi>          <OMA>
  </mrow>                 <OMS cd= "relationl" name= "eq" />
</mrow>                   <OMV name = "A" />
</math>                   <OMA>
                            <OMS cd= "arith1" name= "plus" />      <!--- Java--->
                            <OMV name= "B" />
                            <OMV name= "B" />                      public double A()
                          </OMA>                                  {
                          </OMA>                                  "return B() + C()" ;
                          </OMBJ>                                 }
```

Fig. 8 The equation "A = B + C" is converted to MathML, OpenMath, and a Java method

the root operator template (which is always the "equal" operator) to each leaf symbol. An example of generating XML expression is shown in Fig. 8 for the simple equation "A = B + C" in both MathML and OpenMath format. A Java method for solving this equation is also shown.

3.2 Citrus Water and Nutrient Management System

Ontology-based simulation methodologies were applied to building a model describing water and nutrient balance processes for citrus production called the Citrus Water and Nutrient Management System (CWMS) [24]. To aid growers in water management decision making, a computer-based decision support system was developed to facilitate more efficient use of water and nutrients by basing recommended application rates on site-specific characteristics and local weather data.

CWMS was constructed using the ontology-based simulation environment provided by Lyra. The entire model contains about 700 symbols and 500 equations. We use this example to demonstrate that a relatively complex model, which is now fully validated and implemented with growers, can be constructed in such an environment, and that the process can benefit from the methodology we have presented.

3.2.1 Model Structure

As illustrated in Fig. 9 (and shown formally in Fig. 7), seven concepts are defined for the model structure: block, soil cell, soil profile, soil layer, root distribution, irrigation system, and weather. The basic unit of the model is a restricted area, a cell in a block, representing a single citrus tree and the drainage field surrounding it. A commercial block of citrus consists of many cells as it has

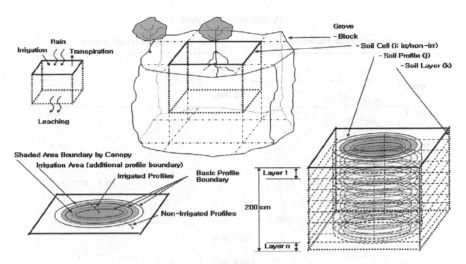

Fig. 9 CWMS soil cell, profile, and layer structure

many trees, but in this model to simplify the simulation process, the model is based on a single cell, and the single tree represented by that cell is characteristic of all the other trees in the block. To build a water balance model, a cell is defined as a cubical soil area surrounding one citrus tree, having a depth of 200 cm from the top of the soil. A cell is further divided into soil profiles within a cell and soil layers within a profile.

A soil cell consists of an area with the tree in the center. The width and length of the cell are described by the distances between trees in the row and trees between two adjacent rows. The cell may have from five to ten soil profile(s) that consist of forty 5-cm-thick soil layers. Each profile is designated as being one of four operational zones (i.e., a nonirrigated & dry-fertilized area, an irrigated & dry-fertilized area, a nonirrigated & non–dry-fertilized area, and an irrigated and non–dry-fertilized area as shown in Fig. 10) according to the irrigation and the dry-fertilized status. The physical and chemical characteristics of each soil layer are determined based on the particular soil series used in the simulation.

Basically, a profile is determined by the distance from the trunk of a tree to three root zone radii (75, 125, and 175 cm). Other profile boundaries are the irrigation diameter and the dry-fertilized area. Depending on the irrigation type (360 degree or less than 360 degree), soil profiles can be divided into irrigated-areas and nonirrigated areas. Irrigation and dry fertilization are assumed to be conducted at a planning area by two working path areas. Therefore, two working path areas between tree rows are always considered as a nonirrigated and non–dry-fertilized area (NINDF). An irrigated & non–dry-fertilized area (INDF) is an irrigated area in the working path area.

Equations for calculating each type's profile number and area is defined at the cell level. Figure 11 shows equations of profile numbers of three types (profile

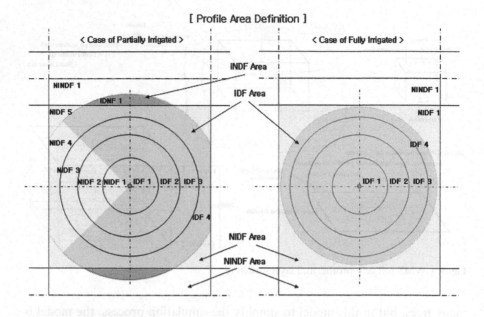

[Profile Area Definition]

< Case of Partially Irrigated >   < Case of Fully Irrigated >

INDF Area
IDF Area

NINDF 1   NINDF 1
IDNF 1
NIDF 5   NIDF 1
NIDF 4   IDF 4
NIDF 3
NIDF 2 NIDF 1 IDF 1 IDF 2 IDF 3   IDF 1 IDF 2 IDF 3
IDF 4

NIDF Area
NINDF Area

[i index]
i = 1 – NIDF : Non Irrigated–Dry Fertilized , i = 2 – IDF : Irrigated–Dry Fertilized
i = 3 – NINDF : Non Irrigated– Non Dry Fertilized , i = 4 – INDF : Irrigated–Non Dry Fertilized

Fig. 10 Soil profile areas irrigation and nutrient supply designations

$$
NPIDF = \begin{cases} 2 & 2 \times RZR_1 < WD \le 2 \times RZR_2 \\ 3 & 2 \times RZR_2 < WD \le 2 \times RZR_3 \\ 4 & 2 \times RZR_3 < WD \\ 1 & \text{otherwise} \end{cases}
$$

$$
NPNIDF = \begin{cases} \begin{cases} 4 & WD = 2 \times RZR_1 \vee WD = 2 \times RZR_2 \vee WD = 2 \times RZR_3 \\ 5 & \text{otherwise} \end{cases} & SpP < 360 \\ \begin{cases} 4 - NPIDF & WD = 2 \times RZR_1 \vee WD = 2 \times RZR_2 \vee WD = 2 \times RZR_3 \\ 5 - NPIDF & \text{otherwise} \end{cases} & \text{otherwise} \end{cases}
$$

$$
NPINDF = \begin{cases} 1 & WD > 2 \times DistToHedgingBoundary \\ 0 & \text{otherwise} \end{cases}
$$

Fig. 11 Equations of profile numbers of three types (profile number of a nonirrigated & non–dry-fertilized area is always 1, and it is defined as a constant)

number of a nonirrigated & non–dry-fertilized area is always 1, and it is defined as a constant). NPIDF, NPNIDF, and NPINDF are respectively symbols of a profile number of an irrigated & dry-fertilized area, a profile number of a nonirrigated & dry-fertilized area, and a profile number of an irrigated and non–dry-fertilized area. RZR is a symbol of the root zone radius matrix, WD is a symbol of the wetting diameter, and SpP is a symbol of the spray pattern.

A soil layer is one vertical element of a soil profile, and the number of soil layers in a soil profile is determined by layer thickness and total depth of the soil profile, whose maximum depth is 200 cm. The soil layers can be grouped, and it is represented as a matrix shown in Fig. 12. Each row is a layer group; the first column is the thickness of layers in a group, the second column is the number of layers in a group, and the third column is a cumulative layer number to that group.

3.2.2 Model Functions

Soil water movement is used as an example of the complex functions of the CWMS model that is implemented by the ontology-based simulation system. The CWMS model is an enhanced tipping bucket model. A pure tipping bucket model assumes that water moves through the entire simulation depth within one time-step. CWMS does not make this assumption and simulates the wetting front movement among layers based on soil water content within and below the wetting front and the layer-specific hydraulic conductivity characteristics to determine the depth of wetting within each time step. Furthermore, because the exact times of irrigation or rainfall events are not inputs (this information is available only at a daily resolution), CWMS assumes that all irrigation, water with nitrogen (N) application, and rainfall occurs at noon giving a maximum of 12 hours to move through the soil on the first day after irrigation, water with N application, or rainfall. The calculated wetting front speed and the time for which the wetting front travels a layer thickness are used to determine the wetting front depth at the end of the day. Water in excess of the soil's drained lower limit continues to drain to deeper soil layers the next day. These soil drainage processes as well as water infiltration, nutrient transformation and transport, and plant uptake are additional functions that can be easily entered into the ontology simulation system as interdependent equations. Additionally, values for soil characteristics (e.g., hydraulic conductivity, root density, and soil temperature) used by these main function equations are entered into the ontology. These soil characteristics change over time and must be recalculated for each layer of each soil profile with every time step.

$$LThickRangeArr = \begin{vmatrix} 5 & 10 & 10 \\ 10 & 6 & 16 \\ 15 & 6 & 22 \end{vmatrix}$$

Fig. 12 Layer thickness matrix

3.2.3 Defining System Symbols

The symbols for block, cell, soil profile, and soil layer are matrices requiring spatial and temporal dimensions. The soil profile also requires input for type and soil profile number dimensions. The soil layer needs four dimensions: the three dimensions of soil profile and one for soil layer number. For example, a symbol, historical soil layer temperature, in a soil layer has four dimensions, and it is described through the EquationEditor (Fig. 13).

Root density is an example of a symbol with a value that changes with distance from the tree trunk and soil depth over time. The spatial distribution of roots within the model is represented as a matrix of root sections within each soil profile, each containing many soil depths (Fig. 14). The three-dimensional distribution is based on root density distribution as a function of tree size [23]. In the root distribution matrix, a column is a root section with a known soil surface area, and a row corresponds with several soil layers within the root section. The first three root sections are described by a circle divided by three radii (75 cm, 125 cm, and 175 cm). The fourth root section is the remainder of the area occupied by the citrus tree. Each root section contains 10 soil layer groups. The first 6 rows use 15-cm soil depth each, and the last 4 rows have soil depths of 30 cm each. Root density within these last four rows is proportional to the sixth soil depth group. RD is a symbol for root density, cv is a symbol of the canopy volume, and RSL is a matrix symbol of the root density regression parameter at lower layer.

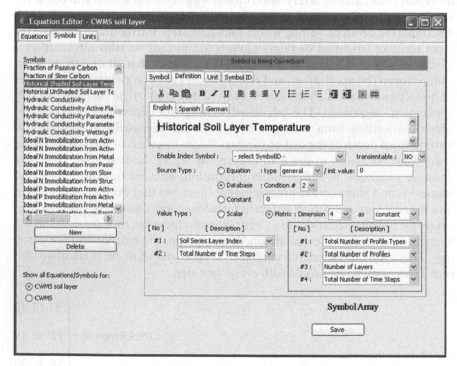

Fig. 13 An example of defining dimension of symbol in soil layer

$$RD = \begin{cases}
\begin{cases}
1.2 + 0.023 \times cv & 0 < cv \le 15 \\
1.1 + 0.043 \times cv - 0.00088 \times cv^2 & 15 < cv \le 35 \\
1.45 + 0.0022 \times cv & 35 < cv \le 50 \\
1.3 + 0.006 \times cv & cv > 50
\end{cases} &
\begin{cases}
0.2 + 0.027 \times cv & 0 < cv \le 20 \\
0.26 + 0.024 \times cv & 20 < cv \le 50 \\
1.1 + 0.007 \times cv & cv > 50
\end{cases} &
\begin{cases}
0.005 \times cv - 0.01 & 2 < cv \le 15 \\
0.02 - 0.0075 \times cv + 0.0007 \times cv^2 & 15 < cv \le 35 \\
0.26 + 0.01 \times cv & 35 < cv \le 50 \\
0.5 + 0.005 \times cv & cv > 50
\end{cases} &
\begin{cases}
0.054 \times cv - 0.8 & 14.8 < cv \le 32 \\
1 - 0.002 \times cv & 32 < cv \le 50 \\
0.55 + 0.007 \times cv & cv \ge 50
\end{cases} \\[2em]

\begin{cases}
0.3 + 0.02 \times cv & 0 < cv \le 15 \\
0.3 + 0.03 \times cv - 0.0007 \times cv^2 & 15 < cv \le 35 \\
0.8 - 0.0026 \times cv & 35 < cv \le 50 \\
0.22 + 0.005 \times cv & cv > 50
\end{cases} &
\begin{cases}
0.1 + 0.02 \times cv & 0 < cv \le 15 \\
0.07 + 0.035 \times cv - 0.0009 \times cv^2 & 15 < cv \le 35 \\
0.19 & 35 < cv \le 50 \\
0.02 \times cv - 0.81 & cv > 50
\end{cases} &
\begin{cases}
0.0053 \times cv - 0.008 & 1.5 < cv \le 15 \\
0.008 \times cv - 0.04 & cv > 15
\end{cases} &
\begin{cases}
0.005 \times cv - 0.125 & cv > 25 \\
0 & \text{otherwise}
\end{cases} \\[2em]

\begin{cases}
0.12 + 0.012 \times cv & 0 < cv \le 15 \\
0.1 + 0.019 \times cv - 0.0004 \times cv^2 & 15 < cv \le 35 \\
0.35 - 0.002 \times cv & 35 < cv \le 50 \\
0.005 \times cv & cv > 50
\end{cases} &
\begin{cases}
0.01 + 0.012 \times cv & 0 < cv \le 15 \\
0.019 \times cv - 0.004 - 0.0004 \times cv^2 & 15 < cv \le 35 \\
0.16 + 0.0005 \times cv & 35 < cv \le 50 \\
0.02 \times cv - 0.8 & cv > 50
\end{cases} &
\begin{cases}
0.002 \times cv - 0.002 & 1 < cv \le 15 \\
0.0043 \times cv - 0.037 & cv > 15
\end{cases} &
\begin{cases}
0.002 \times cv - 0.05 & cv > 25 \\
0 & \text{otherwise}
\end{cases} \\[2em]

\begin{cases}
0.07 + 0.013 \times cv & 0 < cv \le 15 \\
0.06 + 0.016 \times cv - 0.0002 \times cv^2 & 15 < cv \le 35 \\
0.13 + 0.007 \times cv & cv > 35
\end{cases} &
\begin{cases}
0.01 + 0.006 \times cv & 0 < cv \le 15 \\
0.01 \times cv - 0.05 & 15 < cv \le 50 \\
0.2 + 0.005 \times cv & cv > 50
\end{cases} &
\begin{cases}
0.0013 \times cv & 1 < cv \le 15 \\
0.012 \times cv - 0.16 & 15 < cv \le 35 \\
0.09 + 0.005 \times cv & cv > 35
\end{cases} &
\begin{cases}
0.0018 \times cv - 0.045 & cv > 25 \\
0 & \text{otherwise}
\end{cases} \\[2em]

\begin{cases}
0.07 + 0.008 \times cv & 0 < cv \le 20 \\
0.03 + 0.01 \times cv & 20 < cv \le 50 \\
0.23 + 0.006 \times cv & cv > 50
\end{cases} &
\begin{cases}
0.02 + 0.004 \times cv & 0 < cv \le 15 \\
0.015 \times cv - 0.15 & 15 < cv \le 50 \\
0.45 + 0.003 \times cv & cv > 50
\end{cases} &
\begin{cases}
0.001 \times cv - 0.0008 & 0.8 < cv \le 15 \\
0.013 \times cv - 0.18 & 15 < cv \le 35 \\
0.01 \times cv - 0.075 & 35 < cv \le 50 \\
0.22 + 0.004 \times cv & cv > 50
\end{cases} &
\begin{cases}
0.001 \times cv - 0.025 & cv > 25 \\
0 & \text{otherwise}
\end{cases} \\[2em]

\begin{cases}
0.01 \times cv - 0.007 & 0.7 \le cv < 20 \\
0.013 \times cv - 0.08 & 20 \le cv < 50 \\
0.29 + 0.006 \times cv & cv \ge 50
\end{cases} &
\begin{cases}
0.0015 \times cv - 0.002 & 1.4 < cv \le 15 \\
0.01 \times cv - 0.13 & 15 < cv \le 50 \\
0.12 + 0.005 \times cv & cv > 50
\end{cases} &
\begin{cases}
0.0008 \times cv - 0.0009 & 1.2 < cv \le 15 \\
0.009 \times cv - 0.12 & 15 < cv \le 35 \\
0.006 \times cv - 0.005 & cv > 35
\end{cases} &
\begin{cases}
0.0008 \times cv - 0.02 & cv > 25 \\
0 & \text{otherwise}
\end{cases} \\[2em]

RSL_1 \times RD_{61} & RSL_1 \times RD_{62} & RSL_1 \times RD_{63} & RSL_1 \times RD_{64} \\
RSL_2 \times RD_{61} & RSL_2 \times RD_{62} & RSL_2 \times RD_{63} & RSL_2 \times RD_{64} \\
RSL_3 \times RD_{61} & RSL_3 \times RD_{62} & RSL_3 \times RD_{63} & RSL_3 \times RD_{64} \\
RSL_4 \times RD_{61} & RSL_4 \times RD_{62} & RSL_4 \times RD_{63} & RSL_4 \times RD_{64}
\end{cases}$$

Fig. 14 Root density matrix

3.2.4 CWMS Application Implementation

The CWMS model is used to implement a CWMS application program for use by a grower that utilizes crop, soil and weather data. A CWMS application consists of generated simulation source codes and a graphical user interface. Generated simulation code is plugged into the graphical user interface without further coding required.

A graphical user interface allows growers to interact with the system through three phases: the setup, the irrigation scheduling, and reporting. The setup phase, shown in Fig. 15, is for configuring the cell and block information for a particular grove. The irrigation scheduling phase, shown in Fig. 16, provides irrigation scheduling information to growers. By default, it is based on a 14-day simulation followed by a 3-day prediction period (the simulation period can be extended) to provide immediate term recommendations on irrigation rates. We have also run this simulation for full seasons (over 250 days per year, over multiple years) in order to do strategic planning on irrigation scheduling.

Detailed simulation results are provided in the form of daily, monthly, and yearly reports (Fig. 17). The daily report contains each soil layer's root length, water content, nitrate and ammonium content, and soil coefficient. Data can be browsed by selecting a specific data and profile type. The monthly report shows

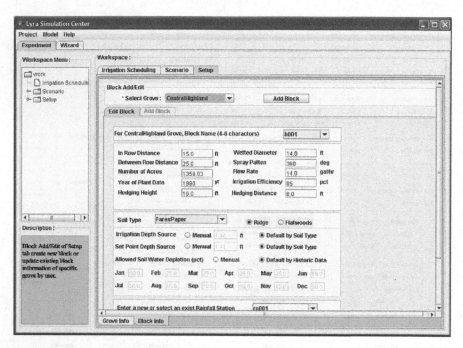

Fig. 15 System configuration phase

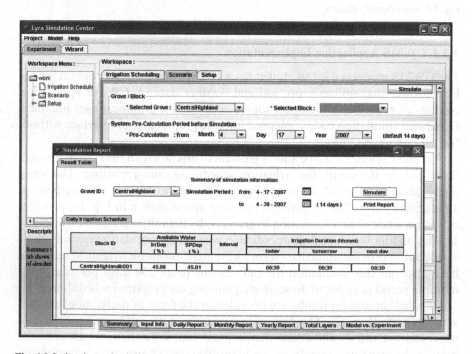

Fig. 16 Irrigation scheduling recommendation

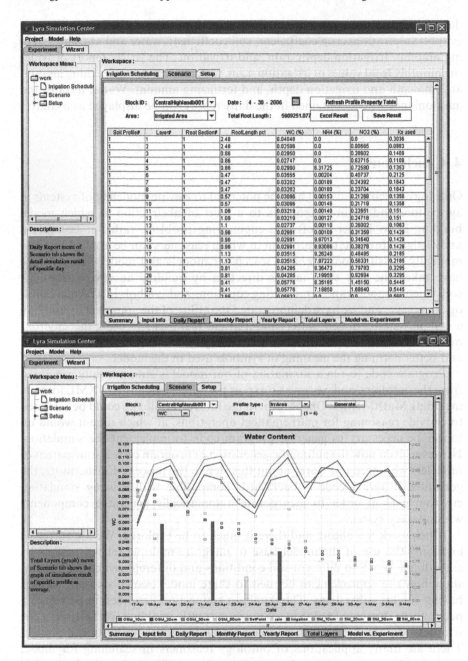

Fig. 17 Simulation result reports: (a) table and (b) graph

data for a particular month including irrigation interval days and duration, evapotranspiration, crop coefficient, soil coefficient, water and nutrient leaching amount, irrigation depth, rain, and irrigation. The yearly report provides the monthly total values of irrigation, rain, water, and nutrient leaching amount at 2-m depth and irrigation depth, and fertilizing amount. Water, nitrate, and ammonium content contained in the daily report can be displayed as a chart.

4 Conclusions

Ontology-based simulation has the potential to elevate modeling of systems in agriculture and natural resources to a level of abstraction at which model building is no longer a software engineering problem, rather it becomes a knowledge representation problem in which reasoners can be applied to automatically classify, compare, and search for models and model elements. We have taken the first steps in this direction by building an environment for constructing models and representing equations and symbols in a formal ontology language, and we have demonstrated the utility of the approach by building a complex model of soil, water, and nutrient management using this environment. Already this approach has the advantage of making models more explicit and better defining the meaning of symbols used in the model.

Program code to run a simulation of our model is automatically generated from the ontology, and we can also export the model equations in XML formats including MathML and OpenMath. In the future, ontologies could be extended to provide reasoning for mathematical operations, at which time it would no longer be necessary to generate program code to implement the simulation. However, even now the automatic generation of program code is transparent to modelers who need not worry about this step in the process. Furthermore, the generated program code is backwards compatible with existing simulation environments in order to incorporate automatically generated components within legacy systems.

Much work lies ahead, and there is much to be explored. We are currently exploring the use of the model base to integrate multiple related models by extending CWMS to different soil conditions and different crops. We hope to show how the approach can be used to share model elements among models having similar subsystems. The database connection described in Section 2.5 needs to be further developed. We are currently building Web services for wrapping databases that can be integrated into models using this approach. There are many applications of ontology reasoners in searching for model components and comparing the structure of two different but similar models that need to be explored. Finally, sharing ontologies among modelers who collaborate on an international level has only just begun. Much discussion is needed to develop standards for sharing and coordinating ontology development at various levels within the modeling community.

References

1. Athanasiadis, I. N., A. E. Rizzoli, M. Donatelli and L. Carlini. 2006. Enriching software model interfaces using ontology-based tools. Proceedings of the iEMSs Third Biennial Meeting, "Summit on Environmental Modelling and Software". International Environmental Modelling and Software Society, Burlington, USA, July 2006.
2. Beck, H. W., L. G. Albrigo and S. Kim. 2004. DISC citrus planning and scheduling program. In: Proceeding of the Seventh International Symposium on Modelling in Fruit Research and Orchard Management, Acta Horticulturae. pp. 25–32.
3. Benjamin, P. C., M. Patki and R. J. Mayer. 2006. Using ontologies for simulation modeling. In: Winter Simulation Conference WSC 2006, Montery, California, USA, December 3–6, 2006. pp. 1151–1159.
4. Cuske, C., T. Dickopp and S. Seedorf. 2005. JOntoRisk: An ontology-based platform for knowledge-based simulation modeling in financial risk management. European Simulation and Modeling Conference 2005. Riga, Latvia, June 1–4, 2005.
5. Design Science. 1996. MathType. http://www.dessci.com/en/products/mathtype.
6. Devanbu, P., R. J. Brachman, P. G. Selfridge and B. W. Ballard. 1990. LaSSIE: A knowledge-based software information system. Proceedings, 12th International Conference on Software Engineering. Nice, France, March 26–30, 1990. pp. 249–261.
7. Donatelli, M., G. Bellocchi and L. Carlini. 2006a. A software component for estimating solar radiation. Environmental Modelling and Software 21(3): 411–416.
8. Donatelli, M., G. Bellocchi and L. Carlini. 2006b. Sharing knowledge via software components: models on reference evapotranspiration. European Journal of Agronomy 24(2): 186–192.
9. Eitzinger, J., M. Trnka, J. Hosch, Z. Zalud and M. Dubrovsky. 2004. Comparison of CERES, WOFOST and SWAP models in simulating soil water content during growing season under different soil conditions. Ecological Modeling 171(3): 223–246.
10. Ewert, F., H. Van Keulen, M. K. Van Ittersum, K. E. Giller, P. A. Leffelaar and R. P. Roetter. 2006. Multi-scale analysis and modelling of natural resource management. In: Proceedings of the iEMSs Third Biannual Meeting, "Summit on Environmental Modelling and Software", International Environmental Modelling and Software Society, Burlington, USA, July, 2006.
11. Fishwick, P. A. and J. A. Miller. 2004. Ontologies for modeling and simulation: issues and approaches. In: Proceeding of 2004 Winter Simulation Conference. Washington, D.C., December 5–8, 2004, pp. 251–256.
12. Forrester, J. W. 1971. World Dynamics. Cambridge, MA: Productivity Press, p. 144.
13. Friedman-Hill, E. 2007. Jess, the rule engine for Java platform. http://herzberg.ca.sandia.gov. Sandia National Laboratories.
14. Haan, C. T., H. P. Johnson and D. L. Brakensiek. 1982. Hydrologic modeling of small watersheds ASAE Monograph No. 5. American Society of Agricultural Engineers, St. Joseph, MO. 553 pp.
15. Islam, A. S. and M. Piasecki. 2008. Ontology based web simulation system for hydrodynamic modeling. Simulation Modelling Practice and Theory 16: 754–767.
16. Ittersum, M. K. V., F. Ewert, T. Heckelei, J. Wery, J. Alkan Olsson, E. Andersen, I. Bezlepkina, F. Brouwer, M. Donatelli, G. Flichman, L. Olsson, A. E. Rizzoli, T. van der Wal, J. E. Wien and J. Wolf. 2007. Integrated assessment of agricultural systems – A component-based framework for the European Union (SEAMLESS). Agricultural Systems 96(1–3): 150–165.
17. Jones, J. W., B. A. Keating and C. H. Porter. 2001. Approaches to modular model development. Agricultural Systems 70(2–3): 421–443.
18. Jurisica, I., J. Mylopoulos and E. Yu. 2004. Ontologies for knowledge management: an information systems perspective. Knowledge and Information Systems 6: 380–401.

19. Lu, H.-Y., C.-T. Lu, M.-L. Wei and L.-F. Chan. 2004. Comparison of different models for nondestructive leaf area estimation in taro. Agronomy Journal 96(2): 448–453.
20. MathML. 2001. http://www.w3.org/Math/.
21. Microsoft. 2003. Microsoft Office Equation Editor. http://office.microsoft.com/en-us/word/HP051902471033.aspx.
22. Miller, J. A., G. T. Baramidze, A. P. Sheth and P. A. Fishwick. 2004. Investigating ontologies for simulation modeling of 37th Annual Simulation Symposium, Arlington, VA, April 18–22, 2004.
23. Morgan, K. T., H. W. Beck, J. M. S. Scholberg and S. Grunwald. 2006. In-season irrigation and nitrogen decision support system for citrus production. In: Proceedings 4th World Congress on Computers in Agriculture and Natural Resources. Orlando, FL, July 2006, pp. 640–654.
24. Morgan, K. T., T. A. Obreza, and J. M. S. Scholberg. 2007. Characterizing orange tree root distribution in space and time. Journal of the American Society for Horticultural Science 132(2): 262–269.
25. OpenMath. 2000. http://www.openmath.org.
26. OWL. 2005. Web Ontology Language Guide. http://www.w3.org/TR/owl-guide.
27. Park, M. and P. A. Fishwick. 2005. Integrating dynamic and geometry model components through ontology-based interface. Simulation 81(12): 795–813.
28. Peart, R. M. and R. B. Curry. 1998. Agricultural Systems Modeling and Simulation. New York: Marcel Dekker.
29. Raubel, M. and W. Kuhn. 2004. Ontology-based task simulation. Spatial Cognition and Computation 4: 15–37.
30. Reddy, V. and V. Anbumozhi. 2004. Development and application of crop simulation models for sustainable natural resource management. In: International Agricultural Engineering Conference. IAEC, Beijing, P.R. China, October 11–14, 2004, p. 10-035A
31. Rizzoli, A. E., M. Donatelli, R. Muetzelfeldt, T. Otjens, M. G. E. Sevensson, F. V. Evert, F. Villa and J. Bolte. 2004. SEAMFRAME, a proposal for an integrated modelling framework for agricultural systems. Proceedings of the 8th ESA Congress, Copenhagen, Denmark, July 11–15, 2004. pp. 331–332.
32. Scholten, H., A. Kassahun, J. C. Refsgaard, T. Kargas, C. Gavardinas and A. J. M. Beulens. 2007. A methodology to support multidisciplinary model-based water management. Environmental Modeling & Software 22(5): 743–759.
33. Simulistics. 2007. Simile. http://www.simulistics.com/products/simile.php.
34. W3C. 2007. SPARQL. http://www.w3.org/TR/rdf-sparql-query.
35. Wolfram. 2007. Mathematica. http://www.wolfram.com/products/mathematica.

Precision Farming, Myth or Reality: Selected Case Studies from Mississippi Cotton Fields

Jeffrey L. Willers, Eric Jallas, James M. McKinion, Michael R. Seal, and Sam Turner

Abstract There is a lot of interest in the concept of precision farming, also called precision agriculture or site-specific management. Although the total acreage managed by these concepts is increasing worldwide each year, there are several limitations and constraints that must be resolved to sustain this increase. These include (1) collecting and managing the large amounts of information necessary to accomplish this micromanagement, (2) building and delivering geo-referenced fine-scale (i.e., change every few meters or less) prescriptions in a timely manner, (3) finding or developing agricultural machines capable of quickly and simultaneously altering the rates of one or more agrichemicals applied to the crop according to a geo-referenced prescription, (4) the need to have personnel stay "current" with advancements in developing technologies and adapting them to agriculture, (5) refining existing and/or creating new analytical theories useful in agriculture within a multidisciplinary, multi-institutional, and multibusiness environment of cooperation, and (6) modification of agricultural practices that enhances environmental conservation and/or stewardship while complying with governmental regulations and facing difficult economic constraints to remain profitable. There are many myths that overshadow the realities and obscure the true possibilities of precision agriculture. Considerations to establish productive linkages between the diverse sources of information and equipment necessary to apply site-specific practices and geographically monitor yield are daunting. It is anticipated that simulation models and other decision support systems will play key roles in integrating tasks involved with precision agriculture. Discovering how to connect models or other software systems to the hardware technologies of precision agriculture, while demonstrating their reliability and managing the flows of information among components, is a major challenge. The close cooperation of the

---

J.L. Willers (✉)
USDA-ARS-Genetics and Precision Agriculture Research Unit, MS, Mississippi State, USA
e-mail: jeffrey.willers@ars.usda.gov

P.J. Papajorgji, P.M. Pardalos (eds.), *Advances in Modeling Agricultural Systems*, DOI 10.1007/978-0-387-75181-8_12, © Springer Science+Business Media, LLC 2009

extension, industrial, production, and research sectors of agriculture will be required to resolve this constraint.

1 Introduction

Production agriculture has experienced dramatic changes since its beginnings, and more changes loom on the horizon [38]. Not long ago, production agriculture was organized around small farms, each with an owner/operator who had intimate knowledge of his farm and its available resources – he knew almost exactly what to expect from each field whenever nature changed a certain way; for example, where certain insects would attack first, where and when soil erosion was most likely, what crop diseases to expect and when, and what yield to expect from different sections of the fields. Economic pressures and technology have changed this figure. Today, large farms dominate most agricultural enterprises [14]. Skills in handling capital investments in equipment, land, labor, and technology have as much or more impact on the success of farm management as agricultural skill and production knowledge. The influences of economies of scale was a major factor in the current dominant management strategy of uniform crop treatments across large fields planted in a single crop, despite the presence of heterogeneous soil and environmental conditions. This type of management results in the simultaneous conditions of over and under supply of the inputs necessary to produce a crop [54]. The requirement for maintaining low, competitive commodity prices coupled with pressures of increased production costs is challenging given farm management strategies based on broadcast (or flat-rate) applications of inputs.

These realities resulted in the birth of precision agriculture, which is a general term for using newer technologies in the management and control of farming activities and agricultural production with the intent to geographically optimize inputs (agri-chemicals, time, materials, etc.) for the purpose of increasing crop yield, quality, and profits. Other names for precision agriculture include *precision farming, site-specific farming, site-specific crop management, prescription farming*, and others [29]. These technologies have the potential to manage large fields as if they are composed of numerous small fields; an interesting shift that parallels the farming practices of the historical small farm. This shift has been driven primarily by the availability of technologies [22] that allow for the exact determination of remote sensing data and the geographic location of variable-rate agricultural machines and harvesters upon the surface of the earth [5, 24]. Site-specific information (i.e., information about a certain piece of land (site) defined at a certain areal size; e.g., 4 m^2) is becoming available at increasingly lower costs for increasingly smaller areas of a farm; thus, providing for possible site-management of smaller parts of the farm to produce a crop of equal or better quality at a lower or equal cost to traditional management practices. Management practices are becoming more specialized and customized for smaller, individually managed units within fields. The use of computers, global positioning systems (GPS), variable-rate equipped farm implements, geographical information systems (GIS), real-time kinetic (RTK) machine guidance systems, remote sensing, simulation, and innovative statistical analysis techniques are increasingly being developed and made available to help provide farm managers with unprecedented levels of information and possibilities. However, there are some barriers and limitations hampering these advances.

The return toward the management of small-scale units that potentially number into the hundreds, thousands, or hundreds of thousands of units on a large farm presents many challenges, some of which are (1) collecting and managing large amounts of information necessary to accomplish this micromanagement, (2) building and delivering geo-referenced fine-scale[1] prescriptions in a manner timely with farming deadlines, (3) finding or developing the machines capable of quickly changing prescriptions of amounts of agrichemicals applied to the crop (every few meters or less), (4) staying "current" with advancements in developing technologies and adapting them to agriculture, (5) refining existing and/or creating new analytical theories that can be used in the "real world" of agriculture, and (6) modifying current practices to enhance environmental conservation and stewardship while complying with governmental regulations and economic constraints and still remain profitable. These challenges provide many opportunities for both difficult and fascinating research [4, 13, 25, 29, 68]. Without a doubt, the agricultural industry is looking to these geospatial technologies as tools to help keep costs under control and production profitable [51].

In this new world of production agriculture, some myths have sprouted up and grown to such a size that they overshadow reality. To separate the myths from reality, our objective is to develop these six target areas into themes by describing four illustrative case scenarios and developing discussions around them. These perspectives have arisen over the years from our precision agriculture experiences in some Mississippi cotton fields. Each of these six challenges is not restricted to any one discipline of the many that are involved in precision agriculture. Also, one of the frustrating aspects is that they are all part reality and part myth. In light of this mixture of myth and reality, we first address one of the main myths behind precision agriculture.

2 Multidisciplinary Teams

It is a myth that experts from multiple disciplines can be easily recruited into a team to find a solution to a problem area in precision agriculture. Whereas the reality is that a multidisciplinary team is actually required [6, 13, 15, 46, 47, 48, 64], these teams are difficult to assemble and maintain. There are

---

[1] A fine-scale map will depict great detail (e.g., 1:5,000), whereas a broad-scale map will not show great detail (e.g., 1:250,000). Map resolution is another important consideration.

many reasons for this, but three main ones are (1) money, (2) level of skill needed, and (3) availability. Money is always a problem in research. The lead-time and justifications required often limit amounts that can be secured for a given project. Often it reverts back to the farmer to provide support for the research in the form of land, crops, treatments, machinery, manpower, and information, thus greatly decreasing some of the capital needed to conduct the research. Seldom are the skill levels needed in this type of team composed of "entry level skills" as precision agricultural problems are not simple or of the type commonly used as examples in textbooks. Once a researcher or technician matures in skill, they become increasingly busy with additional projects. Primarily, because once it is known that they exhibit the necessary skills, others make requests of them, which further limits their availability. Often, only part-time or partial assistance on a problem, or a question-answering session on a "time as available basis," is the best that can be hoped for. Although this conflict for time is real, most skilled personnel who work on precision agriculture–related technologies will often go out of their way to provide assistance.

In this technical environment, a multidisciplinary team also runs into a common problem – communication. Each technical area has its own "language," which is confusing to others because of specialized words, with some common words redefined. This can cause enough frustration that many do not want to work out of their own area of expertise, so it falls to the team leader to learn enough "jargon" to facilitate communication among the team members and build bridges of understanding in order to complete a project. The need to communicate, especially at start up, complicates by many fold the time and energy needed to get a solution to an experiment and have the results transferred to, and then accepted into, the other highly technical fields related to farming.

Another myth related to communication is the thought that the sharing of results can be easily done using maps, charts, equations, or other common forms of data representation. In reality many people – even very well educated people – can have a hard time understanding common forms of data representation due to color blindness, unfamiliarity with the method of representation, use of unfamiliar words or symbols, or other reasons. Often, it is necessary to present the same results in multiple ways to reach different segments of the agricultural audience.

Another myth that creates a common problem for a multidisciplinary team is that of corporate stability and cooperation. Often, if a company has been in business for some years, everyone expects them to continue in business. In the business world of agriculture, that is not a given. Some go out of business. Often, a company is a "leader" in an area or technology, then it is gone or bought out, and new ownership changes what is offered or alters in some way the quality of goods and/or services. Even harder on the team is the myth that any business will cooperate with it (or another business effort) even if they too will benefit. The reality is that interteam cooperation is uncommon. There are trade secrets, rivalries, cost considerations, lack of vision, or lack of patience to overcome conflicts and persist in "pushing through" to the land of promise that lies beyond, if and when true cooperation occurs. Thus, whereas the establishment of a multidisciplinary team is a highly desirable and effective strategy in precision agriculture, the reality is that these teams are hard to create, manage, and coordinate.

3 Precision Agriculture and Information

The backbone of precision agriculture is information. This means a lot of information is collected and processed to create other usable pieces of data to send back to the farmer as a data product that he can use to implement changes in crop management policy. This is a simple idea, but hard to do in practice. Today, technology is growing by leaps and bounds, with the coordination of these diverse technologies into integrated solutions lagging behind, while paradoxically, the demand to apply these solutions to complex problems is growing the fastest. We introduce at this time the first of several case studies to begin to illustrate our concerns.

3.1 Case 1: Simulation and Variable-Rate Nitrogen with Mississippi Delta Cotton, 1998

This study involved a portion of a more than 80-hectare (ha) field near the Mississippi River in Bolivar County, Mississippi. Remote sensing is a technology that monitors and maps sources of variability at the field and subfield levels. Figures 1 and 2 are examples of the type of images that can be collected

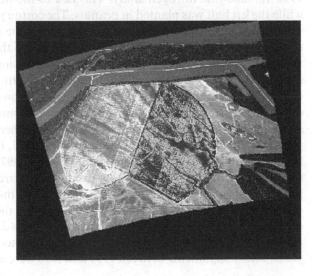

Fig. 1 Multispectral image of ~80-ha cotton field during July 1997 [63]. (Image courtesy ITD Spectral Visions, Stennis Space Center, MS.)

Fig. 2 Multispectral image
of the two ~80-ha cotton
fields during July 1998.
(Image courtesy ITD
Spectral Vision, Stennis
Space Center, MS.)

throughout the growing season by federal, state, university, or commercial
programs. A three-band digital camera system mounted in the belly of a fixed
wing aircraft (Cessna 210 Turbo flown at an average speed of 110 knots)
obtained images from an altitude of 1,829 m above ground level (AGL),
rendering a 1 m × 1 m spatial resolution. The spectral resolution involved
three bands (band 1 = 540 ± 5 nm, band 2 = 695 ± 5 nm, and band 3 = 840
± 5 nm), and the pixel reflectance information was processed with ERDAS
Imagine (Leica Geosystems, Norcross, GA) software. The result of image
processing in this instance grouped the different pixels into 16 categories that
corresponded with plant vigor (see below).

Figure 1 shows a semi-circular field where the right-half was used during the
1998 variable-rate nitrogen study. This half of the field was planted in cotton
while the left half was planted in peanuts. The cotton portion of this aerial view
was processed to show the status of crop during the month of July 1997. It is
evident that the cotton field is not homogenous; therefore, the within field
variability observed on the ground was not surprising. Note the sharp line of
contrast between the darker and lighter areas of cotton, about one-third of the
way toward the right edge. In 1996, peanuts were grown in this nearly 43 ha area
of the 1997 cotton planting and appear to have promoted stronger plant vigor
among those cotton plants in comparison to the other plants during July 1997.

Figure 2 shows the field half as shown in Figure 1, but now includes the west
half of cotton that was planted in peanuts during 1997. Both sides were planted
to cotton during 1998 and although both fields are spatially close and are
similar in agronomic histories for the year 1998, the spatial pattern of crop
vigor is different. As was true for 1997, an effect in the following year due to a
crop rotation cycle between cotton and peanuts was discernible (more lush) for
the west field. Despite changes in image processing techniques between the two
years, color assignments between 1997 and 1998 were arbitrary and differ due to

different choices and selections of color, saturation, and hue in the image processing software [45] and different analysts' preferences [49].

Both fields (Fig. 2) were planted in cotton May 8, 1998, and then uniformly managed except for four transects in the east field where a variable-rate nitrogen application experiment was conducted. Here, we use for the purpose of illustration only the control and the variable-rate nitrogen treatment based on simulation runs by the cotton growth model, GOSSYM [2]. The simulation model was the integrating vehicle for combining diverse technologies into a precision agriculture application. Each of the 72 plots in this example were about 65 m long, where 18 plots, 8 rows wide, were apportioned to each transect that spanned the length of the field. The crop was harvested during 5–9 October 1998. A four-row CASE-IH cotton picker harvested four rows of each plot. At the end of each plot, the harvester was stopped, the seed cotton dumped into a boll buggy equipped with a scale monitor, and the yield of the plot was weighed and recorded. Additionally, during the crop season, weekly observations were performed on subplots of 5 m in length (5 per transect) embedded within 20 of the 72 GOSSYM managed plots. (This study was accomplished prior to the development of a cotton yield monitor, so other innovations had to be improvised by the research team to obtain the plot yields at harvest.) Another variable-rate nitrogen study in the same vicinity and year, but within other strips of the four transects, provides additional details [60].

Several plant attributes (Table 1) measured from these 20 small plots indicated plant density, plant height, and properties of fiber quality, strength, and maturity were highly variable. The variability of plant attributes in Table 1 is mirrored by variability of the imagery. Specifically, the plant density attributes were correlated with the imagery, where the darkest hues represent lower populations of plants of shorter heights (i.e., more soil reflectance), whereas the grayer hues represent higher populations of plants of taller heights (i.e., less

Table 1 Crop attributes as measured from twenty 5-m row samples in a 200-acre section of a Mississippi delta cotton field

	Minimum	Maximum
Plant density, in plt/ha	50,000	100,000
Plant height, in cm	59	142
No. open bolls in 1-m row	33	87
Lint yield, in kg/ha	560	1446
ML in mm	22.3	24.8
Strength in g/tex	28.8	31.5
IM	4.2	5.4
MR	0.83	0.97

Note: Sites (n = 5 sites per transect) were selected within four 8-row transects where simulation-based recommendations for nitrogen on a 1-acre scale were applied in 1998. Site selection was a function of remote sensing reflectance and subjective selection.

Fig. 3 These panels show the on-the-ground variability of the cotton on the same day (1998). (Images courtesy E. Jallas, CIRAD, Montpellier, France.)

soil reflectance). The grayer zones contain cotton plants that are more heavily fruited than the lighter and darkest zones. The cotton is less heavily fruited in the darkest zones. The lighter zones are between the gray and dark extremes. In Fig. 3, the date of collection for the panels was the same day, but differences in cotton maturity are evident among them. All persons are of similar heights. Notice that the individual in the right panel is shown in the background of the center panel.

Using as representative fields those shown in Fig. 1 and Fig. 2, we can ask if it is practical to manage the more than 161 ha on the basis of each individual hectare. In other words, which hectares are similar enough to be grouped together as a management unit? In an attempt to determine this, a k-means cluster analysis was performed on the 209 soil profiles (over 627 samples) taken from this field. The soil data were grouped into 3, 5, 7, 10, 15, and 20 clusters. An analysis using the rule-based system COMAX [33] was performed using soil files representative of the soil types within each distinct cluster. For the entire field, six discrete recommendations were made by COMAX. With the three-cluster analysis, we show in Fig. 4 the spatial separation of groups that relate to soil type (these boundaries are shown as irregular polylines).

With five-cluster and higher groupings, the analyses showed there was little correlation to soil type. Eventually, the process created a patchwork having a more variable spatial organization (Fig. 5) than did the three-cluster map. Can it be inferred that the cluster analysis with three groups is more useful than the analysis with 20 groups? Answering this question is easier *a posteriori*, but is difficult, even impossible, *a priori*. Farmers have access to data, but often to not

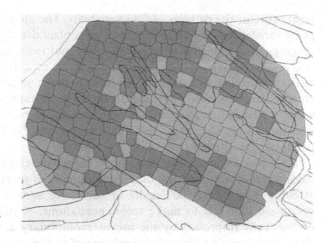

Fig. 4 The k-means clustering results for three clusters. (Image courtesy E. Jallas, ITK, Montpellier, France.)

all of the data that they need. Acquiring data is costly and there is still no general algorithm to determine what represents "good" information for a specific farmer's situation. (However, the approach developed in Case 2 may be one option.) This exploratory, k-means analysis indicated that tremendous gaps exist in our understanding, or determination, of the correct scale for aggregating information to manage a crop at an optimal level.

Farmers do have available tools to manipulate information and provide technical recommendations. One such tool, as already shown, is simulation [2, 18, 19, 20, 30, 31]. Cotton models such as GOSSYM or COTONS have been shown to predict phenological events such as fruit initiations and stresses within ±2 days from the beginning to the end of the season [17, 18]. The model GOSSYM, as shown, was integrated into a GIS [31]. Site-specific information can now drive a crop model thus allowing for variations across a field to be predicted and displayed by GIS software. The display of this information as a visual map can convey to the

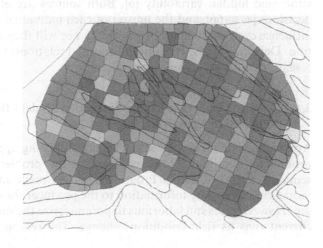

Fig. 5 The k-means clustering results for 20 clusters. (Image courtesy E. Jallas, ITK, Montpellier, France.)

farm operator the patterns of field variability. The model COTONS can simulate plant variability based on emergence timing, plant density within the row, the row spacing, competition for light, soil moisture, nutrients, and genetic characteristics [18, 19]. The site-specific implementation of such models is one way to account for in-field variability and, most importantly, to make an educated guess on how the crop will respond in the future if different management practices or different future weather scenarios are selected for other simulation runs. Geo-referenced simulation models are the major tool that can provide "look ahead" capabilities in precision agriculture.

Rule-based systems in agriculture have been developed to make management recommendations of cotton fields for a number of years [33, 61]. These systems attempt to optimize the results of successive simulations and then use the optimized schedule to make recommendations. A rule-based system can be dramatically improved by the use of evolutionary algorithms. Evolutionary algorithms have no inherent knowledge of cotton growth or management but they can be used to optimize complex, nonlinear problems like the management of a crop. They have, in fact, been used to make management recommendations for inputs such as fertilizer and irrigation [19, 55]. Scientists can use evolutionary algorithms to evolve and expand the expertise of a rule-based system. Implementing newer rule-based systems based on genetic algorithms, rather than a heuristic rule-base, within the environment of GIS could provide a significant next step toward developing site-specific management software.

Earlier it was mentioned that farmers are accustomed to dealing with variability that they can directly observe or easily measure. This case study illustrates that for other characteristics, such as lint quality, how to describe and manage its pattern of variation in the field is not so obvious. Lint quality is one characteristic that determines lint cotton price. Thus, describing the spatial/ temporal variability for characteristics (e.g., soil type, soil moisture, soil nutrient levels, etc.) that determine lint quality has an important economic impact. Farmers must deal with management practices that involve both easily observable and hidden variability [6]. Both sources are affected by management decisions [4, 6, 66], and the impact of each individual finding will not equally influence each component of variability nor will they do so in the same direction. Delimiting these types of impacts and relationships is one of many major research challenges in precision agriculture.

3.1.1 Case 1A: Update of Cotton Simulation Model Efforts in Precision Agriculture

Applications of these cotton models are still being applied to make in-season production decisions. However, some of the problems known in 1998 still remain today (i.e., 2007). Progress has been made in the summarization of model runs to convey information to the producer for in-season decision making. However, it is still laborious to parameterize the simulation model to make current runs as field conditions change. The reason for this is because the promising work to employ rule-based systems and genetic algorithms [19, 33, 55] to automate model parameterization was not completed due to a major redirection and dismantling of the core team. This persistence of an old problem and the interruption of a promising solution demonstrate that multidisciplinary teams need to have a long life and stability.

Another major deficiency is the establishment of linkages by the model to a database structure that stores and allows prompt access to diverse sources of geo-referenced information. This lack of access to a geo-database means that at times, for a commercial farming operation, it takes too long to get the proper data to the model. The runs of the model apply to thousands of hectares with unique decisions being made across them. Thus, the assignment of inputs to the model that match the current state of field variability, the establishment of the run queues, the collation of the output, and then the assembling of the output into formats for presentation to the producer are all steps that need to be prompt. On a commercial cotton farm, some decisions need to be made today, because tomorrow's forecast may call for 5 days of rain to be probably followed by yet another 5–7 days, or longer, waiting for the fields to dry so that the machines can go to work again.

3.2 Case 2: Statistical Analyses of Field-Level Precision Agriculture Experiments

The experiences learned through several experiments (e.g., Case 1 [7, 10]) made it clear to us that new methods of data analyses would have to be developed. One observation that led to this recognition was the fact that when the geo-referenced polygons of the 1998 study plots were placed upon the geo-referenced image by GIS processing, it was obvious that these symmetrical plots (i.e., of the same size and shape) were not congruent with the geospatial variability of the crop reflectance. This realization required that the initial, core group of researchers reach out to additional skill sets in both the GIS and statistical disciplines [64]. As a result, concepts for GIS processing and the design of experiments were linked together to develop a new process for analyzing geo-referenced information collected from the field.

The progress of such forms of cooperation have been previously described [56, 64, 66], but are summarized again here. Once it was recognized that other expertise was necessary, the first obstacle was how to integrate the diverse kinds of geo-referenced information from the cotton field for use in statistical model building. It was here that the GIS specialist provided the first of several key insights. Through the use of macro-language programming, he wrote a script that extracted from both the raster and vector layers the attribute value of each at the coordinate location of each cotton yield monitor record [64]. These compiled attributes were written to a shapefile and then imported into statistical software for subsequent data analysis.

Once the statistical analysis steps began, additional constraints arose that taxed the skills of not only the agriculturalist, or the GIS specialist, but also that of the biometrician. In other words, these problems at the first taxed all of us and clearly required a team effort to overcome. The solution was discovered when we realized that there were two types of experimental units [36, 66]: (1) those involved with traditional management practices and (2) those involved with the site-specific practices. Furthermore, the concept of what comprised an experimental unit, or "plot," no longer entailed an arbitrary, symmetrically sized and shaped area of ground, but these were actually asymmetrical, and differed in size and shape (Fig. 6) [66], especially with respect to the site-specific practices. (A similar fact was learned from the simulation study with the k-means cluster analysis [recall Fig. 5], but we did not fully value what it meant at the time. Here, nearly 9 years later with another tool, we now begin to have some light of understanding. Again, the sequencing of related [yet dissimilar] work demonstrates that multidisciplinary teams need a lengthy longevity to make insightful discoveries.) This was followed by the realization that the information from the various remote sensor systems described geo-referenced topography characteristics [64] that were covariates associated with both the broadcast (i.e., blanket or flat-rate) and site-specific treatments [6, 66]. Similarly, it had to be recognized that the characteristics of the farm machines, including the cotton harvester (as they traveled along their paths of activity), were additional sources of variation that needed to be statistically modeled.

As a part of this realization, it was necessary to once again return to the GIS specialist. Focal processing (of each harvest swath 4 rows wide, within an 8-row-wide cotton cultivar strip) had to be employed to reduce the variation between paired harvest passes of a cultivar strip [64], or a level of the broadcast

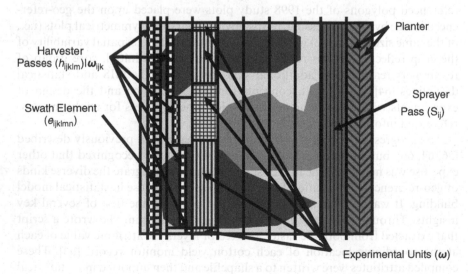

Fig. 6 Representation of asymmetrical experiment units and sources of variation [66]

treatment factor. Focal processing [64] is a neighborhood process in which the output value of the process cell depends on the values of the neighboring cells in the defined focal area. Figures 7 and 8 show the before and after (SAY = Strip Adjusted Yield) results for two paired cotton picker loads collected from 1 of 21 cultivar strips.

Focal processing did not create artifacts but rather reduced the noise present in the yield data [39, 40, 41] to allow the subsequent statistical analysis steps to better identify meaningful effects that described this cotton production system [64, 66].

A statistical model to assess the effect of a one-time, site-specific, variable-rate application on the cotton yield of one or more cotton cultivars (i.e., the broadcast management treatments) is developed as follows [66]. In general, the intersecting geometries of the topography (or productivity or management) zones and the farm equipment's characteristics [66] define the design structure [36] of a site-specific experimental design and its different sizes and kinds of experimental units. The site-specific and traditional practices applied to the

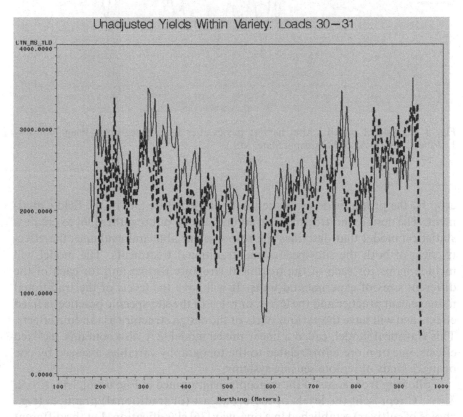

Fig. 7 Example of paired cotton harvest passes prior to focal processing. (Image courtesy J. Willers, USDA-ARS, Mississippi State, MS.)

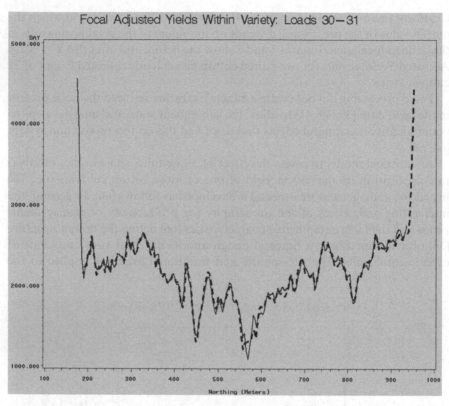

Fig. 8 Example of paired cotton harvest passes after focal processing. (Image courtesy J. Willers, USDA-ARS, Mississippi State, MS.)

crop by the producer describe the treatment structure [36] of the field experiment. The design and treatment structures can be put together [36] to create a statistical model that describes the response variable and evaluates the effectiveness of both the site-specific and traditional treatments. The model will include terms for each of the treatment structure factors and for each of the different sizes of experimental units. It will have the levels of the traditional management practice and the levels, or rates, of the site-specific practice as fixed effects and will have the various parts of the design structure as random effects. This statistical model, called a linear mixed model [27], also contains, as fixed effects, one or more terms related to the topography variables assessed by one or more kinds of remote sensing systems.

Following from some of the concepts diagrammed in Fig. 6 [66], let i = 1, 2, 3, ..., C represent the levels of the broadcast (or traditional) practices (e.g., choice of cultivar) established in a one-way [36] classification. Let the different rates (or levels) of a one-way classification of an agri-chemical (e.g., a plant growth regulator) applied site-specifically to various zones be represented as $r = 1, 2, 3, \ldots, R$. Let $y_{irjklmn}$ represent the yield value of the nth yield monitor reading in the mth harvest pass within the lth planter pass of the kth experimental unit (or replication of a site-specific recommendation) receiving the rth rate in the jth sprayer application pass of the ith broadcast practice. Let the term μ represent the mean yield of the field, let γ_i be the effect of the ith broadcast practice, ρ_r be the effect of the rth level or rate of the agri-chemical, and let $(\gamma\rho)_{ir}$ be the interaction between the broadcast treatments and the site-specific rates (or levels). Define several random effects (see Fig. 6) to be s_{ij} (the jth sprayer pass for the ith level), w_{ijk} (the kth experimental unit within the i, j^{th} combination of the broadcast level and sprayer pass where a particular rate of an agri-chemical was applied), p_{ijkl} (the lth planter pass within the kth experimental unit within the jth sprayer pass for the ith level), h_{ijklm} (the mth harvester pass within the lth planter pass within the kth experimental unit within the jth sprayer pass for the ith level), and the residual, ε_{ijklmn} (the nth yield monitor observation within the mth harvester path within the lth planter pass of the kth experimental unit in the jth sprayer application pass of the ith level).

As the broadcast treatment and site-specific treatment consists of a two-way treatment structure where all combinations of the C levels of the broadcast treatments occur with each of the R levels of rate, it is possible that due to variability in field topography, not all of the site-specific management rates will be applied to all levels of the broadcast treatments in every sprayer path, forming an unbalanced [36] treatment structure. Because the site-specific treatment rates are quantitative, the rate (ρ_r) and rate by broadcast treatment interactions (($\gamma\rho)_{ir}$) terms can be represented by a linear trend in rate having a different slope for each broadcast treatment. Next, let φ_i represent the slope that describes the ith broadcast treatment's response to the site-specific treatment's rates. It is assumed that two or more rates of the site-specific treatment are applied to each broadcast treatment somewhere in the field. In other words, it is not required that two or more rates of the site-specific practice occur in every individual sprayer pass for the ith broadcast treatment.

Therefore, to account for the variability in the yield response described by the topography (or the site characteristics) attribute of each swath element (or yield monitor point), we derive the following mixed, linear model [27, 64, 66] that incorporates linear effects of site characteristics as covariates:

$$y_{ijklmn} = \mu_i + \varphi_i r_{ijk} + \sum_{g=1}^{G} \beta_{gi} X_{gijklmn} + s_{ij} + w_{ijk} + p_{ijkl} + h_{ijklm} + \varepsilon_{ijklmn} \quad (1)$$

where each broadcast practice has its own intercept (μ_i) and its own slope (φ_i) with respect to the site-specific management practice applied at the different locations of the swath elements. The slope β_{gi} is the regression coefficient corresponding with site characteristic value $X_{gijklmn}$ for the ith broadcast practice. Presented in Fig. 9 is a hillshade representation of field elevations acquired by a Light Detection and Ranging (LiDAR) sensor system.

Fig. 9 Hillshade surface of
field elevational relief
determined by LiDAR.
(Image courtesy C.G.
O'Hara [64].)

Drainage, erosion, row patterns, and terrace features are apparent. (Two or
three horizontal streaks known as step-errors are also visible. These arise at
times along margins of overlapping LiDAR strips and are artifacts that
require additional GIS-based processing to remove.) From this elevation
surface, additional site characteristics such as aspect, curvature, flow accumulation, and slope were derived by GIS processing [64]. Information from
imagery was also used as a source of information to define attributes related to
crop vigor and density [64, 66]. Notice as well that with this model, the planter
passes, harvester passes, and the swath elements are nested within the sprayer
passes.

The model of Eq. (1) can include various refinements, such as transformations (squares, reciprocals, logarithms, etc.) of the site characteristic attributes, the construction of cross-products or ratios, and the building of
interaction terms involving the levels of the planned management practices
that could relate to the yield measurements. There are also very large experimental units associated with the traditional practices that are also apportioned by their intersections with the different productivity zones (see the
different regions of the field shown in Fig. 6). Across these zones, various
rates have been assigned to the site-specific experimental units (w_{ijk}) within
them (see more with Case 3).

The random terms s_{ij}, w_{ijk}, p_{ijkl}, h_{ijklm}, and ε_{ijklmn} are assumed to be normally
distributed [66] as:

$$\mathbf{s} = [s_{11}, s_{12}, ..., s_{CS}]^T \sim N(0, \mathbf{Q}_1), \tag{2}$$

$$\mathbf{w}_{ij} = \left[w_{ij1}, w_{ij2}, ..., w_{ijt_{ij}} \right]^{T} \sim N(0, \mathbf{Q}_2), \tag{3}$$

$$\mathbf{p}_{ijk} = \left[p_{ijk1}, p_{ijk2} \right]^{T} \sim N(0, \mathbf{Q}_3), \tag{4}$$

$$\mathbf{h}_{ijkl} = \left[h_{ijkl1}, h_{ijkl2} \right]^{T} \sim N(0, \mathbf{Q}_4), \tag{5}$$

$$\varepsilon_{ijklm} = \left[\varepsilon_{ijklm1}, \varepsilon_{ijklm2}, ..., \varepsilon_{ijklmn_{ijklm}} \right]^{T} \sim N(0, \mathbf{R}_{n_{ijklm}}). \tag{6}$$

In Eq. (3), t_{ij} counts the number of experimental units within the i, j^{th} combination of cultivar and sprayer pass for a total of S passes. The dimensions for $\mathbf{p}_{ijk}$ (the planter) and $\mathbf{h}_{ijkl}$ (the harvester) are fixed at 2 for these two different farm implements. (Similar terms could be defined for a nitrogen applicator, etc.) One-dimensional spatial covariance matrices are represented by the terms $\mathbf{Q}_1$–$\mathbf{Q}_4$ and $\mathbf{R}_{n_{ijklm}}$ and these, and other details, are further described in [66]. An interesting area of inquiry would be to determine the effects upon analysis results if these one-dimensional matrices are, in fact, not normally distributed. Related topics and concepts are discussed in Refs. 35 and 37.

Applications of this type of integration between GIS and statistics are finding interesting interactions, such as RATE*CURVATURE*CULTIVAR that are statistically highly significant (P < 0.0001). Such a three-way interaction indicates that it is necessary to evaluate different combinations of site characteristics in order to recommend the best selection of a cotton cultivar and rates of agri-chemical inputs, such as the amounts of a plant growth regulator, for different locations in the field [66].

From these results obtained from typical Mississippi cotton fields described in Cases 1 and 2, the answer to the question "Is there a need for precision agriculture? " is "yes." When fields are bigger than a few acres, the amount of variability is large enough to justify attempts to manage subareas. The hope is that by identifying and treating each section of the field individually, farmers can geographically avoid nutrient deficiencies, pest, or plant stresses that limit yield across the field. Conversely, the geographic identification of fertile sections can help reduce overapplications of nutrients to locations where they would simply be leached from the soil and cause problems for the environment. Other areas may not be made productive no matter what is done, and these should be ingeniously allocated to other uses.

The overall result should be to increase yields while decreasing inputs at lower costs. These considerations lead to the next question: Do we have the capacity to acquire the information to manage this field variability?

4 Collecting and Managing Information

For producers to take advantage of precision farming technologies, it is necessary to correctly describe variability and its causes, acquire or create recommendations tailored to areas of variability, and finally, apply these recommendations efficiently and economically. The information needed to manage variability can be classified [6] into several broadly defined categories: (1) stable, (2) seasonal, and (3) daily or recurring. These are arbitrary classes established for convenience in handling information; however, if considered in different ways, the same data or information may fit into more than one category – changes in soil pH over time is an example. Examples of agricultural information that is stable over a season include soil type, bulk density, hydraulic conductivity and moisture release change with time, organic matter, elevation, and so forth. Information that is needed only once in a season includes cultivar selection, planting date, residual soil conditions as percent organic matter, residual nitrogen, and initial water content. Finally, information that is needed on a daily or weekly basis includes fertilization scheduling, weather data, pest infestation, pesticide scheduling, crop status, phenology, morphology, and so forth.

There are numerous sources for acquiring site-specific information and not all are dependent on new high-tech methods. For example, soil survey maps are available for most farmland (e.g., for the United States, one link is http://datagateway.nrcs.usda.gov/). However, some of this information may be dated and may not closely match the productivity zones currently in the field, especially after a field has been land formed. Soil sampling is another source of information useful for assessing soil fertility and providing the basis for fertilizer recommendations, however, it is difficult to collect these data and keep costs low, without degrading the usefulness of the information. Therefore, while in the beginning, soil sampling was accomplished using a regular grid of approximately 1 ha spacings, current recommendations are to use directed sampling [11, 26]. Similarly, traditional field sampling to determine a field average for the abundance of a cotton insect pest cannot economically detect locations where insects are present or the plants of the crop are otherwise not healthy. To do most sampling tasks at the level of detail to make a precision agricultural approach feasible would involve tremendous costs in time and labor to "map" problem areas. However, multispectral and hyperspectral images from satellite or fixed wing aircraft are good alternatives for identifying the relative health of plants within a given field (see Figs. 1 and 2) [42, 56, 63, 65]. These maps are useful [12] to the field consultant, because sometimes the pest insect preferentially prefers some cotton habitats over others. In such instances, maps can be used to perform directed scouting for cotton pest control, particularly, for the tarnished plant bug (*Lygus lineolaris* [Palisot de Beauvois]) [63, 65] or other needs [7, 42, 46, 47, 48].

For other commodities, such as forestry or orchards, remote sensing has detected where insects have damaged, or are damaging, the crop [50]; however, in cotton, once such stress caused by insects is detectable by the remote sensor system, it is most often economically[2] too late for the producer [65]. Other instances of where physical damage, such as that caused by hail, can be measured can be found in the literature [42]. These images can address the time and labor issue involved with scouting crops for damage. However, the cost of these images currently remains too expensive, or difficult to acquire, for most large-scale agricultural needs [16]. But, progress is being made [56], including some commercial interests (e.g., www.gointime.com).

As gains are made by service providers of remotely sensed information, it will mean that farmers need to be alert and demand from suppliers or vendors of information for precision farming purposes their strict conformity to standards that ensure that the data are of high quality. Vendors, or extension services, should provide training to ensure that producers and consultants know the limits and the proper application of these data products [1, 9, 23].

How much cost [25, 28] is the farmer willing to spend for new data? For example, soil sampling done on a 1-ha grid has been demonstrated to be inadequate. Because it is expensive to also determine soil hydrology characteristics, farmers are reluctant to sample even a single field on a fine-resolution grid to obtain the data needed to run a detailed plant simulation model. Similarly, for insect pests, traditional sampling using a drop cloth may typically involve sampling for insects at six to eight random locations in a 163-ha field. The problem is that insects do not attack plants in a random fashion. Traditional sampling at the necessary resolution for site-specific management would be too time consuming and expensive. For insect scouting using whole plant, or terminal counts, a common recommendation is 100 plants for every 20 ha. Following this guideline for the two fields portrayed in Fig. 2 would require a sample of 1000 plants. Very few if any farmers can afford to pay crop consultants to spend the time necessary to collect pest samples at this intensity. Clearly, there is a need to develop new observation and sampling techniques based on remote sensing data [42, 56, 65].

Correlation between soil characteristics, field variability, and yield is not always easy to detect or interpret [6]. For instance, yield variation would be expected to follow soil type boundaries, but that is often not the case. For an extended study, it was reported [21] that many locations had consistently high or consistently low yields; however, for most locations the variability was inconsistent from year to year and temporal behavior of the yields was inconsistent. Geographical information systems are excellent for viewing yield distribution maps [39, 40], but in many cases these maps only show what has already been known. The key is to develop concepts that can promptly link the information from a series of maps (or databases) of diverse nature, precision, and scale to management practices that optimize crop production at the lowest

---

[2] Perhaps more colleagues in the remote sensing industry should see the landmark work by Stern et al. [58].

cost and effort. Considerable research is directed toward this goal, but it remains a major problem to solve.

Our final question to be examined is if the information is available, then "Do we have the tools to manipulate this information and thus provide farmers with useful recommendations?"

5 Precision Farming Equipment

The following quote is worth mentioning: ". . . implementation of precision farming systems is not inexpensive. Costs will decrease as the technology develops, but it now (i.e., 1998) requires $15,000–$20,000 to purchase a yield monitor, GPS receivers, computers, software, and variable controllers for application equipment. There may also be increased costs for sampling and data management, analysis, and interpretation" [43]. History teaches that mass production makes the manufacturing process of end products easier and less expensive. As precision farming becomes more widely used, equipment manufacturers will start to produce more variable-rate (VR) equipment. Everyone involved with precision farming expects that in the end, costs will decrease, as has happened with GPS receivers, now available for below $100. In precision agriculture, we can realistically expect this to happen for some of the electronic equipment because of the economy of scale. Today, we can purchase a 3-GHz microcomputer with terabytes of storage and gigabytes of RAM for less money than we could purchase an AT class PC with a few hundred kilobytes of RAM and almost no storage in the mid-1980s. Computers are becoming more popular, with the result that more capacity is being produced while the cost per unit continues to decline. However, this trend may not be as evident with the larger farm equipment, or sophisticated software. Remember that the number of farmers worldwide continues to decline and the size of farms is increasing. This may cause a decreased demand due to the fewer number of farms, thus reducing the number of units produced and raising the per unit cost. It will be necessary to revise current marketing tactics or government policy involving the sale of precision agriculture equipment and software.

Complexity of equipment (e.g., [3]) can add even more difficulty in the requirements to operate, maintain, repair, and replace failing parts. This sophistication increases the chance that a single part will fail, rendering the entire unit inoperable or unpredictable until it is repaired. For example, we had one summer when after collecting soil hydrology data, soil fertility data, running numerous simulations, developing site-specific prescriptions, and finally transferring all the necessary data to the variable-rate controller on the fertilizer applicator, an injector nozzle became clogged when the nitrogen was being applied. Only three of the four rows fertilized received the prescribed application. In this case, the equipment designed to manage the variability introduced another source of variation. When these types of problems arise, many producers (or consultants) quickly lose their enthusiasm for precision agriculture. The unfortunate reality is that many research or business operations do not acknowledge this small fact and take it into account as part of their protocol (there are fortunately a few that do).

5.1 Case 3: Development of Geo-referenced Site-Specific Prescriptions

Presented in Fig. 10 are six major boxes that are linearly interconnected and describe tasks necessary to build a site-specific insecticide prescription in cotton. Each box contains several subtasks that must be accomplished before the final goal is obtained, which is the application of a site-specific pesticide application for a cotton pest [7, 10, 56]. Although other developers of precision agricultural prescriptions may have different workflows (e.g., [8]), all of these schemes have a common quality. Geo-referenced information has to be acquired, processed, linked to field data, converted into a decision by someone,

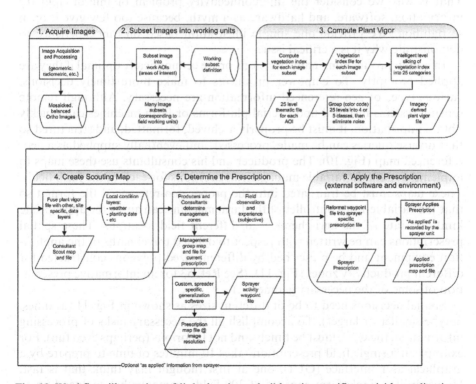

Fig. 10 Workflow illustration of linked tasks to build a site-specific pesticide application. (Image courtesy A. Zusmanis, Leica Geosystems, Norcross, GA, and J. Willers, USDA-ARS, Mississippi State, MS; unpublished.)

transformed into a map, and ultimately converted into a format compatible for use by a variable-rate controller. The variable-rate controller ideally can log geo-referenced "as-applied" data to confirm that the decision was correctly implemented and later used for statistical analyses or refinements of other decisions later in the production season.

Currently, to do the tasks between Box 1 and Box 6 requires at least two or more different software applications, the transfer of information from among two or more storage devices, and the use of labor and time within two or more different organizations to accomplish one or more of these tasks. The involvement of personnel from the farm itself is also necessary. Whereas some integration of these steps is being accomplished by the development of standards [62], neither we nor other developers of software or hardware have yet succeeded in creating the interconnectivity necessary to link all of the steps together in transparencies that are simplistic, of low cost, and do not reduce the time required by the ultimate user, which is either the producer or his consultant. This reality is the ignored myth of precision agriculture and it is a major problem. It appears to us, given the numerous papers describing precision agricultural applications, that few [3, 62] are working on this difficult task. That is why we consider the interconnectivity problem of interdisciplinary information, software, and hardware as a myth, because too few give it open recognition. We provide one simple illustration based on our experience to demonstrate why it is a critical issue.

In cotton production, many decisions prove economically optimal if they are applied promptly. To empower producers to make more timely decisions, various types of geo-referenced information are necessary. Although remote sensing systems installed upon aerial platforms or farm machines can quickly obtain information, it must be properly archived, formatted, and structured so that diverse queries can be made, processed, and eventually supplied as a geo-referenced map (Fig. 10). The producer and his consultants use these maps to implement spatially variable management decisions for each of the farm fields. These maps must be translated into the proper geo-referenced file format so that the variable-rate controller of the farm machine can apply the appropriate kind and rate of the agri-chemical at different field locations. These spatial prescriptions can be written with respect to defined travel paths of the application equipment and be prescribed by different-sized polygons constructed for different productivity zones (Fig. 11). (See Ref. 53 for an informative presentation of some of the necessary geometry.)

Spatial decisions need to be of different sizes as shown in Fig. 11 (at times, maybe smaller or larger). To accomplish all the necessary tasks of processing information, the work must be timely and not laborious (perhaps even fun). For example, if a single field prescription takes 10 minutes of time to prepare by a graphical user interface (GUI), one at first thought may think that is fast. However, if a single farm consists of 300 fields, someone has to spend up to 50 hours just in processing time alone. This is too costly, too slow, and too laborious and leaves no other time to fix the fence, feed the livestock, repair the

Fig. 11 Diagram showing
the sprayer passes with
different sizes of application
polygons. (Image courtesy
J. Willers, USDA-ARS,
Mississippi State, MS, and
A. Zusmanis and A. Leason,
Leica Geosystems,
Norcross, GA;
unpublished.)

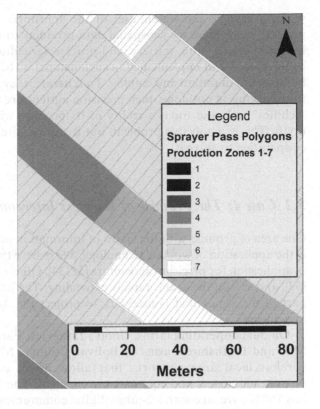

tractor, or visit with family and friends. For some, there would be no time to
hunt or fish. The crux of the problem is that current capabilities for the archival,
retrieval, and processing of hundreds of gigabytes of geo-referenced agricultural information in the form of dozens of data files for each field are not
adequate. Future requirements are expected to involve larger number of files,
involving thousands to tens of thousands of gigabytes.

Therefore, we believe it will be necessary to develop a manageable geo-
referenced database (data warehouse) and decision support system that is
accessible to farmers and other agricultural decision makers. We believe that
because agriculture is critically dependent upon the timeliness of management
operations, this database will need a design quite different from those currently in use for other applications, such as urban planning. This system
should be linked to easy-to-use and *locally* controlled, wireless, high-speed,
local area network of software tools and decision aids to move digital information collected on the farm. Especially, there must be wireless links to the
farm machines and the operators who drive them. There similarly must be a
return link from the remote machine (e.g., 10 km away from the farm office)
back to the data warehouse. This farm-centric data warehouse must also have

wireless links to the warehouses of the appropriate service providers who provide supplementary geospatial data products to the farm. These alternative methods to store, retrieve, and process agricultural information have to be developed and implemented in such manner as to not be too costly or too laborious to maintain and employ. To reiterate, they must be simple and easy to use. If this is not done, then precision agriculture will suffer from its main Achilles' heel, and today's reality of its promise will fade into tomorrow's myth. This is a solvable problem, but it will require an uncommon spirit of cooperation to solve it.

5.2 Case 4: The Promise of Wireless Interconnectivity

One area of promise with the myths of information acquisition and processing is the application of wireless technology. We believe that wireless is an essential coapplication for precision agriculture [32, 34]. Others have also recognized this [67], and in fact, the area is rapidly expanding. The justification is simple – it is part of the armor that is necessary to protect the Achilles' heel of precision agriculture.

On our cooperating farms, Good's Longview Farms in Noxubee County, MS, and Perthshire Farms in Bolivar County, MS, we have established wireless local area networks that allow direct communication between farm headquarters and equipment in the field. The Good's Longview Farm has 100% coverage with a 2-mbps digital communication network. Approximately 80% of Perthshire Farms has been covered. Additionally, for Perthshire Farms, digital communication has been provided between farm headquarters and two commercial gins operated by the farm, one on-site and another 4.8 km away (see also [59]). Working with a commercial Internet supplier, we have established in Noxubee County a wireless wide area network that provides non–line-of-sight coverage to anyone within 8 km of a 91-m-tall microwave tower located on Prairie Point Road, 9.7 km east of Macon, MS. Line-of-sight coverage extends up to 64.4 km from the tower base station, which provides 360° coverage and 2-mbps up and down connection speed to end users.

These wireless networks are there and in place. However, they have been upgraded several times as the hardware has made advances. A consequence is that the supporting software for the wireless network also changes as the hardware has changed. This is problematic in light of the connectivity among the various junction points along the pathway of information flow (Fig. 10). Will it be necessary with each wireless hardware/software upgrade to revise the other software gateways in order to maintain the flow of information throughout the enterprise? How troublesome and costly this will be if such is the case. We hope that much forethought by hardware and software developers at all points along the linkages will concurrently occur.

5.3 Dollars and Sense

Earlier we described how equipment malfunctions increase field variability. Now we ask, "Does it make economic sense to invest in the new technology if we are hamstrung by problems due to these malfunctions? " Equipment manufacturers have been working on methods and designs to help reduce mechanical failure, but they still exist and will continue to exist. Another issue that has been troublesome for even conventional equipment is calibration. It is probably not unreasonable to expect that a variable-rate equipped machine may be even more difficult to keep calibrated. If the variable-rate equipment cannot be depended on to apply the prescribed amount, then we are in the same situation we were in 10 years ago: we can identify and even predict the variability in the field, but we still cannot manage it promptly or properly. There is no gain in net return on the investments spent to detect, categorize, and manage the variability in the field if the equipment employed to do the task is not more dependable than current implements or able to match the spatial resolution provided by current remote sensing systems.

There are other considerations to precision farming. Even with these obvious possible problems, a major benefit is that some variable-rate application equipment can keep track of what was actually applied. This is very important information to have when considering additional applications later in the season, but do these application records benefit us any other way? Suppose, for example, someone says your spray plane sprayed his or her lawn. Can you prove it didn't? With geo-referenced spray records, you can. Suppose new environmental regulations say you cannot apply herbicide within 30 m of a seasonal waterway. Accurate application data may be the only method producers have to prove compliance. It may also be used to show noncompliance. There may be a day when federal regulations require all applications be done with variable-rate or geo-referenced equipment. However, the use of such regulations without the concurrent ability to easily manage information will be burdensome upon the people involved with agriculture.

At present, precision farming is expensive [25, 28]. Producers will continue to invest in these endeavors only as long as they can see a cost advantage or simplification of their workload – either this year or very soon [1,44]. This trend will continue as long as there is a substantial net return for selling the crop.

We could say more, but there are already many other excellent discussions on the problems and promise of precision agriculture [6, 29, 38, 44, 51, 52, 57, 68]. The main themes presented in the introduction have now been discussed from the perspective of the cotton production system. Other commodity systems should be very similar to the perspectives we have described. This is a good parallelism that means the motivation to do better is there, because the benefits of precision agriculture apply to many commodities. And its tools apply to more than just agriculture. We learned after the devastation of Hurricane Katrina on the Mississippi Gulf Coast that many pathways for the acquisition and flow of information also arise with and apply to disaster recovery operations or other civic responsibilities. Therefore, the market opportunities for precision agriculture are there, if noteworthy problems can be resolved in a cooperative and commonsense way, for the benefit of all participants of these technologies.

6 Conclusions

The research focus of precision agriculture needs to be more than the development and application of technology and should involve the demonstration that the available information from sensor systems can be managed to match the increasing capabilities of hardware and software to solve production problems on commercial farms. Tools such as variable-rate equipment and controllers, yield monitors, agriculturally oriented GIS software, and geo-referenced crop models are available and are now being used. However, these tools are at an early stage of development when balanced against what is ultimately required. In fact, we do not yet know what to do with a lot of the data available or how much of what information is important, or how much of the field variation is manageable or acceptable. It is an interesting contrast that, despite the recent improvements in technological tools, further refinements and capabilities are needed to allow for better site-specific management. For now, at least, some progress is being seen in identifying significant patterns in the data and interactions.

As shown by the case study illustrations, it would appear that over the years, some of the main problems of precision agriculture have not changed. We could have included other case scenarios for the precision agriculture enterprise; however, the four that were presented are representative of the domains of our expertise. Therefore, the fact that other scenarios were not developed in this work indicates that we too lack several key linkages to other skill sets. Thus, we too are caught in the gap between the myth and the reality of the promise of precision agriculture.

In summary, precision agriculture has many myths, but its reality is becoming clearer. Without the timely management of information at an affordable cost, the promise of precision agriculture will remain less than par; however, if that task is skillfully mastered, then yields should increase while costs and environmental disturbances should decrease. Thus, although the promises of site-specific farming have not been fully actualized, they are on the horizon. But for now, the business, extension, production, and research elements of the agricultural community have a lot to accomplish. By analogy, we conclude that despite the glamour and sophistication of the current technology, precision agriculture is probably at the equivalent point in the development of today's aircraft as were the Wright Brothers at Kitty Hawk. Time will tell. Time may also tell more than ever expected.

Acknowledgments We are thankful for the support of Kenneth Hood, Perthshire Farms, Gunnison, MS, and Paul Good, Good's Longview Farm, Macon, MS, for permission to work in their fields. Thanks are also expressed to Mr. Dan Woodard for providing the variable-rate applicator and writing the final nitrogen prescription that were based on the simulation model recommendations described in the Case 1 scenario. Thanks are expressed to Mr. Ronald E. Britton and Dr. Martin Wubben, USDA-ARS, Mississippi State, MS, and Dr. Wes Burger, Wildlife and Fisheries, Mississippi State, MS, for their assistance during the preparation of the manuscript.

References

1. Adrian, A. M., Norwood, S. H., Mask, P. L. 2005. Producers' perceptions and attitudes toward precision agriculture technologies. Computers and Electronics in Agriculture 48: 256–271.
2. Baker, D. N., Lambert, J. R., McKinion, J. M. 1983. GOSSYM: A Simulator of Cotton Growth and Yield. South Carolina Experiment Station Technical Bulletin 1089.
3. Batte, M. T., Ehsani, M. R. 2006. The economics of precision guidance with auto-boom control for farmer-owned agricultural sprayers. Computers and Electronics in Agriculture 53: 28–44.
4. Bongiovanni, R., Lowenberg-Deboer, J. 2004. Precision agriculture and sustainability. Precision Agriculture 5: 359–387.
5. Bugayevskiy, L. M., Snyder, J. P. 1995. Map Projections. A Reference Manual. Taylor and Francis, Ltd., Philadelphia, PA.
6. Bullock, D. S., Kitchen, N., Bullock, D. G. 2007. Multidisciplinary teams: A necessity for research in precision agriculture systems. Crop Science 47: 1765–1769.
7. Campenella, R. 2000. Testing components toward a remote-sensing-based decision support system for cotton production. Photogrammetric Engineering & Remote Sensing 66(10): 1219–1227.
8. Cox, S. 2002. Information technology: the global key to precision agriculture and sustainability. Computers and Electronics in Agriculture 36: 93–111.
9. Daberkow, S. G., McBride, W. D. 2003. Farm and operator characteristics affecting the awareness and adoption of precision agriculture. Precision Agriculture 4: 163–177.
10. Dupont, J. K., Campenella, R., Seal, M. R., Willers, J. L., Hood, K. B. 2000. Spatially variable insecticide applications through remote sensing. In: 2000 Proceedings of the Beltwide Cotton Conferences, edited by P. Duggar and D. Richter. National Cotton Council, Memphis, TN. Vol. 2: 426–429.
11. Fitzgerald, G. J., Lesch, S. M., Barnes, E. M., Luckett, W. E. 2006. Directed sampling using remote sensing with a response surface sampling design for site-specific agriculture. Computers and Electronics in Agriculture 53: 98–112.
12. Fleischer, S. J., Blom, P. E., Weisz, R. 1999. Sampling in precision IPM: when the objective is a map. Phytopathology 89: 1112–1118.
13. Fountas, S., Wulfsohn, D., Blackmore, B. S., Jacobsen, H. L., Pedersen, S. M. 2006. A model of decision-making and information flows for information-intensive agriculture. Agricultural Systems 87: 192–210.
14. Gliessman, S. T. 2000. Agroecology: Ecological Processes in Sustainable Agriculture. Lewis Publishers, an imprint of CRC Press, Boca Raton, FL.
15. Heermann, D. F., Hoeting, J., Thompson, S. E., Duke, H. R., Westfall, D. G., Buchleiter, G. W., Westra, P., Peairs, F. B., Fleming, K. 2002. Interdisciplinary irrigated precision farming research. Precision Agriculture 3: 47–61.
16. Hergert, G. W. 1998. A futuristic view of soil and plant analysis and nutrient recommendations. Communications in Soil Science and Plant Analysis 29: 1441–1454.

17. Hodges, H. F., Whisler, F. D., Bridges, S. M., Reddy, K. R., McKinion, J. M. 1998. Simulation in crop management:GOSSYM/COMAX. In: R. M. Peart, R. B. Curry. Agricultural Systems Modeling and Simulation. Marcel Dekker, Inc, New York.
18. Jallas, E. 1998. Improved model-based decision support by modeling cotton variability and using evolutionary algorithms. Ph.D. Dissertation. Mississippi State University.
19. Jallas, E., Sequeira, R., Boggess, J. E. 1998. Evolutionary Algorithms for Knowledge and Model-based Decision Support. IFAC Artificial Intelligence in Agriculture. Chiba, Japan.
20. Jallas, E., Sequeira, R. A., Martin, P., Turner, S., Cretenet, M. 1998. COTONS, a Cotton Simulation Model for the Next Century. Second World Cotton Research Conference, Athens, September 1998.
21. Jaynes, D. B., Colvin, T. S. 1997. Spatiotemporal variability of corn and soybean yield. Agronomy Journal 89:30–37.
22. Kennedy, M. 1996. The Global Positioning System and GIS: An Introduction. Ann Arbor Press, Chelsea, MI.
23. Kitchen, N. R., Snyder, C. J., Franzen, D. W., Wiebold, W. J. 2002. Educational needs of precision agriculture. Precision Agriculture 3: 341–351.
24. Langley, R. B. 1998. The UTM Grid System. GPS World, February, pp. 46–50.
25. Lavergne, C. B. 2004. Factors determining adoption or non-adoption of precision agriculture by producers across the cotton belt. Master of Science Thesis. Texas A & M University.
26. Lesch, S. M. 2005. Sensor-directed response surface sampling designs for characterizing spatial variation in soil properties. Computers and Electronics in Agriculture 46: 153–179.
27. Littell, R. C., Milliken, G. A., Stroup, W. W., Wolfinger, R. D., Schabenberger, O., Stepanski, E. 2006. SAS® System for Mixed Models, 2nd ed. SAS Institute Inc., Cary, NC.
28. Martin, S. W., Cooke, Jr., F. 2002. Summary of precision farming practices and perceptions of Mississippi Cotton Producers: Results from the 2001 Southern precision farming survey. In: 2002 Proceedings of the Beltwide Cotton Conferences, edited by P. Duggar and D. Richter. National Cotton Council, Memphis, TN [non-paginated CD].
29. McBratney, A., Whelan, B., Ancev, T., Bouma, J. 2005. Future directions of precision agriculture. Precision Agriculture 6: 7–23.
30. McCauley, J. D. 1999. Simulation of cotton production for precision farming. Precision Agriculture 1: 81–94.
31. Mckinion, J. M., Jenkins, J. N., Akins, D., Turner, S. B., Willers, J. L., Jallas, E., Whisler, F. D. 2001. Analysis of a precision agriculture approach to cotton production. Computers and Electronics in Agriculture 32: 213–228.
32. McKinion, J. M., Jenkins, J. N., Willers, J. L. 2007. Wide area wireless network for supporting precision agriculture. In: 2007 Proceedings of the Beltwide Cotton Conferences, edited by P. Duggar and D. Richter. National Cotton Council, Memphis, TN [non-paginated CD].
33. McKinion, J. M., Lemmon, H. 1985. Expert systems for agriculture. Computers and Electronics in Agriculture 1(1): 31–40.
34. McKinion, J. M., Turner, S. B., Willers, J. L., Read, J. J., Jenkins, J. N., McDade, J. 2004. Wireless technology and satellite internet access for high-speed whole farm connectivity in precision agriculture. Agricultural Systems 81: 201–212.
35. Milliken, G. A. 2003. Multilevel designs and their analyses. Paper 263–228. Proceedings Twenty-Eighth Annual SAS® Users Group International Conference. SAS Institute, Inc., Cary, NC.
36. Milliken, G. A., Johnson, D. E., 1992. Analysis of Messy Data, Vol. 1. Designed Experiments. Chapman and Hall/CRC, New York.
37. Milliken, G. A., Johnson, D. E., 2002. Analysis of Messy Data, Vol. 3. Analysis of Covariance. Chapman and Hall/CRC, New York.

38. National Research Council. 1997. Precision Agriculture in the 21st Century: Geospatial and Information Technologies in Crop Management. Washington, DC, National Research Council, National Academy Press. Available at: http://books.nap.edu/openbook/0309058937/html/R1.html (verified 22 December 2007).

39. Pierce, F. J., Anderson, N. W., Colvin, T. S., Schueller, J. K., Humburg, D. S., McLaughlin, N. B. 1997. Yield mapping. In: Pierce, F. J., Sadler, E. J. (Eds.), The State of Site-Specific Management for Agriculture. American Society of Agronomy, Crop Science Society of America, and Soil Science Society of America, Publishers, Madision, WI.

40. Ping, J. L, Dobermann, A. 2003. Creating spatially contiguous yield classes for site-specific management. Agronomy Journal 95: 1121–1131.

41. Ping, J. L., Dobermann, A., 2005. Processing of yield map data. Precision Agriculture 6: 193–212.

42. Pinter, P. J., Jr., Hatfield, J. L., Schepers, J. S., Barnes, E. M., Moran, M. S., Daughtry, C. S. T., Upchurch, D. R. 2003. Remote sensing for crop management. Photogrammetric Engineering and Remote Sensing 69: 647–664.

43. Pioneer Hi-Bred International, Inc. 1998. Precision Farming Offers Opportunities and Challenges. Available at: http://www.pioneer.com /xweb/usa/technology/precise.htm (unverifiable 22 December 2007).

44. Plant, R. E. 2001. Site-specific management: the application of information technology to crop production. Computers and Electronics in Agriculture 30: 9–29.

45. Pouncey, R., Swanson, K., Hart, K. (Ed.). 1999. ERDAS Field Guide, 5th ed. ERDAS, Atlanta, GA.

46. Reed, J. 2001a. Dollars in and dollars out, Part 1. Cotton Farming 45(7): 26–39.

47. Reed, J. 2001b. Dollars in and dollars out, Part 2. Cotton Farming 45(8): 27–29.

48. Reed, J. 2001c. Dollars in and dollars out, Part 3. Cotton Farming 45(9): 6–8.

49. Richards, J. A., Jia, X. 1999. Remote sensing Digital Image Analysis. An Introduction, 3rd ed. Springer-Verlag, Berlin, Germany.

50. Riley, J. R. 1989. Remote sensing in entomology. Annual Review of Entomology 34: 247–271.

51. Robinson, E. 2007. GPS, GIS, VR, and remote sensing technologies continuing to evolve. Delta Farm Press 64(49): 12.

52. Rural Advancement Foundation International. 1997. Inch by inch, Row by Row ... What will precision farming sow. RAFI Communique. April/May 1997.

53. Rudnicki, M., Meyer, T. H. 2007. Methods to convert local sampling coordinates into Geographic Information System/Global Positioning Systems (GIS/GPS)-compatible coordinate systems. Northern Journal of Applied Forestry 24(3): 233–238.

54. Schnug, E., Panten, K. and Haneklaus, S. 1998. Soil sampling and nutrient recommendations – the future. Communications in Soil Science and Plant Analysis 29(11–14): 1455–1462.

55. Sequeira, R. A., Olson, R. L., Willers, J. L., Mckinion, J. M. 1994. Automating the parameterization of mathematical models using genetic algorithms. Computers and Electronics in Agriculture 11: 265–290.

56. Shaw, D. R., Willers, J. L. 2006. Improving pest management with remote sensing. Outlooks on Pest Management 17(5): 197–201.

57. Sonka, S. T. 1985. Information management in farm production. Computers and Electronics in Agriculture 1: 75–85.

58. Stern, V. M., Smith, R. F., van den Bosch, R., Hagen, K. S. 1959. The integrated control concept. Hilgardia 29(2):81–101.

59. Thomasson, J. A., Ge, Y., Sui, R. 2006. Integrating cotton quality information between gin and farm. In: 2006 Proceedings of the Beltwide Cotton Conferences, edited by P. Duggar and D. Richter. National Cotton Council, Memphis, TN [unpaginated CD].

60. Thompson, J. M., Varco, J. J., Seal, M. R. 1999. Formulating decision support factors for variable rate nitrogen fertilization. In: 1999 Proceedings of the Beltwide Cotton Conferences, edited by P. Duggar and D. Richter. National Cotton Council, Memphis, TN. Vol. 2: 1283–1285.

61. Wagner, T. L., Williams, M. R., Willers, J. L., Akins, D. C., Olson, R. L., McKinion, J. M. 1995. Knowledge Base for rbWHIMS: An Expert System for Managing Cotton Arthropod Pests in the Midsouth. Mississippi Agricultural and Forestry Experiment Station Technical Bulletin 205.

62. Wei, J., Zhang, N., Wang, N., Lenhert, D., Neilsen, Mizuno, M. 2005. Use of the "smart transducer" concept and IEEE 1451 standards in system integration for precision agriculture. Computers and Electronics in Agriculture 48: 245–255.

63. Willers, J. L., Seal, M. R., Luttrell, R. G. 1999. Remote Sensing, line-intercept sampling for tarnished plant bugs (Heteroptera: Miridae) in Mid-south cotton. Journal of Cotton Science 3: 160–170.

64. Willers, J. L., Milliken, G. A., O'Hara, C. G., Jenkins, J. N. 2004. Information technologies and the design and analysis of site-specific experiments within commercial fields. In: G. Milliken (Ed.), 16th Applied Statistics in Agriculture Conference, 25–28 April, 2004, Manhattan, KS, pp. 41–73. [Available in PDF format from J.L.W.]

65. Willers, J. L., Jenkins, J. N., Ladner, W. L., Gerard, P. D., Boykin, D. L., Hood, K. B., McKibben, P. L., Samson, S. A., Bethel, M. M. 2005. Site-specific approaches to cotton insect control. Sampling and remote sensing analysis techniques. Precision Agriculture 6: 431–452.

66. Willers, J. L., Milliken, G. A., Jenkins, J. N., O'Hara, C. G., Gerard, P. D., Reynolds, D. B., Boykin, D. L., Good, P. V., Hood, K. B. 2008. Defining the experimental unit for the design and analysis of site-specific experiments in commercial cotton fields. Agricultural Systems 96: 237–249.

67. Vellidis, G., Garrick, V., Pocknee, S., Perry, C., Kvien, C., Tucker, M. How wireless will change agriculture. In: J. V. Stafford (Ed.), Precision agriculture '07. Wageningen Academic Publishers, Wageningen, The Netherlands, pp. 57–68.

68. Zhang, N., Wang, M., Wang, N. 2002. Precision agriculture-a world overview. Computers and Electronics in Agriculture 36: 113–13.

Rural Development Through Input–Output Modeling

Konstadinos Mattas, Efstratios Loizou, and Vangelis Tzouvelekas

Abstract Input–output (I-O) models developed in the late 1930s and ever since have been applied extensively. Though the contribution of I-O models in depicting economic transactions was recognized early on, computational constraints have limited their use. This is mainly because of huge data requirements, difficulties in computational handling, and lack of software developed and adjusted for I-O analysis. Today, I-O analysis can be applied extensively in regional and local economies and can provide valuable information on growth and investment priorities, sectoral interrelationships, and policy impacts. I-O analysis has been employed in research on both agriculture and rural development to evaluate the importance of agricultural activities, the interdependence among agriculture and the rest of the economy, the intensity of the rate of growth, and the impacts of policy interventions.

In addition, I-O analysis can identify the most dynamic sectors in a regional economy and record the financial flows among sectors, as well as record the origin (domestic or foreign) of each sector's inputs. I-O analysis also provides a breakdown of each sector's output, sales to other industries, and to final demand (consumption, gross fixed capital formation, and exports). The basic advantage of the I-O analysis is its ability to model the economic conditions for an entire region. Simultaneously, I-O analysis can assess the economy-wide impacts (both direct and indirect) of changes in one sector. Only I-O analysis can capture indirect effects and pursued impacts due to general structural disturbances.

Today, I-O analysis is within the reach of any decision maker in a rural area; it can be very easily used as long as user-friendly software is available.[1] A glance

---

[1] Schaffer [44] refers some sources of models used in impact analysis (IMPLAN, RIMS, etc.). See also the works of Johnson and Otto [21] and Jordan and Brooks [22].

K. Mattas (✉)
Department of Agricultural Economics, Aristotle University of Thessaloniki,
Thessaloniki, Greece
e-mail: mattas@auth.gr

P.J. Papajorgji, P.M. Pardalos (eds.), *Advances in Modeling Agricultural Systems*, 273
DOI 10.1007/978-0-387-75181-8_13, © Springer Science+Business Media, LLC 2009

at I-O applications reveals that this analytical tool for rural development has been used only in certain cases and only by experts. Thus, practitioners in development and in agriculture are either not aware of the benefits from the application of this tool or they are not able to perform the analysis and interpret the results.

The purpose of this chapter is to offer a quick theoretical background of the I-O analysis and to illustrate, following simple steps, the application of I-O analysis at the regional level. Moreover, a simple-to-use code (which can run under the GAUSS software) is given; this code can be adjusted to the needs of the user. To make the presentation clear and understandable, a real example of rural development analysis is presented for a Greek region.

1 Input–Output Models and Applications in Rural Development and Agriculture

Ov er the past 30 years, several studies have dealt with various rural development issues. Extensive reviews of such studies can be found in Refs. 24, 32, 34. In reading through these applications, a practitioner or a policymaker can easily discover that I-O analysis provides a useful contribution for studying a variety of real problems in rural development.

Harrison-Mayfield [13, 14] examined the link of agriculture with all other sectors in a rural economy; the author provided some indicative statements that highlight the role of agriculture in a rural economy. Using I-O analysis, the author discovered strong interconnections between agriculture and other sectors of the rural economy. Additionally, the author pointed out the role of agriculture in supporting rural employment indirectly.

Sharma et al. [45] examined the role of agriculture in Hawaii's economy and argued for the suitability of the final demand–based approach in identifying the role of agriculture in an economy from output-based and hypothetical extraction approaches. Leones et al. [23] also examined the role of agriculture and its definition in I-O analysis (i.e., the sectors that constitute agriculture). Mattas and Shrestha [29] and Tzouvelekas and Mattas [49] examined the role agriculture and food sectors are playing in the Greek economy by employing the final demand–based approach. Mattas and Pagoulatos [28] examined the extent to which investment stimulates an economy's total gross output and growth; they also assessed the contribution of investment in agriculture to the whole Greek economy. Additionally, the impacts of agricultural exports were examined by Bairak and Hughes [2], while the relationship among agriculture, the rural economy, and the urban economy was examined by Hughes and Litz [18].

The effectiveness of farm policies for the social and economic development of rural areas, and specifically the assessment of the impacts of agricultural support payments on the output of agriculture and on the whole economy, was examined by Doyle et al. [7]. Roberts [43] examined the impact of changes in milk quotas on employment in the United Kingdom. She developed a modified Leontief model based on Social Accounting Matrix (SAM) approach in order to assess the impacts of milk quotas by exogenizing agriculture. Agriculture's importance through exogenization has also been studied by Papadas and Dahl, [40], using Make and Use matrices. Tiller et al. [48] examined the economic impacts of the quota buyout system in the U.S. tobacco sector on output, value added, and employment.

On the other hand, the validity of I-O analysis in assessing the role of a sector in an economy and evaluating the impacts of various policy measures is also criticized for several reasons. Among others, Midmore and Harrison-Mayfield, [35] supported the uniqueness of I-O analysis in showing how exogenous changes in an industry affect the whole economy; yet, they also presented a number of shortcomings of the method. Midmore [33] and Midmore et al. [36] also provided a detailed criticism of the effectiveness of this approach.

I-O multipliers have also been criticized, specifically, for following an upward bias. The backward and forward linkages, as proposed by Rasmussen [42] and Hirschman [16], are used in impact analysis. However, although they are widely applied, scholars' works raise important doubts about the way that the I-O linkages evaluate sectoral interdependences [1, 6, 15, 17, 30, 37, 39, 41, 46, 47].

In addition to the problems related to the I-O multipliers and the basic assumption of I-O analysis, regionalization techniques are another important issue. The problems arising when constructing a nonsurvey or a hybrid regional I-O table from a national one are examined by a considerable number of studies in the literature. Such issues are extensively analyzed in studies pertaining to regional I-O models and regionalization techniques (see, among others, [5, 27, 38]).

2 Case Application

To examine the role of agriculture in rural development, the Greek (NUTS 2) region of Anatoliki Makedonia and Thraki (AMT) was selected as a case application. In this rural region, agriculture (especially the cultivation of tobacco) plays a dominant role. In this case, we examine what would be the employment, output, and income effects caused by a change in tobacco regime. This policy change would be expected to cause significant impacts on the region's economy and development. The region of AMT[2] is taken as a case in point because tobacco constitutes an important activity for the region. Tobacco's contribution in terms of income and employment is essential for the region's social and economic sustainability and development.

---

[2] The analysis for Anatoliki Makedonia and Thraki region was taken as example in the present study, from a previous work of Mattas et al. [26], in which can be found more details; moreover, analysis of the tobacco sector in Greece can be found in Mattas et al. [25].

Agriculture is a dominant sector in the region both in gross value added and in employment. According to data published by the National Statistical Service of Greece (NSSG), more than 35% of the region's total labor force is involved in rural activities. Moreover, a significant part of the industry's labor force is employed in sectors connected indirectly with the agricultural sector (such as food, tobacco, and cotton processing). As for tobacco, 25% of the country's growers are located in the region of AMT.

The current study, by employing I-O analysis, aims to examine the potential economic and social impacts from a policy scenario that assumes fully decoupled payments. This scenario is in line with the decisions of the EU Commission; it includes fully decoupled tobacco premiums and assumes that the support will be based on a lamp-sum payment according to the average support of the period 2000–2002. Moreover, along with the reduction in tobacco premiums, it is assumed that tobacco market prices will be increased by 50% due to an expected reduction in leaf tobacco supply.

To investigate the possible impacts of the above scenario, a regional I-O model was developed. In doing so, the results of a linear programming model were used to identify the potential changes in the crop mix on tobacco-producing farms and to indicate alternative profitable cultivations. The results of the linear programming model were fed into the I-O model as a final demand change (for details, see [26]). Hence, the I-O model, using the I-O multipliers, provides sectoral and total changes for the entire region in total output, household income, and employment.

2.1 Methodology and Data

An I-O model identifies the major sectors in an economy and the financial flows among them. By portraying the total economy with a number of sectors, I-O models provide an effective tool for sectoral analysis. The basic advantage of I-O analysis is its ability to examine a regional or a national economy as a whole. I-O models can simultaneously assess both the direct and the indirect impacts from changes in one sector on all other sectors of the economy. Therefore, the approach provides evidence on the size of the effects (due to changes in the tobacco regime) on total employment, income, and gross output by employing a final demand–based approach.

The well-known Leontief demand driven I-O model in its general form can be presented in the following equation[3]:

$$X = AX + Y \tag{1}$$

[3] The mathematical presentation of the simple Leontief model can be found easily in the literature; for a thorough presentation, see [38].

where A is the matrix of technical coefficients, X is the vector of sectoral output, and Y is the vector of sectoral final demand components. Solving for final demand Y:

$$X - AX = Y \Rightarrow (I - A)X = Y \qquad (2)$$

and then solving system (2) for total output X:

$$X = (I - A)^{-1}Y \qquad (3)$$

where I is an identity matrix. The solution to system (3) constitutes the basic solution of the Leontief I-O system. Any exogenous changes in the final demand vector, Y, induce changes in the total gross output, X, of the economy, which can be assessed by Eq. (3). The matrix $(I - A)^{-1}$ is the so-called Leontief inverse, or the matrix of interdependence coefficients. Each element of that matrix indicates the total (direct and indirect) requirements of sector i per unit of final demand for the output of sector j.

Because a regional table was not available for the region under examination, a regionalization technique is followed to construct it; in order to do that, the Greek national I-O table was used. The latest available table was the 1998 symmetric I-O table for the Greek economy published by the National Accounts Department of the Greek Statistical Service (NSSG). The table initially consisted of 30 economic sectors, classified at a two-digit classification level according to the Standard Industrial Classification (SIC) system. The classification scheme is suitable to depict the region, and few modifications were performed, thereby avoiding any aggregation or disaggregation of the national table (which would violate the I-O hypotheses and create biases). For the purposes of analysis, only agriculture was disaggregated into five sectors: tobacco, wheat, maize, cotton, and other agricultural activities. The four primary crops are recorded as separate sectors in the final classification scheme.

2.2 Regionalization Technique

Various regionalization techniques are presented in the literature to counter the lack of survey regional I-O tables. These techniques can be grouped into purely nonsurvey techniques and partial survey (or hybrid) techniques. In the current study, the hybrid GRIT technique (generation of regional I-O tables) – developed initially by Jensen et al. [19] – was used for the regionalization of the national I-O table. It is a formalized, nonsurvey method of compilation with facility for the user to insert superior data at any stage of the compilation procedure. Most I-O studies that apply the GRIT technique estimate interindustry flows by using an employment-based location quotient (LQ) for the corresponding elements of the national direct requirement matrix.[4]

In the current study, the GRIT was applied by employing an LQ developed by Flegg et al. [12], known as FLQ, which was modified by Flegg and Webber [8].[5] In contrast with other LQs, such as the simple location quotient (SLQ) and the cross industry location quotient (CILQ), the FLQ provides an alternative way of estimating the relevant trade coefficients. Trade coefficients measure the proportion of any given commodity supplied from the region's production (that is, they measure the degree of self-sufficiency of a region). Those trade coefficients depend on three variables: (1) the relative size of the supplying sector, (2) the relative size of the purchasing sector, and (3) the relative size of the region. The most commonly used LQs do not take into account all three of the above-mentioned variables. Specifically, the SLQ takes into account only the first and the third, whereas CILQ takes into account only the first two. All the three variables are considered by the FLQ; this is its main advantage.

To identify the significance of tobacco cultivation and the tobacco processing sectors for the regional economy, I-O multipliers are estimated. High sectoral multipliers, in terms of output, income, and employment, reveal the ability of a sector to induce multiple impacts within the whole economy due to an exogenous change. Multipliers measure the change in the entire economy caused by changing a sector's output, income, or employment by only one unit.

After estimating the I-O multipliers, the scenario analysis is performed to examine the impacts from the change in tobacco policy. The impacts are measured in terms of output, income, and employment losses or gains for the national and regional economy. Direct or indirect losses or gains are traced out and measured in total.[6] In this way, the most affected sectors of the economy can be identified, and both relative and absolute changes can be measured.

3 The Computational Procedure

The regionalization technique is described step-by-step, as the main objective of this work is to demonstrate the I-O modeling procedure and the assessment of possible impacts at a regional level. The application of the GRIT technique, as it is presented below, was incorporated in the GAUSS mathematical computer package. GAUSS allows for easy data-handling and matrix manipulation, and the software module can be monitored step-by-step.

---

[4] Detailed discussion and application of various regionalization techniques, including GRIT, can be found in Refs. 19, 20, 27, 38, 49.

[5] The original proposal of FLQ by Flegg et al. [10] was introduced in 1995. Since then, a dialogue was opened in the literature [3, 8, 9, 10, 11, 31] and the original version was improved and modified to the version appearing in Ref. 8.

[6] A detailed description of the procedure for calculating I-O multipliers can be found among others in Refs. 27, 38.

In the sections that follow, an attempt is made to present the procedure to use I-O analysis in a verbal form, with a theoretical background, while staying in line with the GAUSS code steps as they are presented in the Appendix. To make the procedure easier to understand, it is developed and presented in distinct sections (called phases) and subsections (called steps). In addition, phases and steps reflect the various methodological stages.

PHASE 1: LOAD BASIC TABLES AND DATA IN TXT FORMAT

First, the necessary data for the regionalization procedure and the calculation of the multipliers have to be loaded. Such data are as follows:

- The transactions matrix, the primary inputs, and the final demand matrices are loaded (source: the national I-O table). In this example, the transactions matrix consists of 34 sectors and the primary inputs (imports, compensation of employees, and other primary inputs) and final demand (consumption, exports and other final demand) of three categories.
- The national sectoral output taken from the national I-O table.
- The national and regional sectoral employment vectors (source: the National Statistical Service of Greece).

PHASE 2: COMPUTATION OF THE LOCATION QUOTIENTS

In this phase, two indices are estimated (the SLQ and the FLQ) using the regional and national employment vectors. Both are estimated for use in the next steps of the GRIT regionalization procedure.

PHASE 3: APPLICATION OF THE GRIT TECHNIQUE STEPS[7]

Step 3.1: Sectoral Aggregation

This step is an optional one and takes place only when it is needed. Aggregation and disaggregation for the purpose of impact analysis takes place in consequent steps. Here, aggregation is enforced due to I-O and employment data noncompatibility. Specifically, sectoral aggregation becomes necessary in cases where the aggregation scheme of the available employment[8] data is less than that of the national I-O table. The national I-O table is aggregated in such a way that the number of sectors in the table must correspond with the number of sectors for which employment data exists. In this case, employment data are available in at least the same classification level (34 sectors) as that of the national I-O table; thus, there is no need for any further aggregation.

---

[7] The mathematical presentation of the GRIT can be found in Ref. 27, where each step's equations are described in detail.

[8] Employment is the measure used for the calculation of the LQs; it is the most commonly used measure due to its availability.

Step 3.2: Reallocation of International Trade

Though the treatment of international trade in a regionalization procedure is a contentious issue in the literature, in this example, the reallocation of international trade back to the national table is adopted [19]. There are two cases where there is no need to reallocate international trade. The first case is when the national transactions table is given with total flows (i.e., imports are included in the intermediate transactions); in such situations, this step is skipped. The second case is when an import matrix is available; when that occurs, imports are redistributed to the original sectors and there is no need to follow the procedure followed by the program. When the national transactions matrix is expressed only in domestic flows, the imports row is reallocated proportionally to the relevant rows of the sectors in the national table. In the current study, following Jensen et al. [19], imports are redistributed only to the secondary sectors of the matrix, as imports are generally made up of secondary products.

As mentioned before, different views on the issue of international trade treatment exist. Some people retain imports in the primary inputs and consider them as regional imports after their regionalization.

Step 3.3: Computation of the National Direct Requirements Matrix

The regionalization procedure is applied using the direct requirements matrix, so this matrix must be calculated. In doing this, some additional adjustments need to be pursued. The first adjustment uses either a net table or a gross table. Those who prefer to use a net table suggest that all intrasectoral flows (the main diagonal elements) of the national transactions matrix should be eliminated. This is because intrasectoral flows include interregional trade. By maintaining these flows within the table when deriving the regional table, the regional intermediate purchases might be overestimated. On the other hand, a retort in the literature suggests that important information is lost by eliminating the on-diagonal elements, so a gross table is a better choice. To have a more accurate regional I-O table in the case where the main diagonal is eliminated, a small-scale survey can be performed to identify some of the eliminated coefficients.

The second adjustment that might be needed is the elimination of the non-existent sectors in the region; this involves the deletion of sectors that are present in the national table but do not exist in the regional economy, which is especially common if the region is a small one. Then, the column coefficients of the nonexisting sectors are added to the exports vector, while the row coefficients should be added to the imports vector. In the current case, all national sectors are present in the region as well; therefore, such adjustment does not appear in the GAUSS code (Step 3.3) in the Appendix.

Step 3.4: Computation of the Regional Direct Requirements Matrix

This step is crucial in the regionalization procedure because it is when the core matrix (direct requirements matrix) of the regional table is derived. To derive the regional direct requirements matrix, the GRIT technique suggests the application of a location quotient; in the present study the FLQ[9] is employed. With the application of the FLQ, we attempt to split the national technical coefficients into regional purchase and import coefficients. A LQ measures the relative importance of an industry in a region compared with its importance nationally. The computation of the FLQ was performed in Phase 2 (see Step 2.1).

The procedure to regionalize the national direct requirements matrix is as follows: The FLQ matrix is multiplied with the national direct requirements matrix. If the FLQ between any two purchasing and selling sectors is greater than zero but less than one $(0 < FLQ_{ij} < 1)$, it is assumed that regional output is insufficient to satisfy the local demand and that imports will be required to make up the deficiency. In this case, the respective technical coefficient of the national direct requirements matrix will overestimate the regional interindustry transactions and will have to be reduced. To do this, the national coefficients are multiplied by the relevant FLQ_{ij} $(a_{ij}^R = FLQ_{ij} \cdot a_{ij}^N)$, where α is the direct coefficient, and R and N indicate regional and national coefficients, respectively. The residual is added to the relevant national import coefficient to yield an enhanced regional import coefficient $(m_{ij}^R = a_{ij}^N(1 - FLQ_{ij}))$, where m is the import coefficient.

In the case where the FLQ_{ij} is greater than one $(FLQ_{ij} > 1)$, it is assumed that the regional sector's i supply is sufficient to meet the purchasing sector's j demand, and the national coefficient is considered to be the same as the regional one. In fact, all resulting regional coefficients will take values between their national value and zero.

Step 3.5: Sectoral Aggregation – Definition of the Regional Classification Scheme

After all of the above adjustments, the classification scheme of the regional I-O table has to be defined. Until this stage, it is assumed that the economic structure (number of sectors) is the same in the region as in the country; usually this is not true for small regions. Therefore, the dimensions of the regional I-O tables should be adjusted (aggregated/disaggregated) in such a way that reflects adequately the economic conditions in the region. Usually small and minor sectors (i.e., those with low employment or output levels) are aggregated, though high aggregation of the national table leads to a loss of information and creates problems with the product mix among the sectors. In this particular application, the regional classification level follows the same classification as the national. Thus, in the current case, this procedure was skipped.

The regional Leontief and transactions matrices and the regional imports and total output vectors are calculated. A key variable is the regional sectoral total output vector, which is needed to derive the regional transactions matrix (in monetary flows) from the national direct requirements matrix. If the regional output vector is not available, it can be derived by regionalizing the national output vector. Thus, the national output vector is multiplied by the sectoral

---

[9] The use of the FLQ was supported above; for a more comprehensive analysis, see Ref. 27.

ratios of regional to national employment (or gross domestic product). Certainly, the best approach is to acquire the regional sectoral output data from any secondary data sources.

In a case where any aggregation of the output vector is needed, the procedure is the same as the one described in Phase 1. To perform the aggregation, the regional direct requirements matrix as well as the regional import coefficients vector are adjusted using regional employment weights.

At this stage, the user can insert any superior data available in order to improve the derived regional matrix and the other data. Such superior data usually come either from surveys or from other secondary sources. The collection of such data should focus on the important sectors of the regional economy, that is, on the sectors with high direct coefficients that more strongly affect the multipliers. A discussion on the issue can be found in the original GRIT study and in many other recent studies in the literature.

Step 3.6: Computation of the Complete Regional I-O Table

In the last step of the regionalization procedure, the complete regional I-O table is calculated in symmetric form. During the computation sequence, adjustments related to final demand and primary inputs categories are needed to balance the table.

The final demand (FD) in the current study consists of three components (consumption, exports, and other final demand) that have to be estimated. Initially, the total regional final demand is calculated as the difference between total sectoral regional output and total sectoral intermediate sales. The other FD component is calculated as a residual, while consumption and exports are regionalized using the sectoral shares of regional versus national employment. The two vectors are further adjusted with the use of the SLQ (see the first commands of the code in Step 3.6).

Specifically, the regionalization of consumption and exports is done following a two-stage procedure. In the first stage, the vectors are regionalized (as mentioned above) by multiplying them by the employment ratios. In the second stage, the SLQs calculated in Phase 2 are used to further adjust the values of consumption and exports. The procedure is as follows:

- If the computed SLQ of a sector is less than one (SLQ < 1), then the respective value in the final demand vector (in this case consumption and exports) is reduced further by multiplying it by the SLQ.
- If the computed SLQ of a sector is greater than one (SLQ > 1), then it is assumed that the sector is well represented in the region and thus no further adjustment is needed.

The primary inputs in the current study also consist of three components: compensation of employees, imports, and other primary inputs. The primary inputs (and final demand) of an I-O table might consist of many other sub-components, depending on how detailed the I-O table is. The regionalization of primary inputs is performed following the two-stage procedure (as described above in the case of final demand). The other primary inputs component is derived residually.

Using the transactions matrix estimated above, along with the primary inputs and the final demand, the regional symmetric I-O table can be constructed. The regional I-O table can then be used to form an I-O model and examine impacts from exogenous policy changes. In order to do this, I-O multipliers and other backward and forward linkages are estimated.

PHASE 4: COMPUTATION OF I-O MULTIPLIERS

Having constructed the regional I-O table, the procedure then shifts to the calculation of various I-O multipliers[10] and indices, which are very meaningful tools in rural development analysis. In this particular work, only the most common [16, 42] (output, household income, and employment) multipliers are estimated. These multipliers are calculated using the estimated Leontief inverse matrix and the sectoral employment and household income vectors (for the case of employment and income multipliers). In addition, the direct coefficients (backward and forward or alternatively as Chenery and Watanabe [4] linkages) are calculated using the direct requirements matrix. The calculation of the Chenery and Watanabe indices is shown in Step 4.1 in the GAUSS code. The calculation of the Rasmusen [42] and Hirschman [16] multipliers is described in Step 4.2 of the code.

4 Input–Output Multipliers and Impact Analysis Results

Following the computational procedure presented above, the I-O multipliers were estimated. An I-O model has the ability to identify the important sectors of an economy at a regional or a national level. Key sectors are identified in terms of multipliers; the higher the multiplier, the stronger is the ability of the corresponding sector to create multiple impacts in the economy. The computed sectoral multipliers and total effects are used in the impact analysis to estimate the direct and indirect impacts from the policy change in the tobacco regime. The calculated multipliers are shown in Table 1.

Using the calculated multipliers, the impacts in terms of output, employment, and household income are shown below. These are negative effects, because the fully decoupled tobacco premiums will induce farmers to abandon in large degree the tobacco cultivation; this will have direct, negative consequences for employment, output, and income.

From the mathematical programming model, the changes in the crop mix on tobacco-producing farms in the AMT region were estimated. Changes will be radical; tobacco cultivation is found to decline by 77% in the region. An estimated 48.9% of the tobacco-producing farms will be forced out of

---

[10] A detailed presentation of the I-O multipliers and linkages can be found in, among others, Refs. 27 and 38.

Table 1 Sectoral regional multipliers for Anatoliki Makedonia and Thraki

Sectors	OM	R	Type I EM	R	Simple EM	R	Type I IM	R	Simple IM	R
Agriculture	1.1444	22	1.1406	21	0.1356	8	1.2092	16	0.0492	30
Tobacco	1.0392	33	1.0117	34	0.3619	2	1.0166	31	0.1359	15
Wheat	1.0540	29	1.0150	32	0.4394	1	1.0776	26	0.0438	31
Maize	1.0335	34	1.0155	31	0.2637	4	1.0481	28	0.0426	34
Cotton	1.0459	31	1.0187	30	0.3006	3	1.0659	27	0.0434	32
Forestry	1.0824	27	1.0254	29	0.2065	5	1.0149	32	0.4325	3
Fishing	1.1841	19	1.1616	20	0.0655	18	1.1828	20	0.1278	17
Mining and quarrying	1.0538	30	1.0822	24	0.0216	34	1.0382	29	0.1362	14
Food and beverages	1.3450	10	2.6775	4	0.0636	20	1.2520	13	0.1047	22
Tobacco processing	1.2451	14	3.6628	1	0.0958	11	1.3658	10	0.1105	21
Textile and clothing	1.2650	12	1.4085	9	0.0563	23	1.2103	14	0.1630	11
Leather and leather products	1.4545	5	1.4392	8	0.0552	24	1.6240	5	0.0968	24
Wood and wood products	1.2853	11	1.2679	13	0.1054	9	1.3573	11	0.1579	12
Paper and publishing	1.4009	6	1.3765	11	0.0671	16	1.4411	8	0.1151	19
Petroleum products	1.8600	2	3.3293	2	0.0279	32	1.7336	4	0.1143	20
Chemical products	1.3873	7	1.7319	7	0.0219	33	1.5049	7	0.0908	27
Rubber and plastic products	1.1757	22	1.1403	20	0.0367	28	1.1176	23	0.1350	16
Nonmetal mineral products	1.1956	17	1.1658	17	0.0440	26	1.1440	21	0.2175	5
Basic metals and products	1.3579	8	1.3823	10	0.0510	25	1.4237	9	0.1158	18
Machinery	2.0484	1	2.8932	3	0.0581	22	3.5816	1	0.0945	25
Electrical equipment	1.5502	4	2.1975	5	0.0294	31	2.4901	3	0.0671	29
Transport equipment	1.6332	3	2.1686	6	0.0347	30	2.5550	2	0.0726	28
Other manufacturing	1.2613	13	1.2230	15	0.0825	13	1.2093	15	0.1648	10

Table 1 (continued)

Sectors	OM	R	Type I EM	R	Simple EM	R	Type I IM	R	Simple IM	R
Electricity, gas, and water	1.2193	15	1.2484	14	0.0387	27	1.2016	18	0.1487	13
Construction	1.2141	16	1.1720	18	0.0632	21	1.1966	19	0.1674	8
Trade	1.1114	25	1.0623	26	0.0989	10	1.1194	22	0.1014	23
Hotels and restaurants	1.1887	18	1.2199	16	0.0641	19	1.2044	17	0.0928	26
Transport and communication	1.1529	21	1.1142	23	0.0796	14	1.0936	24	0.1814	7
Financial intermediation	1.3540	9	1.3638	12	0.0659	17	1.3379	12	0.1656	9
Real estate and business activities	1.1222	24	1.2115	17	0.0365	29	1.5211	6	0.0433	33
Public administration and defense	1.0721	28	1.0445	28	0.0895	12	1.0119	33	0.6450	1
Education	1.0393	32	1.0128	33	0.1383	7	1.0064	34	0.6166	2
Health	1.1022	26	1.0479	27	0.0792	15	1.0333	30	0.2471	4
Other services	1.1281	23	1.0728	25	0.1393	6	1.0785	25	0.2167	6

R, the sectoral ranking; OM, output multipliers; EM, employment multipliers; IM, income multipliers.

Table 2 Impacts due to policy changes in the AMT region

	Employment		Income		Output	
	Persons	(%)	000 €	(%)	000 €	(%)
Tobacco Cultivation						
Total effect[*]	10,960	(5.03)	12,086	(0.90)	92,609	(1.02)
Direct	10,818	(4.96)	11,866	(0.88)	88,751	(0.98)
Indirect	142	(0.07)	220	(0.02)	3,859	(0.04)
Tobacco Processing						
Total effect[*]	1,569	(0.72)	14,150	(1.05)	174,986	(1.93)
Direct	1,481	(0.68)	13,443	(1.00)	171,569	(1.89)
Indirect	88	(0.04)	707	(0.05)	3,417	(0.04)
Total Change	12,529	(5.75)	26,236	(1.95)	267,595	(2.95)

[*]Values in parentheses are the relevant shares in the total regional economy's employment, income, and output levels.

farming.[11] After incorporating the results as final demand changes to the I-O model, the potential employment, income, and output impacts were estimated. The results of the impact analysis are presented in Table 2.

The total reduction of the regional employment rate will be 5.75%. This underlines the regional significance of tobacco cultivation. Specifically, an estimated 12,529 persons will be unemployed after the change in the tobacco policy regime in the region. The majority of this change (86.3%, or 10,818 persons) stems from tobacco cultivation and refers to tobacco growers that will be forced out of farming. In the tobacco processing sector, 1,481 persons (11.8% of the total) will lose their jobs. Finally, the indirect employment effects caused by both primary and processing sectors are low, as they account only for the 1.8% of the total employment reduction. The corresponding reductions in household income and output levels can be seen in the table.

5 Conclusions

The design and implementation of policies related to agriculture and rural regions has always been a challenge. Such policies affect a considerable number of people as well as many issues related to rural activities and, thus, policy designers keep them high in their agendas. Successful planning of such policies requires *ex ante* evaluation of their potential impacts. Given this notion, an attempt was made in this work to assess the potential impacts of a policy change in a rural economy. Specifically, to examine potential impacts of the tobacco regime reform, I-O modeling was used. Despite various shortcomings, the specific tool that we employed provides an adequate solution for such kind of studies. That is, our I-O model can adequately examine the potential impacts for a whole economy in

---

[11] For more details about the mathematical programming model and the results, see Ref. 26.

the form of a general equilibrium rather than partial impacts on a sector or on a specific activity.

Although I-O analysis is relatively simple in its application, the limited existence of relevant computational tools prevents its wider application. The existence of such computational tools is much more limited for studies dealing with regional models than for studies of national models. Thus, the primary objective of the current work was to present the basics of a regionalization methodology and to develop a code in a general form that is relatively simple to apply in such studies. The GRIT regionalization technique was the one that was analyzed step-by-step through the derivation of the regional symmetric table. Moreover, the calculation of the well-known I-O multipliers has also been demonstrated. Finally, though the mathematical presentation[12] of the regionalization technique is not present, the code developed might be helpful to those interested in performing such studies.

Appendix: The Code of the GAUSS Computer Package

```
/* Gauss program for regionalizing the National I-O table for Anatoliki Makedonia and Thraki region (1998, 34 sectors), using the GRIT technique, and estimating the I-O multipliers */
```

/* PHASE 1: LOAD BASIC TABLES AND DATA IN TXT FORMAT */

```
/*format /mat /on /mb1 /ros 16,8;*/
format /rds 12,5;
outwidth 240;

m = 34;
/* set the dimension (number of sectors) in the national transactions table */
n = 3;
/* set the dimension (number of categories) of the final demand and final payments matrix */

loadm TrNt[ ] = c:\gauss\kapnos\reg\TrNt.txt; TrNt = reshape(TrNt,m,m);    /* Load National Transactions Matrix mxm*/
loadm FDNt[ ] = c:\gauss\kapnos\reg\FDNt.txt; FDNt = reshape(FDNt,m,n);    /* Load National Final Demand Matrix mxn */
loadm FPNt[ ] = c:\gauss\kapnos\reg\PrdnNt.txt;
FPNt = reshape(FPNt,m,n);
/* Load National Final Payments Matrix for Greece mxn */
    loadm EmpNt[ ] = c:\gauss\kapnos\reg\EmpNt.txt;
    PrdnNt = reshape(PrdnNt,m,1);
```

---

[12] As mentioned, the mathematical presentation of the GRIT technique and the calculation of the relevant linkage coefficients can be found in Ref. 41.

```
          /* Load National Sectoral Output mx1 */
loadm EmpNt[ ] = c:\gauss\kapnos\reg\EmpNt.txt;    EmpNt = reshape(EmpNt,m,1);
          /* Load National Employment Vector m sectors */
loadm EmpRg[] = c:\gauss\kapnos\reg\EmpRg.txt;    EmpRg = reshape(EmpRg,m,1);
          /* Load Regional Employment Vector m sectors */
```

/* PHASE 2: COMPUTATION OF THE LOCATION QUOTIENTS */

/* Step 2.1: Computation of the Employment-Based CILQ Type (the FLQ) */

```
EmpRatio = EmpRg./EmpNt;
C1 = reshape(Empratio',m,m);
CILQ0 = (EmpRatio./C1)*0.34181;
 /* CILQ Matrix before adjustment*/

/* Replace the values of CILQ which are greater than 1 with 1 */
CILQ = zeros(m,m);
          i = 0;
             do while i<m;
                i = i + 1;
                j = 0;
                do while j<m;
                j = j + 1;
                   if CILQ0[i,j] > 1;
                      CILQ[i,j] = 1;
                   else;
                         CILQ[i,j] = CILQ0[i,j];
                   endif;
          endo;
endo;
```

/* Step 2.2: Estimation of the Employment-Based SLQ */

```
SL0 = (EmpRg./sumc(EmpRg))./(EmpNt./Sumc(EmpNt));
SL1 = reshape(SL0,m,m);
SLF = zeros(m,m);
          i = 0;
             do while i<m;
                i = i + 1;
                j = 0;
                do while j<m;
                j = j + 1;
                   if SL1[i,j] > 1;
                      SLF[i,j] = 1;
                   else;
```

```
        SLF[i,j] = SL1[i,j];
      endif;
    endo;
endo;
```

/* PHASE 3: APPLICATION OF THE GRIT TECHNIQUE STEPS */

/* Step 3.1: Sectoral Aggregation (No Need for Aggregation in the Current Study)*/

/* Step 3.2: Reallocation of International Trade */

```
TrNt1 = TrNt[8:25,.];
/* Partitioned Transactions with Secondary Sectors */
d1 = ones(1,rows(TrNt1))*TrNt1;
/*primary sectors (1-7), sec(8-25), ter (26-34)*/
d2 = d1 + FpNt[.,2]';
TrNt2 = TrNt1*diagrv(zeros(m,m),d2)*inv(diagrv(zeros(m,m),d1));
/* National Transactions SubMatrix Adjusted for International Trade (mxn) */
TrNt3 = TrNt[1:7,.]|TrNt2|TrNt[26:34,.];
/* National Transactions Matrix Adjusted for International Trade (nxn) */
```

/* Step 3.3: Computation of the National Matrices (There Are No Nonexistent Sectors) */

```
DRMNt = (TrNt3)*inv((diagrv((zeros(m,m)),(PrdnNt))));
/*National Direct Requirements Matrix mxm */
```

/* Step 3.4: Computation of Regional Matrices */

```
DRMRg = CILQ.*DRMNt;
/* Regional Direct Requirements Matrix mxm */
ImpRgCf1 = (sumc(drmnt-drmrg));
/*Regional Imports Coefficients vector adjusted for CILQ mx1*/
```

Print; print "Direct Requirements Regional for AMT"; print DRMRg;

/* Step 3.5: Sectoral Aggregation – Definition of the Regional Classification Scheme*/

```
C3 = eye(m);C4 = C3-DRMRg;
LEONTIEFRg = inv(C4);
/* Regional Leontief Inverse Matrix mxm */
PrdnRg = (PrdnNt).*(EmpRg./EmpNt);
/* Regional Sectoral Output mx1 */
TransRg = DRMRg*diagrv(zeros(m,m),PrdnRg);
/* Regional Transactions Matrix */
ImpRgf = ImpRgCf1.*PrdnRg;
```

/* Regional Imports Vector mx1, from coefficients to values */

Print; print "Leontief Regional for AMT"; print LEONTIEFRg;
Print; print "Regional Sectoral Output"; print PrdnRg;

format /rds 14,3;
Print; print "Regional Transactions Matrix for AMT"; print TransRg;

/* Step 3.6: Computation of the Complete Regional Input–Output Table */

/*Final Demand and Primary Inputs Computation */

/ Regionalization of the Household Income Using Employment Ratios and an*
*Employment-Based SLQ */*

```
SLQH0 = (Emprg./sumc(EmpRg))./(EmpNt./sumc(EmpNt));
r0 = (EMPRg)./(EMPNt); r1 = FpNt;
SLQHd = zeros(m,1);
 i = 0;
  do while i<m;
  i = i+1;
    if SLQH0[i,1] > 1;
      SLQHd[i,1] = r0[i,1] ;
    else;
      SLQHd[i,1] = r0[i,1]*SLQH0[i,1];
    endif;
endo;

HousRg = r1[.,1].*SLQHd;
/* Regional Household Income Vector mx1 */
IPRg = (ones(1,m))*TransRg;
/* Regional Sectoral Intermediate Purchases 1xm */
OPRg = PrdnRg-sumc(TransRg)-HousRg-ImpRgf;
/* Regional Other Payments Vector mx1 */
FpRg = HousRg + ImpRGf + OPRg;
/* Regional Final Payments mx3 */
/*PrdnRg-(OPRg + HousRg + ImpRGf + IPRg'); */

Print; print "Regional Final Payments for AMT"; print FpRg;

ISRg = sumc(TransRg');
/* Regional Sectoral Intermediate Sales kx1 */
FDi = PrdnRg-ISRg;

format /rds 16,5;
```

/* *Regionalization of Private Consumption Vector Using Employment Ratios and an Employment-Based SLQ* */

```
SLQF0 = (EmpRg./sumc(EmpRg))./(EmpNt./sumc(EmpNt));
r3 = (EmpRg)./(EmpNt); r2 = FdNt;
SLQC = zeros(m,1);
  i = 0;
    do while i<m;
    i = i+1;
      if SLQF0[i,1] > 1;
        SLQC[i,1] = r3[i,1] ;
      else;
        SLQC[i,1] = r3[i,1]*SLQF0[i,1];
      endif;
endo;
```

```
ConsRg = r2[.,1].*SLQC;
/* Regional Sectoral Private Consumption kx1 */
```

/* *Regionalization of Exports Vector Using Employment Ratios and an Employment-Based SLQ* */

```
SLQE = zeros(m,1);
  i = 0;
    do while i<m;
    i = i+1;
      if SLQF0[i,1] > 1;
        SLQE[i,1] = r3[i,1] ;
      else;
        SLQE[i,1] = 0;
      endif;
endo;
```

```
ExpRg = r2[.,2].*SLQE;
/* Regional Sectoral Exports kx1 */
OFDRg = PrdnRg-ISRg-ExpRg-ConsRg;
/* Regional Other Final Demand kx1 */
FDRg = ConsRg + ExpRg + OFDRg;
/* Regional Final Demand kx3 */
```

```
Print; print "Regional Final Demand for AMT"; print FDRg;
```

```
" Regional Final Demand ";
?;
" Int Sales Consumption Exports Other FD Output ";
```

ISRg~ConsRg~ExpRg~OFDRg~PrdnRg;
?;
?;

" Regional Final Payments ";
?;
" Int Purchases Households Imports   Other FP Output ";

IPRg'~HousRg~ImpRGf~OPRg~PrdnRg;

/* PHASE 4: COMPUTATION OF THE I-O MULTIPLIERS */

```
b1 = ones(1,m);
FDShare = FDRg/sumc(FDRg);
FPShare = FPRg/sumc(FPRg);
SecId = seqa(1,1,m);
```

/* Step 4.1: Chenery and Watanabe Linkage Indices (or Direct Coefficients) */

```
CH1 = B1*DRMRg;    /* Output Backward linkages */
CH2 = DRMRg*B1';   /* Output Forward Linkages */

print;print"Chenery and Watanabe Linkages";
print; print" Sector Backward Forward";
print SecId~CH1'~CH2;
print "Total " OCH1~OCH2;
```

/* Step 4.2: Rasmussen and Hirschman Linkage Indices (or I-O Multipliers)*/

```
B2 = B1*LeontiefRg;
/* Output Backward linkages or OUTPUT MULTIPLIERS */
B4 = HousRg./PrdnRg;
/* Income technical coefficients or DIRECT INCOME EFFECT */
B5 = B4'*LeontiefRg;
/* Simple Income Multiplier or TOTAL INCOME EFFECTS */
B6 = B5'./B4;
/* TYPE I INCOME MULTIPLIERS */
B7 = EmpRg./PrdnRg;
/* Employment technical coefficients or DIRECT EMPLOYMENT EFFECT */
B8 = B7'*LeontiefRg;
/* Simple Employment Multiplier or TOTAL EMPLOYMENT EFFECTS */
B9 = B8'./B7;
/* TYPE I EMPLOYMENT MULTIPLIERS */
B3 = LeontiefRg*B1';
```

```
/* Output Forward Linkages */
B10 = LeontiefRg*B4;
/* Income Forward Linkages */
B11 = LeontiefRg*B7;
/* Employment Forward Linkages */

?;
print;print"Rasmussen and Hirschman Linkages";

print; print"Backward";
print;
print" Sector Output Income Employment Type I Income Type I Employment";
print SecId~B2'~B5'~B8'~B6~B9;
print " Type I            " OB2~OB5~OB8~OB6~OB9;

print; print "direct effects";
print; print "B4 Direct Income Effect, B7 Direct Employment Effect";
print SecId~B4~B7;

?;
print; print"Forward";
print; print" Sector Output Income Employment ";
print SecId~B3~B10~B11;
print " Type I            " OB3~OB10~OB11;


CLOSEALL;
END;
```

References

1. Augustinovics M. (1970). Methods of international and intertemporal measures of structures. In: Carter A. P. and Brody A. (eds.), *Contributions to Input-Output Analysis*. Amsterdam: North Holland.
2. Bairak R. and D. Hughes (1996). Evaluating the impacts of agricultural exports on a regional economy. *Journal of Agricultural and Applied Economics*. 28: 393–407.
3. Brand S. (1997). On the appropriate use of location quotients in generating regional input-output tables: a comment. *Regional Studies*. 31: 79–94.
4. Chenery H. B. and T. Watanabe (1958). International comparisons of the structure of production. *Econometrica*. 26: 487–521.
5. Dewahurst J. H. (1992). Using RAS technique as a test of hybrid methods of regional input-output table updating. *Regional Studies*. 26: 81–91.
6 Dietzenbacher E. and J. A. Van der Linden (1997). Sectoral and spatial linkages in the ec production structure. *Journal of Regional Science*. 37: 235–257.

7. Doyle C., M. Mithcell and Topp K. (1997). Effectiveness of farm policies on social and economic development in rural areas. *European Review of Agricultural Economics*. 24: 530–546.
8. Flegg T. A., and C. D. Webber (2000). Regional Size, regional specialization and the FLQ formula. *Regional Studies*. 34: 563–569.
9. Flegg T. A. and C. D. Webber (1996). Using location quotients to estimate regional input-output coefficients and multipliers. *Local Econ. Quart* 4: 58–86.
10. Flegg T. A. and C. D. Webber (1996). The FLQ formula for generating regional input-output tables: an application and reformation. Working Papers in Economics, 17, University of the West of England, Bristol.
11. Flegg T. A. and C. D. Webber (1997). On the appropriate use of location quotients in generating regional input-output tables: Reply. *Regional Studies* 31: 795–805.
12. Flegg, T. A., C. D. Webber and M. V. Elliot (1995). On the appropriate use of location quotients in generating regional input-output tables. *Regional Studies*. 29: 547–561.
13. Harrison-Mayfield L. (1993). The impact of the agricultural industry on the rural economy tracking the spatial distribution of the farm inputs and outputs. *Journal of Rural Studies* 9: 81–88.
14. Harrison-Mayfield L. (1996). Agriculture's links with the rural economy: an input-output approach. In Midmore P. and Harrison-Mayfield L. (eds.), *Rural Economic Modeling: An Input-Output Approach* Wallingford: CAB International.
15. Heimler A. (1991). Linkages and vertical integration in the Chinese economy. *Review of Economics and Statistics*. 69: 261–267.
16. Hirschman A. O. (1958). *The Strategy of Economic Development*. New Haven: Yale University Press.
17. Hughes D. (2003). Policy uses of economic multiplier and impact analysis. *Choices*. Second quarter: 25–29.
18. Hughes D. and V. Litz (1996). Rural-urban economic linkages for agriculture and food processing in the Munroe, Louisiana, functional economic area. *Journal of Agricultural and Applied Economics*. 28: 337–355.
19. Jensen, R. C., T. D. Mandeville and N. D. Karunaratne (1979). *Regional Economic Planning: Generation of Regional Input-Output Analysis*. London: Croom Helm.
20. Johns, P. M. and P. M. K. Leat (1987). The application of modified GRIT input-output procedures to rural development analysis in Grampian region. *Journal of Agricultural Economics*. 38: 245–256.
21. Johnson T. G. and D. M. Otto. (1993). *Microcomputer-Based Input-Output Modeling: Applications to Economic Development*. Westview Press, Boulder, CO.
22. Jordan J. L. and R. Brooks (1984). IO/EAM: an input-output economic assessment model. *Southern Journal of Agricultural Economics*. 16: 145–149.
23. Leones J., G. Schulter and G. Goldman (1994). Redefining agriculture in interindustry analysis. *American Journal of Agricultural Economics*. 76: 1123–1129.
24. Mattas, K. (2005). Measuring policy impacts on rural regions. *New Medit*, (Editorial), 4: 2.
25. Mattas, K., C. Fotopoulos, V. Tzouvelekas, S. Loizou and K. Polymeros (1999). The dynamics of crop sectors in regional development: the case of tobacco. *International Advances in Economic Research*. 5: 255–268.
26. Mattas K., E. Loizou, S. Rozakis and V. Tzouvelekas (2006a). Agricultural modelling: an input-output approach. In: Ferretti F. (eds.), *Leaves and Cigarettes: Modelling The Tobacco Industry: With Applications to Italy and Greece*. Franco Angeli Milano Italy.
27. Mattas K., S. Loizou, V. Tzouvelekas, M. Tsakiri and Bonfiglio A. (2006b). Deriving regional I-O tables and multipliers. In: Bonfiglio A., Esposti R. and Sotte F. (eds.), *Rural Balkans and EU Integration: An Input-Output Approach*. Franco Angeli Milano Italy.
28. Mattas K. and A. Pagoulatos (1990). Determining differential sectoral impacts of investments. *European Review of Agricultural Economics*. 17: 495–502.
29. Mattas K. and C. Shrestha (1989). The food sector and economic growth. *Food Policy*. 14: 67–72.

30. Mattas K. and C. Shrestha (1991). A new approach to determining sectoral priorities in an economy: input-output elasticities. *Applied Economics*. 23: 247–254.
31. McCann P. and J. Dewhurst (1998). Regional size, industrial location and input-output expenditure coefficients. *Regional Studies*. 32: 435–444.
32. Midmore P. (1991). Input-output and agriculture: a review. In: Midmore P. (eds.), *Input-Output Models in the Agricultural Sector*. Aldershot: Gower: 1–20.
33. Midmore P. (1993). Input-output forecasting of regional agricultural policy impacts. *Journal of Agricultural Economics*. 44: 284–300.
34. Midmore P. and L. Harrison-Mayfield (1996). *Rural Economic Modeling: An Input-Output Approach*. Wallingford: CAB International.
35. Midmore P. and L. Harrison-Mayfield (1996). Rural economic modeling: multi-sectoral approaches. In: Midmore P. and Harrison-Mayfield L. (eds.), *Rural Economic Modeling: An Input-Output Approach*. Wallingford: CAB International.
36. Midmore P., R. Medcalfe and L. Harrisson-Mayfield (1997). Regional input-output analysis and agriculture. *Cahiers d'Economie et Sociologie Rurales*. 42–43: 8–31.
37. Midmore P., M. Munday and A. Roberts (2006). Assessing industry linkages using regional input–output tables. *Regional Studies*. 40: 329–343.
38. Miller R. E. and P. D. Blair (1985). *Input-Output Analysis: Foundations and Extensions*. Prentice Hall, Englewood Cliffs, NJ.
39. Oosterhaven J. and D. Stelder (2002). Net multipliers avoid exaggerating impacts: with a bi-regional illustration for the Dutch transportation sector. *Journal of Regional Science*. 42: 533–543.
40. Papadas C. and D. Dahl (1999). Supply-Driven input-output multipliers. *Journal of Agricultural Economics*. 50: 269–285.
41. Pasinetti L. (1973). The notion of vertical integration in economic analysis. *Metroeconomics*. 25: 1–29.
42. Rasmussen P. N. (1956). *Studies in Intersectoral Relations*. Amsterdam: North-Holland.
43. Roberts D. (1994). A modified Leontief model for analyzing the impact of milk quotas on the wider economy. *Journal of Agricultural Economics*. 45: 90–101.
44. Schaffer W. (1999). Regional impact models. In: Loveridge, S. (ed.), *The Web Book of Regional Science*. Morgantown, WV: Regional Research Institute, West Virginia University. Available at www.rri.wvu.edu/regscweb.htm.
45. Sharma K., P. Leung and S. Nakamoto (1999). Accounting for the linkages of agriculture in Hawaii's economy with an Input-Output model: a final demand-based approach. *The Annals of Regional Science*. 33: 123–140.
46. Skolka J. (1986). Input-output multipliers and linkages. 8th International Conference on Input-Output Techniques, Sapporo, Japan, July 28 to August 2.
47. Sraffa P. (1960). *Production of Commodities by Means of Consumption*. Cambridge: Cambridge University Press.
48. Tiller K., B. English and J. Menard (2004). Tobacco buyout legislation: economic impacts in the Southeast. Paper presented at the Southern Agricultural Economics Association (SAEA), Tulsa, Oklahoma, February 17, 2004.
49. Tzouvelekas V. and K. Mattas (1999). Tourism and agro-food as a growth stimulus to a rural economy: the Mediterranean island of Crete. *Journal of Applied Input-Output Analysis*. 5: 69–81.

Modeling in Nutrient Sensing for Agricultural and Environmental Applications

Won Suk Lee, Ismail Bogrekci, and Min Min

Abstract This chapter describes applications of modeling in nutrient prediction, such as nitrogen (N) for citrus production and phosphorus (P) for agricultural and environmental purposes. Heavy reliance on agricultural chemicals has raised many environmental and economic concerns. Some of the environmental concerns include the presence of agricultural chemicals in groundwater and eutrophication in lakes due to excessive nutrients. To prevent groundwater contamination or eutrophication in lakes, excess use of chemicals should be avoided. Timely and efficient supplies of nutrients for agricultural production are also essential for high yield and profit.

Nitrogen is an essential nutrient for growing crops and is also a concern in maintaining a healthy environment. It is well-known that excess P entering a lake from surrounding agricultural fields causes many problems, such as periodic algal blooms and displacement of native ecosystems. Currently, N and P concentrations are measured from samples obtained in agricultural fields through standard laboratory analysis procedures, which are very time consuming, costly, and labor intensive.

Real-time sensing systems using N and P prediction models will enable cost-effective nutrient detection, which will greatly decrease the time and labor required for monitoring nutrient levels in crops and in tributaries of lakes. Citrus tissue samples are acquired from commercial groves at different times of the year and at different stages of growth. Soil samples are obtained from different locations in drainage basins of lakes. Reflectance spectra of samples are measured in the ultraviolet, visible, and near-infrared regions. Nutrient concentrations in the samples are correlated with the absorbance of the same samples.

Prediction models are developed using different statistical methods, such as stepwise multiple linear regression (SMLR) and partial least squares (PLS) regression. Then, N and P concentrations in unknown samples are determined nondestructively from reflectance spectra of the samples. Such prediction could

W.S. Lee (✉)
Department of Agricultural and Biological Engineering, University of Florida,
Gainesville, FL, USA
e-mail: wslee@ufl.edu

P.J. Papajorgji, P.M. Pardalos (eds.), *Advances in Modeling Agricultural Systems*,
DOI 10.1007/978-0-387-75181-8_14, © Springer Science+Business Media, LLC 2009

be used to better assess the effectiveness of best management practices for fertilizers. The sensor systems are combined with a differential Global Positioning System (DGPS) receiver, and they can generate a nutrient concentration map of the entire citrus grove or lake drainage basin under investigation.

1 Introduction

Soil nitrogen (N) and phosphorus (P) are important nutrients for crop production, however excessive nutrients oftentimes cause environmental concerns. Based on the U.S. EPA's report, nutrients were among the two leading causes of water quality degradation in various water bodies throughout the nation [21]. Agriculture has been found to be the primary source of non-point contamination of both surface and groundwater quality [32]. Farmers often apply higher rates of fertilizers as they perceive such a practice as low-cost insurance against crop failure.

Nitrate and phosphates have been reported to be the primary nutrients of concern affecting surface and groundwater quality. Excessive nitrates in groundwater (when used as a drinking water source) can cause human health problems, and excessive nitrates and/or phosphates can result in eutrophication of surface water and imbalances of native ecosystems. The EPA currently has a maximum contaminant level (MCL) of 10 mg/L (as nitrogen) for nitrate in drinking water [33]. Potential health effects from ingestion of contaminated water are listed as, "Infants below the age of six months who drink water containing nitrate in excess of the MCL could become seriously ill and, if untreated, may die. Symptoms include shortness of breath and blue-baby syndrome."

Eutrophication caused by high concentrations of nutrients has become a major environmental concern in recent years. The element most commonly associated with eutrophication of freshwater ponds and lakes is phosphorus [14, 23]. Excessive algal growth reduces the amount of dissolved oxygen available to other aquatic life. This reduction of oxygen, as well as the development of toxic algal blooms, decreases biotic diversity in and around water bodies, thus effectively destroying an environmental niche and making restoration of the area extremely difficult [14]. Nitrogen is the most important and essential nutrient among agricultural chemicals for growing crops, and it is also the nutrient element that causes the most concern for maintaining a healthy environment. N is an integral part of chlorophyll, which is the primary absorber of light energy needed for photosynthesis. If N is used properly in conjunction with other necessary soil nutrients, it can speed crop growth and increase yields.

In the past century, Lake Okeechobee in Florida has been significantly affected by agriculture and water management practices when most of the land surrounding the lake was used for agricultural purposes. Over a long period of time, storm water and irrigation runoff from agricultural land uses in the basin caused large quantities of phosphorus to be released into the lake.

As a consequence, a large increase in the lake's phosphorus level occurred, which affected the water quality and ecological health in the lake [1, 11]. Animal waste from dairy farms in the watershed was the most important source contributing to high phosphorus loads in the lake [1, 21].

Currently, N and P concentrations are measured from samples obtained in agricultural fields through standard laboratory analysis procedures, which are very time consuming, costly, and labor intensive [12, 24]. Therefore, a real-time sensor system is critically needed that could detect P and N concentrations faster, easier, and for less cost.

One practical technique to assess soil and plant nutrient status is near-infrared (NIR) spectroscopy due to its conceivable use for rapid and nondestructive determination of the concentration of certain constituents in a sample while providing significant labor and cost savings. NIR spectroscopy has been extensively used for many agricultural applications, such as water and nutrient stress sensing for agricultural crops [2, 5, 25, 26, 28, 30] and weed detection [34]. The electromagnetic spectrum of a soil and leaf sample holds information about the sample such as water content, nutrients, particle size, disease, and so forth.

This chapter describes modeling applications in nutrient prediction, such as predicting concentrations of N for citrus production and of P for agricultural and environmental purposes. Real-time sensing systems using N and P prediction models will enable cost-effective nutrient detection, which will greatly decrease the time and labor required for monitoring nutrient levels in crops and in lake tributaries.

2 Statistical Modeling

In this section, we will examine two of the most commonly used modeling methods for nutrient sensing; that is, partial least squares regression and stepwise multiple linear regression.

2.1 Partial Least Squares Regression Analysis

The partial least squares (PLS) analysis procedure was developed first in chemometrics to evaluate multivariate data [17, 37], and it was an extension of the econometric path modeling established during the 1970s by Wold [36]. PLS regression has gained importance in many fields of chemistry, such as analytical, physical, and clinical chemistry, and in industrial process control [13].

The procedure is widely used in multivariate calibration [18], because another goal of the procedure is to take into account the variation in the predictors and improve the prediction of new observations during the validation process [22]. After the unknown structure of the data is detected with a PLS model, the model can be used to predict new values for a new data set from the same population. The PLS procedure is also a good alternative to the traditional multiple regression analysis and principal component regression, because the model parameters obtained by PLS are more robust, which means the parameters do not change extensively when new calibration samples are submitted to the model [13].

PLS regression is a technique that generalizes and combines features of principal component regression (PCR) and multiple regressions. It is particularly useful when trying to predict a set of dependent variables from a very large set of independent variables [15]. PLS regression is a method for constructing predictive models that can handle many independent variables, even when they are highly correlated. It can also relate the set of independent variables to a set of multiple dependent variables. The purpose of this method is to predict response variables and not necessarily to understand the underlying relationship between the variables [31]. In situations where there are many independent variables explaining a dependent factor, PLS can be an appropriate multivariate technique to use [29].

PLS is a two-step process: first, reduction of the matrix-dimension or determination of the number of relevant components [15], and second, identification of latent structure models in the data matrix [16]. In contrast with PCR, which chooses factors that explain the maximum variance in predictor variables without considering the response variables, the PLS method balances the two objectives, seeking the factors that explain both response and predictor variations [22]. Techniques implemented in the PLS procedure work by extracting successive linear combinations of the predictors and optimally explaining response variation and predictor variation.

The ordinary least squares procedure is not applicable when one or more columns of the matrix data can be expressed as a linear combination of other data columns of the same matrix, a condition called *collinearity*. In such cases of high collinearity, unbiased predictors have high variance that generally leads to unrealistic regression coefficients [4]. The technique uses steps to find the best model, using the residual error of each step as an input of the next [4, 27]. An optimal number of factors is generally obtained by a cross-validation procedure, where the calibration set is submitted to "leave-one-out" comparisons until the predicted residual sum of squares (PRESS) is at a minimum [29]. At this point, the model has an optimal number of factors. A smaller PRESS value indicates a better model prediction.

To identify important wavelengths for predicting nutrient concentrations from reflectance spectra, X-loadings and X-weights are usually used [10]. In PLS, X-loadings represent the common variations in the spectral data, and X-weights represent the changes in the spectra that correspond with the regression constituents. Wavelengths having high loadings or weights could be considered as important wavelengths for developing prediction models. X-weights (w) and X-loadings (p) are calculated using the following equations:

$$t = X_0 w \tag{1}$$

$$p' = (t't)^{-1} t' X_0 \tag{2}$$

where t is a score vector, and X_0 is the centered and scaled matrix of predictors.

However, because X-loadings and X-weights do not directly reflect the relationship between predictors and responses, the B-matrix, from the traditional regression equation $Y = XB$, is used to give the accumulated picture of the most important wavelengths. A B-matrix can be calculated from the PLS loadings and weights, Eq. (3). Wavelengths with a high B value contribute more to a calibration model and could be considered to be important wavelengths:

$$B = w(p'w)^{-1} q' \tag{3}$$

where w is X-weight, p is X-loading, and q is Y-weight. The number of factors to be extracted is set by the user.

2.2 Stepwise Multiple Linear Regression

Multiple linear regression is a procedure used to develop a model between two or more explanatory variables and a response variable by fitting a linear equation to observed data. In multiple linear regressions, many methods are used to select the best model. Stepwise multiple linear regression (SMLR) is one of the most widely used methods for selecting the best subset of independent variables. There are three main selection methods; that is, forward selection, backward elimination, and stepwise selection.

In the forward selection method, the model starts with only the constant, and variables are added to the model one at a time. At each step, each variable that is not already in the model is tested for inclusion in the model. The most significant of the tested variables is identified and added to the model, as long as its p value is below a user-specified level. The selection process continues until no remaining variables have F statistic p-values below the specified significance level. A variable remains in the model once it has been added in the model [3].

Backward elimination starts with a model that includes all variables. Each variable in the current model is tested with the F statistic, and any variable with a p value larger than a user-specified significance level is removed from the model. This step is continued until all variables remaining in the model have p values higher than the specified significance level. A variable cannot be added again once it is removed from the model.

Stepwise selection is an improved version of forward regression that permits reexamination at every step of the variables incorporated in the model in previous steps. Each forward selection step, with a user-defined significance level (α), can be followed by one or more backward elimination steps with a significance level of 0.1. The stepwise selection process terminates if no further variables can be added to the model or if the variable just entered into the model is the only variable removed in the subsequent backward elimination.

SMLR has been reported by Card et al. [9] to have good capabilities for wavelength selection. In SMLR, overfitting could be a problem, because too many wavelengths might be selected by the stepwise procedure. In the examples in this chapter, the number of wavelengths for the calibration model was selected based on the model that yielded the best coefficient of determination (R^2) for the validation data set. The "Stepwise" option in the SAS REG procedure was used to select wavelengths by SMLR.

2.3 Prediction Models

Using the statistical methods explained above, prediction models are developed. Usually the whole data set can be divided into two parts: calibration (or training) and validation data sets. For the calibration data set, one-half or two-thirds of the whole data set can be used. In both cases, the remaining data will be used for validation purposes.

In our examples, we use two different cases for dividing the data set: one example uses half for calibration and half for validation, and the other uses two-thirds for calibration and one-third for validation. Note that the prediction model should be robust enough to be used for other data sets. In order to increase robustness of the prediction models, a wide range of data points should be included in the calibration data set.

Now let's examine two different application examples, one for P sensing for soil and the other for N prediction for citrus leaves, as illustrated in [6, 7, 8] and [18, 19].

3 Modeling Application – Example 1: Phosphorus Sensing for Soil

3.1 Soil Sampling and Reflectance Measurement

Soil samples were obtained in November 2002 from two dairy farms (Mac Arthur and Palaez Ranch) and in June 2003 from three farms (Candler Farm, Palaez Ranch South, and Woody Larson) located in the Lake Okeechobee basins in Okeechobee County, Florida. The sampling locations were selected using the experience and expert knowledge of the University of Florida, Institute of Food and Agricultural Sciences (IFAS) extension service officers on the current P status of the Lake Okeechobee basin, focusing on obtaining soil samples with a wide range of P concentrations from very low to high levels. From the soil surface at a depth of 0–15 cm, a total of 68 and 150 soil samples were collected in 2002 and 2003, respectively. Each sample location was recorded with a differential Global Positioning System (GPS) receiver (March-II, Corvallis Microtechnology Inc., Corvallis, Oregon). Each soil sample weighed about 200–400 g. The collected soil samples were preserved on wet ice until delivery to a laboratory to measure reflectance. Soil samples were oven-dried at 104°C for 24 hours. The dried soil samples were ground and sieved using an 0.6-mm sieve (mesh number 30, U.S. standards). The soil samples were sent to a laboratory for chemical analysis of P concentrations. All soil samples were analyzed for Mehlich-1 extractable P. Soil samples were dried and ground in order to remove the effect of water content and particle size of the soil samples on reflectance measurements. For the purposes of this discussion, the samples obtained from the field were named *wet* samples, and the oven-dried samples were named *dry* samples.

A spectrophotometer (Cary 500 Scan UV-VIS-NIR, Varian Inc., Walnut Creek, California) equipped with an integrating sphere (DRA-CA-5500, Lab-sphere North Sutton, NH) was used to collect spectral reflectance data for each soil sample. About 30 g of soil sample was placed in sample holders. A set of two sample holders (one for a reference and one for a sample) was used. The reference was used to collect baseline (background) data, and the sample cell was used to hold the soil samples for reflectance measurements. Reflectance was measured for each soil sample in 400–2500 nm with an increment of 1 nm. The spectral signature for each soil sample was collected using a baseline correction mode. Reflectance of the soil samples was measured before and after drying.

3.2 Data Analysis

Reflectances of all samples were converted into absorbance before further analysis in order to find the relationship between P concentrations and the absorption of light at different wavelengths using Beer–Lambert's law [35]. The data was filtered using a Savitzky–Golay polynomial convolution filter to remove the noise in the signal. The data set was divided into two groups to have a calibration set and validation set. For each year, two-thirds of the data set was randomly selected and used for calibration, and the remaining one-third was used for validation.

A correlation coefficient spectrum was calculated for each wavelength to find highly correlated wavelength bands using the SAS CORR procedure. In order to develop a prediction model for P concentrations in the soil samples, PLS with an option of cross-validation and SMLR were applied to the data sets using the SAS PLS and REG procedures with a stepwise selection option, respectively. The standard error of prediction (SEP) for the wet and dry soil samples was computed using the following equation in order to observe the prediction performance of the calibration models obtained from the PLS regression:

$$SEP(\%) = \sqrt{\frac{1}{n-1}\sum_{i=1}^{n}(e_i - \bar{e})^2} \qquad (4)$$

where n = number of samples, p = number of independent variables in calibration model, e_i = difference between actual concentration and predicted concentration in the ith sample, and $\bar{e}$ = mean of e_i.

3.3 Results and Discussion

Table 1 shows chemical analysis results for the samples obtained in five different locations. The results show that a wide range of P concentrations were obtained so that a robust prediction model could be developed.

Figure 1 shows a typical absorbance curve of the soil samples. Two water absorption bands at 1450 nm and 1940 nm are very distinct. Figure 2 shows

Table 1 Mehlich-1 P analysis results of soil samples

Year	Field	Number of Samples	Range (mg/kg)	Mean (mg/kg)	Standard Deviation (mg/kg)	Coefficient of Variation (%)
2002	Mac Arthur	40	2–1391	267.3	380.9	142.5
	Palaez Ranch	28	2–7	3.2	1.4	41.2
2003	Candler	60	14–2701	1329.3	889.1	66.9
	Palaez Ranch South	60	4–174	28.6	31.1	108.7
	Woody Larson	30	4–159	43.7	42.0	96.1

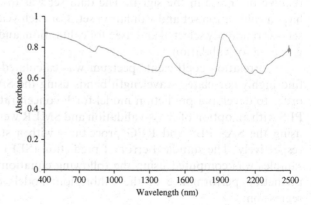

Fig. 1 Absorbance of a soil sample

Fig. 2 Correlation
coefficients between
absorbance and P
concentrations at each
wavelength for the samples
obtained in 2002

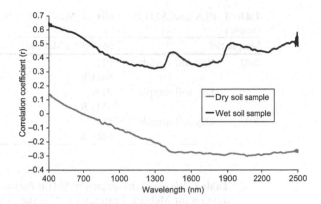

correlation coefficients between P concentrations and absorbance at each wavelength for the samples obtained in 2002. Wet soil samples show higher correlation coefficients than do dry samples.

In Table 2, the percent variation explained by factors extracted from the PLS procedure is shown for the wet soil samples: 99.9% of the spectra can be explained by the calibration model, and 70.0% of P concentrations can be presented by the calibration model with five factors for the wet soil samples. This step indicates that the calibration model could be successfully used for validation.

Tables 3 and 4 show the results from PLS and SMLR modeling. Generally, the PLS method predicted better (higher R^2 values) than did SMLR for all data sets. For the 2002 data set, dry samples yielded higher R^2 values than did wet samples, due to the fact that the effects of water and particle size were removed from the samples. The SMLR method selected wavelengths of 400, 414, 467, 933, 940, 941, and 1666 nm for the wet samples and 408, 416, 2065, 2068, 2069, 2494, and 2496 nm for the dry samples. However, the R^2 values were very low (0.38 and 0.10), implying that the prediction models might have been significantly overfitted to the calibration data set.

Figure 3 shows the relationship between the predicted and actual P concentrations for the wet soil samples in the 2002 validation data set. The prediction model for the wet soil samples showed that the prediction model

Table 2 Percent variation explained by PLS factors for the wet soil samples in the calibration data set

Number of Extracted Factors	Model Effects		Dependent Variables	
	Current (%)	Total (%)	Current (%)	Total (%)
1	89.4	89.4	23.8	23.8
2	6.4	95.9	15.4	39.2
3	3.0	98.9	7.1	46.3
4	0.9	99.8	6.9	53.2
5	0.06	99.9	16.8	70.0

Table 3 PLS and SMLR results of Mehlich-1 extractable P for soil samples

Sample Set		Statistical Method	R^2 for Validation
2002	Wet soil sample	PLS	0.78
		SMLR	0.38
	Dry soil sample	PLS	0.89
		SMLR	0.10
2003	Dry soil sample	PLS	0.91
		SMLR	0.85

Table 4 Wavelengths selected by SMLR for the calibration data set for Mehlich-1 extractable P for the 2002 samples

Sample Set	Selected Wavelengths (nm)
Wet soil sample	400, 414, 467, 933, 940, 941, 1666
Dry soil sample	408, 416, 2065, 2068, 2069, 2494, 2496

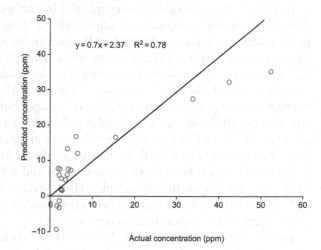

Fig. 3 PLS prediction results of P concentrations for the wet soil samples in the 2002 validation set

underestimated P concentrations from 33 to 52 mg/kg. For the dry soil samples in the validation data set, the R^2 value between predicted and actual P concentrations was computed to be 0.89. The calibration model for the dry soil samples was found to be underestimating the P concentrations in the validation data set from 15 to 52 mg/kg. The standard error of prediction (SEP) was 6.71 mg/kg for the wet soil samples, and the SEP was 4.93 mg/kg for the dry soil sample in the validation data sets. Figure 4 shows the validation results for the 2003 data set. This prediction model produced the highest R^2 value among all analyses.

Figure 5 shows actual and PLS predicted P concentration maps in Candler Farm. The predicted map is not exactly the same as the actual map, however it

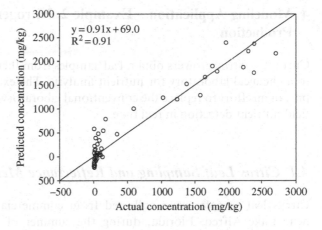

Fig. 4 PLS prediction results of P concentrations for the dry soil samples in the 2003 validation set

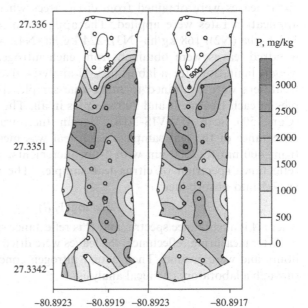

Fig. 5 Maps of actual (*left*) and PLS predicted (*right*) phosphorus concentration in Candler Farm

seems to follow the trend of different P concentrations at different locations in the field.

The overall correlation coefficients for the dry soil samples were lower than those of the wet soil samples. In addition, stepwise multiple linear regression results showed that the R^2 value for the wet soil samples was higher than that of the dry soil samples in the validation data sets. However, the PLS procedure produced higher R^2 values for both the wet and dry soil samples for the validation data sets. The SEP values for the PLS procedure validation for the dry and wet soil samples seem to be reasonable to determine the low to high P areas in a field.

4 Modeling Application – Example 2: Nitrogen Sensing for Citrus Production

Currently, citrus growers obtain leaf samples from their groves and send them to a chemical laboratory for nutrient analysis. The example shown below is a part of an effort to replace the conventional laboratory leaf analysis with an in-field nutrient detection in real time.

4.1 Citrus Leaf Sampling and Reflectance Measurement

Citrus leaf samples were collected from commercial citrus groves located near Lake Alfred, Florida, during the summer of 2002. A total of 1000 citrus leaves were obtained from citrus trees where five different nitrogen application rates were applied. The application rates were 0 kg/ha (N1), 112 kg/ha (N2), 168 kg/ha (N3), 224 kg/ha (N4), and 280 kg/ha (N5). Two hundred leaves were obtained from each nitrogen treatment. Because of weight limitations for a laboratory N analysis, five leaves from the same N treatment were combined to make one sample. Thus, there were 40 samples in each treatment and 200 samples in all. The same spectrophotometer (Cary 500 Scan UV-VIS-NIR, Varian Inc.) was used to measure the reflectance of the leaf samples. Each leaf was measured for its reflectance from 400 nm to 2500 nm with 1-nm increments. Figure 6 shows a typical reflectance spectrum of citrus leaf samples. The spectra, then, were converted into absorbance:

$$A = \log(1/R) \qquad (4)$$

where A is absorbance spectra, and R is reflectance spectra.

After measuring reflectance, the leaves were dried in an oven at 104C for 24 hours and were ground. Their actual nitrogen concentrations were obtained through a laboratory chemical analysis.

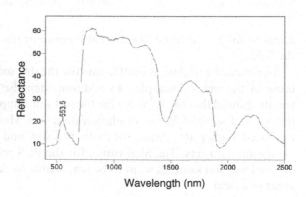

Fig. 6 Reflectance of a citrus leaf sample

4.2 Data Analysis

Stepwise multiple linear regression (SMLR) and PLS analyses were conducted to determine important wavelengths and develop prediction models. The whole data set was separated into two parts. The training data set included 100 samples (20 samples from each N treatment), which were a half of a total of 200 samples. The B-matrix, X-loadings, and X-weights were calculated with the training data set. The remaining 100 samples were used as a validation data set.

To find the important wavelengths, the simplest method is to compute correlation coefficients between absorbance spectra at each wavelength with the actual N concentration of the samples. Wavelength regions showing high correlation are regions that should be selected, while regions showing low or no correlation should be ignored. The SAS CORR procedure was used to calculate correlation coefficients (r).

To develop prediction models, the first method (Method I) used was to apply SMLR to the entire spectral region (415–2485 nm). Considering that wavelengths having high correlation coefficients (r) with N concentration may contribute more to prediction models (Fig. 7), a second method (Method II) was used, which involved applying SMLR only to spectral regions where $|r| > 0.5$. Because spectral reflectance was measured with 1-nm increments, the number of variables (2071 wavelengths) was much larger than the number of samples in the training data set (100 samples), and the resulting high collinearity could make the stepwise procedure unstable. To reduce collinearity, absorbance values of every 20 wavelengths were averaged into one variable, and the number of variables was reduced from 2071 to 103. Then, a third method (Method III) was used, which involved applying SMLR to these 103 variables of the averaged data set with a reduced spectral resolution of 20 nm. The SAS REG procedure was used to conduct the SMLR.

Partial least squares (PLS) regression was conducted, and X-loadings and X-weights were calculated using the SAS PLS procedure. Wavelengths having

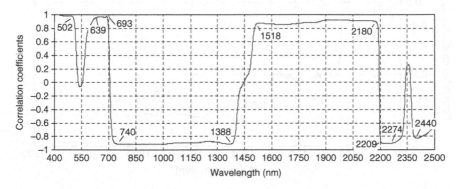

Fig. 7 Example correlation coefficient spectrum between absorbance at 415 nm and absorbance at all other wavelengths

high loadings or weights could be considered important wavelengths for developing prediction models. The SAS PLS procedure was applied to the entire spectral region (415–2485 nm), with the maximum number of PLS factors set to 15. A cross-validation option (CVTEST) in PLS was used to select the optimum number of factors.

4.3 Results and Discussion

In Table 5, the results of the N concentration analysis of the citrus leaf samples are shown. The nitrogen concentration in citrus leaves showed a trend similar to the actual N application rates applied to the groves where the samples were collected.

Figure 8 shows the correlation coefficient spectrum between absorbance and actual N concentration of the calibration samples at each wavelength. Absorbances in some regions were highly correlated with the N concentration. The wavelength range near 550 nm, a well-known N absorption band [5, 30], showed a peak correlation coefficient of 0.51. The highest correlation coefficient in the visible range, 0.56, was seen at 707 nm.

Table 5 Results of nitrogen concentration analysis of the samples from five different nitrogen application rates

N treatment	N1	N2	N3	N4	N5
Actual N application (kg/ha)	0	112	168	224	280
Number of samples	40	40	40	40	40
Average (%)	2.32	2.81	2.83	2.97	3.10
Standard deviation (%)	0.14	0.10	0.13	0.14	0.14
Min (%)	1.99	2.59	2.48	2.56	2.78
Max (%)	2.57	3.02	3.15	3.37	3.38

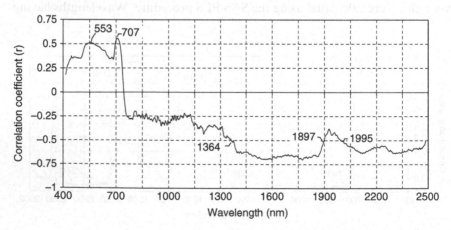

Fig. 8 Correlation coefficients between absorbance at each wavelength and leaf nitrogen concentration of the calibration data set

Table 6 shows the analysis results from the SMLR and PLS procedures. The results show a good relationship between the actual N concentration and the predicted value for both the training and validation data sets. Method I selected 10 wavelengths with an R^2 of 0.816 for the validation data set. Method II used wavelength ranges where $|r| > 0.5$, which were 535–572 nm, 696–721 nm, 1364–1897 nm, and 1995–2485 nm, and it selected five wavelengths. The R^2 of Method II for the validation data set was 0.743. Method III selected nine 20-nm-wide averaged wavebands and showed the best R^2 of the three SMLR methods, because the averaged data set reduced the collinearity and made the SMLR analysis more reliable. The R^2 of Method III for the validation data set was 0.839. The SEP was also the lowest among the three different SMLR prediction models. Table 6 also shows the results obtained with the PLS procedure, resulting in an R^2 of 0.793 and 0.132% for SEP for the validation data set.

Table 7 shows the number of factors, the corresponding percent variation of the variables, and the PRESS achieved by the PLS regression. The smallest error was found at the seventh factor with a PRESS value of 0.44 ($p > 0.1$). These seven factors accounted for 98.8% of the model variation and 88.9% of the variation of the dependent variables.

Table 6 Results of SMLR and PLS analysis

Method	Selected Wavelengths (nm)	R^2 Calibration	Validation	SEP (%)
SMLR I	448, 669, 719, 1377, 1773, 1793, 1834, 2000, 2101, 2231	0.916	0.816	0.126
SMLR II	719, 1773, 1810, 2101, 2231	0.864	0.743	0.145
SMLR III	695–714, 815–834, 1335–1354, 1375–1394, 1555–1574, 2195–2214, 2215–2234, 2375–2394, 2455–2474	0.914	0.839	0.115
PLS	415–2485	0.886	0.793	0.132

Table 7 Percent variation explained by the seven factors in the PLS regression

Number of Extracted Factors	Model Effects Current (%)	Total (%)	Dependent Variables Current (%)	Total (%)	Root Mean PRESS
1	51.7	51.7	52.1	52.1	0.77
2	13.7	65.4	7.5	59.6	0.74
3	7.4	72.7	5.8	65.4	0.69
4	23.0	95.7	1.7	67.1	0.68
5	2.1	97.9	7.0	74.2	0.61
6	0.6	98.4	11.6	85.8	0.47
7	0.4	98.8	2.6	88.9	0.44

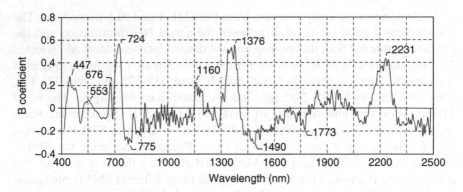

Fig. 9 B-matrix determined from the training data set using PLS with 7 factors

Figure 9 illustrates the B coefficients for each wavelength calculated by the PLS procedure with seven factors. Wavelengths of 447 nm, 676 nm, 724 nm, 775 nm, 1160 nm, 1376 nm, 1490 nm, and 2231 nm had $|B| > 0.2$ and therefore were considered to contribute more information to the calibration model. Wavelengths at 448 nm, 669 nm, 719 nm, 1377 nm, 1773 nm, and 2231 nm determined in SMLR Method I matched with the peaks in the B-matrix very well, an additional indication of their possible importance. Wavelengths of 1377 nm, 1773 nm, and 2231 nm have not been reported in previous research, which mostly concentrated on wavelengths in the 400–1100 nm range.

Comparing the SMLR and PLS procedures, both worked very well in this example application. SMLR worked especially well for the averaged data set that had low collinearity. Models developed using SMLR used fewer wavelengths (5–10) than those developed using PLS. The best prediction results, generated by SMLR Method III, are shown in Fig. 10. The PLS procedure

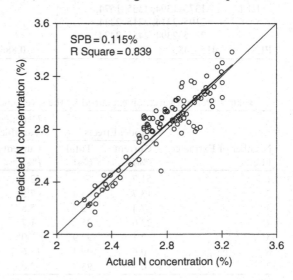

Fig. 10 N concentration prediction using SMLR Method III with $R^2 = 0.839$

appears to be superior to the SMLR procedure for full spectrum analysis due to its ability to compress data. The seven factors from the PLS regression were fewer than the 10 wavelengths selected by SMLR Method I, and the R^2 (0.828) generated by the PLS analysis was better than that of SMLR Method I (0.816).

5 Conclusions

In this chapter, we have examined two different statistical modeling methods with two example applications related to agricultural systems. The P sensing example showed that SMLR and PLS analyses could be used successfully for developing prediction models of P concentration using spectral reflectance, indicating promising results toward the development of a P-sensor system using visible and NIR spectroscopy.

The results from the N prediction modeling example also illustrated that an SMLR or PLS algorithm could be used to develop good prediction models for N detection for citrus production. The N prediction model could be used for on-the-go variable rate fertilizer application. Overall, the prediction models can be used to develop in-field nutrient sensing systems to replace standard laboratory analysis procedures and to maintain the quality of the environment by applying only necessary amounts of fertilizers at different locations.

References

1. Allen, L., Jr. 1987. Dairy-sitting criteria and other options for waste water management on high-table soils. Soil & Crop Science Society 47: 108–127.
2. Bausch, W.C., Diker, K., Goetz, A. F. H., Curtis, B. 1998. Hyperspectral characteristics of nitrogen deficient corn. ASAE Paper No. 983061. St. Joseph, MI: ASAE.
3. Beal, D.J. 2005. SAS code to select the best multiple linear regression model for multi-variate data using information criteria. In: Proceedings of Southeast SAS Users Group Conference. SESUG, Portsmouth, VA.
4. Björkström, A., Sundberg, R. 1999. A generalized view on continuum regression. Scandinavian Journal of Statistics 26: 17–30.
5. Blackmer, T.M., Schepers, J.S., Varvel, G.E. 1994. Light reflectance compared with other nitrogen stress measurements in corn leaves. Agronomy Journal 86: 934–938.
6. Bogrekci, I., Lee, W.S., Herrera, J. 2003. Assessment of P-concentrations in the Lake Okeechobee drainage basins with spectroscopic reflectance of VIS and NIR. ASAE paper No. 031139. St. Joseph, MI: ASAE.
7. Bogrekci, I., Lee, W.S., Herrera, J.P. 2004. Spectral signatures for the Lake Okeechobee soils using UV-VIS-NIR spectroscopy and prediction phosphorus concentrations. ASAE Paper No. 041076. St. Joseph, MI: ASAE.
8. Bogrekci, I., Lee, W.S. 2005. Spectral phosphorus mapping using diffuse reflectance of soils and grass. Biosystems Engineering 91(3): 305–312.
9. Card, D.H., Peterson, D.L., Matson, P.A., Aber, J.D. 1988. Prediction of leaf chemistry by the use of visible and near infrared reflectance spectroscopy. Remote Sensing of Environment 26: 123–147.

10. Esbensen, K.H. 2002. Multivariabte Data Analysis in Practice. 5th ed. Woodbridge, NJ: CAMO Technologies.
11. Federico, A., Dickson, K., Kratzer, C., Davis, F. 1981. Lake Okeechobee water quality studies and eutrophication assessment. Tech. Publ. 81-2. S. Florida Water Management District., West Palm Beach, FL.
12. Gain, S. 1997. An optimized network for phosphorus load monitoring for Lake Okeechobee, Florida. USGS Report 97-4011.
13. Geladi, P., Kowalski, B.R. 1986. Partial least squares regression: a tutorial. Analytica Chimica Acta 185: 1–17.
14. Harper, D. 1992. Eutrophication of Freshwater. Principles, Problems and Restoration. New York: Chapman and Hall.
15. Helland, I.S. 2001. Some theoretical aspects of partial least squares regression. Chemometrics and Intelligent Laboratory Systems 58: 97–107.
16. Lingærde, O.L., Christophersen, N. 2000. Shrinkage structure of partial least squares. Scandinavian Journal of Statistics 27: 459–473.
17. Martens, H. 2001. Reliable and relevant modeling of real world data: a personal account of the development of PLS regression. Chemometrics and Intelligent Laboratory Systems 58: 85–95.
18. Martens, H., Næs, T. 1989. Multivariate Calibration. Chichester, England: John Wiley & Sons.
19. Min, M., Lee, W.S. 2003. Spectral-based nitrogen sensing for citrus. ASAE paper No. 031137. St. Joseph, MI: ASAE.
20. Min, M., Lee, W.S. 2005. Determination of significant wavelengths and prediction of nitrogen content for citrus. Translation ASAE 48(2): 455–461.
21. Nair, V., Graetz, D., Portier, K. 1995. Forms of phosphorus in soil profiles from dairies of south Florida. Soil Science Society of America 59: 1244–1249.
22. SAS Institute Inc. 1999. SAS/STAT Users Guide. The PLS Procedure. Version 8. Cary, NC: SAS Institute, pp. 2691–2734.
23. Sharpley, A. 1999. Agricultural phosphorus and eutrophication. Environmental Quality 24: 947–951.
24. South Florida Water Management District (SFWMD). 2003. Lake Okeechobee. Available at: www.sfwmd.gov. Accessed August 2003.
25. Stafford, J.V., Weaving, G.S., Lowe, J.C. 1989a. A portable infra-red moisture meter for agricultural and food materials: Part 1, Instrument development. Journal of Agricultural Engineering Research 43: 45–56.
26. Stafford, J.V., Bull, C.R., Weaving, G.S. 1989b. A portable infra-red moisture meter for agricultural and food materials: Part 2, Field evaluation on grass. Journal of Agricultural Engineering Research 43: 57–66.
27. Stoica, P., Söderström, T. 1998. Partial least squares: a first-order analysis. Scandinavian Journal of Statistics 25:17–24.
28. Sudduth, K.A., Hummel, J.W. 1996. Geographic operating range evaluation of a NIR soil sensor. Translation ASAE 39(5): 1599–1604.
29. Sundberg, R. 1999. Multivariate calibration – Direct and indirect regression methodology. Scandinavian Journal of Statistics 26: 161–207.
30. Thomas, J.R., Oerther, G.F. 1972. Estimating nitrogen content of sweet pepper leaves by reflectance measurements. Agronomy Journal 64: 11–13.
31. Tobias, R.D. 2000. TS-509: An introduction to partial least squares regression. Cary, NC: SAS Institute Inc.
32. U.S. EPA. 1994. National water quality inventory. 1992 Report to Congress. US EPA 841-R-94-001. Office of Water. Washington, DC: U.S. Government Printing Office.
33. U.S. EPA. 2006. List of drinking water contaminants & MCLs. Available at: www.epa.gov/safewater/mcl.html#1. Accessed 15 August 2006.

34. Wang, N., Zhang, N., Peterson, D.E., Dowell, F.E. 2000. Testing of a spectral-based weed sensor. ASAE Paper No. 003127. St. Joseph, MI: ASAE.
35. Williams, P., Norris, K. 2001. Near-Infrared Technology in the Agricultural and Food Industries. 2nd ed. St. Paul, MI: American Association of Cereal Chemists, Inc.
36. Wold, H. 1975. Soft modeling by latent variables: the nonlinear iterative partial least squares approach. In Perspectives in Probability and Statistics. Papers in Honor of M.S. Barlett. J. Gani, ed. New York: Academic Press, pp. 117–142.
37. Wold, S., Martens, H., Wold, H. 1983. The multivariate calibration method in chemistry solved by the PLS method. In: Proceedings of the Conference in Matrix Pencils. A. Ruhe and B. Kagstrom, eds. Heidelberg, Germany: Springer-Verlag, pp. 286–293.

Estimation of Land Surface Parameters Through Modeling Inversion of Earth Observation Optical Data

Guido D'Urso, Susana Gomez, Francesco Vuolo, and Luigi Dini

Abstract Earth observation (EO) optical data represent one of the main sources of information in the retrieval of land surface parameters (i.e., leaf area index and surface albedo). These parameters are widely used in research and applications in agriculture for improving water and land resources management, especially in the field of precision farming, to monitor crop status, predict crop yield, detect disease and insect infestations, and support the management of farming tasks. During recent years, the technical capabilities of airborne and satellite remote sensing imagery have been improved to include hyperspectral and multiangular observations. In parallel with the advancement of observation techniques, there has been an important development in the study of the interaction of solar radiation with Earth's surface. This process can be described by using canopy reflectance models of different complexity, which can also be used in operative applications in the field of agricultural water and land management. As such, enhanced EO data and canopy reflectance models can be combined together to reduce the empiricism of traditional methods based on simplified approaches and to increase the estimation accuracy.

In this chapter, the application of numerical inversion techniques to a canopy reflectance model is investigated both in the spectral and angular domains. An example of a case study is reported, concerning the estimation of leaf area index in an agricultural site; multidirectional and hyperspectral data, acquired by means of the Compact High Resolution Imager (CHRIS) onboard the Project for On-Board Autonomy (PROBA) platform of the European Space Agency, have been used for the numerical inversion of the canopy reflectance model.

1 Introduction

The biosphere is one of the main components of the earth's system as it regulates exchanges of energy and mass fluxes at the soil, vegetation, and atmosphere level. The vegetation cover, through its active role in the carbon and water cycles,

G. D'Urso (✉)
Department Agricultural Engineering and Agronomy, University of Naples
Federico II, Naples, Italy
e-mail: durso@unina.it

P.J. Papajorgji, P.M. Pardalos (eds.), *Advances in Modeling Agricultural Systems*,
DOI 10.1007/978-0-387-75181-8_15, © Springer Science+Business Media, LLC 2009

largely influences climatic and hydrological processes, which can be modeled at different spatial scales. At continental scales, the interaction between land surface and atmosphere are described by means of global circulation models (GCMs), which can be applied to analyze and predict the impact of land surface processes on climate dynamics. At regional scales, hydrological models are widely used to evaluate the production of run-off after precipitation events, the recharge of aquifers, the amount of soil evaporation and transpiration, and other processes related to the management of soil and water resources. In the context of agricultural water management, the exchange of water in the soil–plant–atmosphere continuum is the basis for the assessment of crop water use, which can be used for improving the allocation of water resources in irrigated areas. The increase of water scarcity – especially for agriculture, which accounts for more than 70% of freshwater consumption from lakes, rivers, and groundwater – requires a sustainable management of this limited resource. The development of canopy also plays a critical role on a much smaller scale, as in precision farming, not only to determine crop water use, but also to reduce the environmental impact of fertilization and to predict crop yield.

All the mentioned processes and phenomena are influenced by the status of vegetation, which can be described by biophysical parameters related to plant morphology, such as the leaf area index (indicated by the acronym LAI and expressing the amount of foliage in m^2 per unit soil surface), the leaf angle distribution (LAD), the surface albedo, the height and the roughness of vegetated surfaces [38]. For example, the currently most used approach for estimating crop water requirement, also suggested by the Food and Agriculture Organization (FAO) [13], is based on the knowledge of LAI, surface albedo, and crop height. These biophysical parameters can be determined from *in situ* measurements in the case of very localized studies (i.e., at the scale of 1–10 hectares); however, when repetitive measurements are needed over larger heterogeneous areas, field campaigns do not represent a feasible solution in most cases.

Remote sensing from space – or Earth observation (EO) – has shown its potential in the detection of biophysical parameters of vegetation [40], in particular LAI, which is the most relevant parameter in many land surface processes. Several studies have provided evidence that the value of LAI, as well as the leaf optical properties, their spatial distribution and orientation with respect to the illumination and viewing angles, affects the way the vegetated surface interacts with solar radiation (i.e., the surface reflectance that we measure from E.O. platforms) [32, 5]. It follows that we can analyze EO data in the solar spectrum (400–1000 nm) to estimate the vegetation parameters of the observed surface through a model of canopy reflectance.

A canopy reflectance (CR) model describes the transfer and interaction of electromagnetic radiation inside the canopy based on physical laws. A CR model can be used in the direct and inverse modes. In direct mode, the canopy reflectance is simulated based on a set of equations describing the transfer of solar radiation within the canopy and the specific set of (biophysical) input parameters. In inverse mode, canopy reflectance from EO data is used to compute these biophysical parameters, solving what is known as an inverse parameter identification optimization problem.

In recent years, there have been two main developments: (1) the availability of new generation of sensors, with enhanced spectral and spatial resolution and angular viewing possibilities, to measure surface reflectance; (2) the detailed knowledge of the reflectance behavior of complex vegetated surfaces, through a comprehensive set of CR models, which can take into account the anisotropic reflectance behavior of vegetation canopies.

Simultaneous directional observations today available from experimental satellite platforms, such as the CHRIS/PROBA [6], can better characterize the anisotropic reflectance of vegetation canopies, thus avoiding inaccuracies due to single-angle parameter estimation approaches provided by former satellites.

Thus, in exploiting the unique and innovative hyperspectral and multiangular information content of CHRIS/PROBA imagery, the aim of this research is to estimate LAI by inverting a CR model and to present some numerical results obtained with this methodology in an agricultural area in Spain.

2 Statement of the Problem: EO Data and CR Modeling

Much progress has been achieved from the pioneer intuition of Monteith, who wrote in the Annual Report of Rothamsted Experimental Station in 1957, the year of Sputnik, the following sentence: ". . . measurements of reflection coefficients may give useful estimates of leaf growth without destructive sampling . . . ."

For more than two decades, starting from the first EO satellite, the Landsat-1, up to the latest very-high-resolution sensors (2.8 m), a feasible approach to estimate vegetation parameters, and in particular LAI, has been based on empirical relationships using as data the nadir-viewing measurements of canopy reflectance in the red and infrared bands. Formulas for different vegetation indexes (VIs) have been defined based on these measurements from Landsat-like type of satellites, and it has been evidenced that the amount of vegetation, and thus the LAI parameter, is correlated with the value of such VIs [4, 9, 11].

However, this approach does not consider other parameters, except LAI, to influence the spectral response of canopy, such as the canopy structure, the leaf spatial distribution, row orientation and spacing, leaf and soil optical properties, and the sun–target–sensor geometry. As a result, the canopy reflectance has to be considered as anisotropic, and hence we cannot observe the canopy from one angle only, like in the case of VIs, as when observing an ideally homogeneous reflecting surface (Lambertian assumption) [23]. Moreover, VIs, being based only on red and infrared wavelength regions (bands), have too limited spectral information to fully explain how different foliage constituents (i.e., pigments, leaf water content, chlorophyll) influence the canopy reflectance.

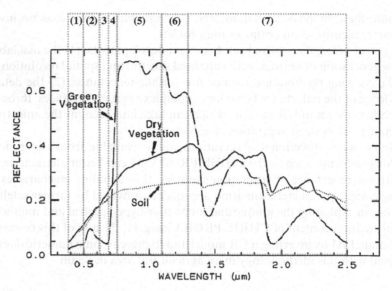

Fig. 1 Typical spectrum reflectance of vegetation

If we consider the typical reflectance spectrum of vegetation (Fig. 1), we can distinguish at least five primary and two transition waveband regions in the interval between 0.4 and 2.5 μm, where differences in leaf optical properties (scattering and absorption) and the background optical properties have an important role in determining the canopy spectral reflectance [40]. These waveband regions are (1) 0.4–0.5 μm, where strong spectral absorption by the chlorophyll and the carotenoids occurs; (2) 0.5–0.62 μm, where we find the maximum reflectance in the visible range (i.e., why green vegetation to our eyes appears "green"); (3) 0.62–0.7 μm, where strong chlorophyll absorption occurs; (4) 0.70–0.74 μm, where strong absorption ceases; (5) 0.74–1.1 μm, where minimal absorption occurs and the leaf scattering mechanisms result in high levels of spectral reflectance; (6) 1.1–1.3 μm, where the liquid-water coefficients of absorption increase from close to 0 at 1.1 μm to values of 4 at 1.3 μm; and (7) 1.3–2.5 μm, where absorption by liquid water occurs.

Concluding, near-nadir-looking VIs are a simple means to obtain an initial approximation of LAI, but they are too limited both in the spectral and angular domain to be used for an accurate estimation of LAI [43].

As we have seen above, the reflectance of vegetation canopies is a complex function in these two domains, defined as *bidirectional reflectance distribution functions* (BRDFs) [35]. The BRDF can be mathematically described by using CR models, which can be applied in *inverse mode* for estimating the LAI; to this end, multispectral (or even hyperspectral) data with simultaneous observations at different viewing angles are needed.

On one hand, the canopy reflectance model allows a higher validity because there are less restrictive assumptions compared with the empirical relations, and it allows one to consider the anisotropy of the surface. However, any model aiming to reproduce real processes needs to make simplifications, and in this sense, they are not totally accurate.

On the other hand, we need not only a sufficient sampling on the surface BRDF to have enough data but also good-quality data meaning accurate sensor calibrations and corrections for the presence of atmospheric effects.

Also, the inversion optimization algorithms [25] have to be both fast and robust enough to avoid problems in the minimization of the error function and may have to use regularization techniques to get stable solutions.

In the following sections, we will address these three aspects of the problem.

3 The PROSPECT–SAILH Canopy Reflectance Model

Since the first CR model was formulated by Monsi [33], a large number of physically based CR models with increasing levels of complexity have been developed to reproduce the BRDF of vegetated surfaces. Recent reviews have been given (see [8, 26], following [16, 34]). The most complex approach considers a three-dimensional geometric description of the distribution and orientation of canopy elements; although computationally intensive, these models accurately simulate within-canopy spatial heterogeneity (e.g., organ size distributions, leaf clumping, gaps) and scene-scale heterogeneity (e.g., topography) that other models must either neglect or approximate with quasi-empirical formulations. Several models have been developed (see [17]), but their high computational demand limits their applicability for inversion purposes.

Referring to the objective of this work, we have considered the *PSH model*, which is the combination of two widely used algorithms: PROSPECT [24] for the simulation of individual leaf reflectance and SAILH [41, 42] for the simulation of the canopy BRDF.

The combination of the PROSPECT and SAILH models has been selected on the basis of the results of the RAMI (radiation transfer model intercomparison) experiment [37], which carried out a benchmarking of different CR modeling techniques. Figure 2 shows the schematization of the coupled PROSPECT and SAILH models, integrating the required input information.

The PROSPECT model [24] provides the leaf hemispherical reflectance and transmittance (ρ, τ) to the SAILH model as a function of:

- The leaf structural parameter (N) [34], which mainly influences the near-infrared spectral region; it typically ranges between 1 and 2.5.
- The leaf chlorophyll a + b concentration (Chl_{a+b}) ($\mu g/cm^2$), which affects the reflectance and transmittance in the visible (400–700 nm).
- The equivalent water thickness (C_w) (g/cm^2), which takes into account light absorption by leaf water content in the middle infrared (1100–2500 nm).
- The dry matter content (C_m) (g/cm^2), which is responsible for light absorption between 800 and 2500 nm.

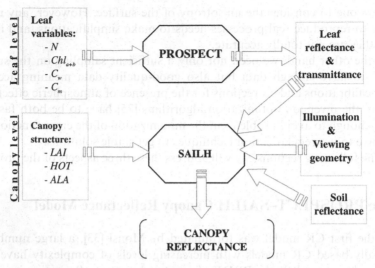

Fig. 2 Schematization of the direct mode of PROSPECT and SAILH models (PSH model)

For details on the PROSPECT model, we refer to the cited literature.

The SAILH (Scattering Arbitrary Inclined Leaves) model [41, 42] assumes the canopy as a horizontal, homogenous, and infinitely extended vegetation layer (turbid medium), made up of leaves randomly distributed within the canopy. The radiative transfer equation is solved by the four-stream approximation method: ascending and descending fluxes of direct and diffuse radiation are considered. The SAILH model requires the following parameters:

- The leaf reflectance and transmittance (ρ, τ), provided by PROSPECT;
- The leaf area index (LAI);
- The average leaf angle (ALA);
- The illumination and observation geometric parameters, such as the solar zenith angle, the view zenith angles, and the azimuth angle between Sun and observer;
- The "hot-spot" parameter (Hot), representing the increase of reflectance when illumination and observation angles are coincident;
- The fraction of diffuse radiation;
- The soil hemispherical reflectance (ρ_{soil}).

The SAILH model is basically represented by a system of four differential equations, relating the variation of radiant fluxes along the vertical direction in a canopy to LAI:

$$\frac{1}{LAI}\frac{d}{dx}\begin{pmatrix} E_s \\ \mathbf{E}^- \\ \mathbf{E}^+ \\ E_o \end{pmatrix} = \begin{pmatrix} k & & & \\ -\mathbf{s}' & \mathbf{A} & -\mathbf{B} & \\ \mathbf{s} & \mathbf{B} & -\mathbf{A} & \\ w & \mathbf{v}^T & \mathbf{v}'^T & -K \end{pmatrix}\begin{pmatrix} E_s \\ \mathbf{E}^- \\ \mathbf{E}^+ \\ E_o \end{pmatrix} \tag{1}$$

In Eq. (1), x is the optical height coordinate, usually taken to vary between -1 and 0 (at the top of the canopy), for the differentiation of radiant fluxes; that is, the direct solar incoming flux E_s, the vector of diffuse incoming solar radiation $\mathbf{E}^-$, the vector of diffuse outcoming (from the canopy) radiation $\mathbf{E}^+$, and the radiant flux in the observer (the sensor) direction E_o. The matrix on the second member of Eq. (1) contains both in scalar and in matrix form the extinction and scattering coefficients, which are related to the other parameters introduced above.

Two fundamental criteria have been considered to choose PSH in this study: (i) simplicity (i.e., the possibility to have a rather good representation of the canopy reflectance using a relatively small amount of input parameters as well as limited computational requirements), and (ii) reliability, as the SAILH model has been successfully tested for a large set of crops, among them corn [29] and sugar beet [1], which are also considered in the case study presented herein.

4 Experimental Data Acquisition

4.1 Test-Site Description and Ground Measurements

The ESA Spectra Barrax Campaigns (SPARC) [31] were carried out in Barrax (N30°3′, W2°6′), an agriculture test area situated within La Mancha region in the south of Spain, from July 12 to 14, 2003, and from July 14 to 16, 2004.

Agricultural remote sensing research has been concentrated in this area for many years thanks to its flat topography (differences in elevation range up to 2 m only) and to the presence of large and uniform vegetation fields (e.g., alfalfa, corn, potato, sugar beet, onion, and garlic; Fig. 3) with a wide range of LAI

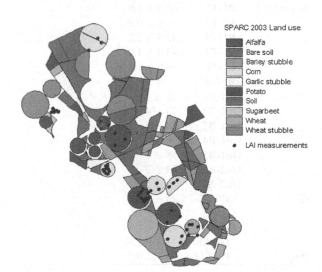

Fig. 3 Land-use classification during SPARC-2003 and LAI measurement points

from 0.5 up to 6.5. The groundwater table is located approximately 20–30 m below the land surface. The typical climatic conditions of the Mediterranean area can be found here with precipitation concentrated during spring and autumn and hot, dry summers. The annual average rainfall is about 400 mm.

During the SPARC campaign, a large amount of ground measurements were collected in the Barrax study area, covering leaf water content C_w, leaf biomass C_m, chlorophyll content Chl_{a+b}, and LAI [14]. Measurements of dry matter and water content were carried out on three samples per each elementary sampling unit (ESU). The leaf chlorophyll content was measured with the Chlorophyll Content Meter 200 (Opti-Science, USA), performing relative measurements, calibrated in the laboratory at a later stage [15]. Dry matter and water content range values obtained from field data campaign were respectively 40–190 and 100–800 g/m^2. Chlorophyll range values obtained from field crop measurements were between 20 and 50 $\mu g/cm^2$. A summary of the biophysical parameters measured for the characterization of the different crops during SPARC-2003 is shown in Table 1.

For this study, LAI measurements were made by means of the digital analyzer LAI-2000 (LI-COR Biosciences, USA) [28]. Each single LAI value considered in this study was the result of the average of 24 measurements taken randomly within an ESU area of approximately 700 m^2. The resulting mean LAI value was 3.01, the minimum was 0.6, and the maximum was 6.3.

Table 1 Biophysical characterization of crops during SPARC-2003

Crop	ESU	ChlorophyllMean Value(μg cm^{-2})	Crop MeanValues(μg cm^{-2})
Corn	C1-A1	48.9 ± 0.5	50.6 ± 0.8
	C1-A2	51.6 ± 0.5	
	C1-A3	50.4 ± 0.4	
Sugar beat	B1-T1	44.9 ± 1.0	44.3 ± 1.4
	B1-T2	48.6 ± 0.9	
	B1-T3	42.5 ± 1.2	
	B1-T4	43.4 ± 1.0	
	B1-T5	38.8 ± 1.0	
	B1-T6	47.4 ± 0.8	
Onion	On1-B1	23 ± 3	20 ± 2
	On1-A1	23 ± 4	
	On1-B2	20 ± 3	
	On1-B4	16 ± 3	
	On1-B5	18 ± 3	
	On2-A1	22 ± 3	
	On2-A2	6 ± 2	
Garlic	G1-A7	20 ± 2	15 ± 2
	G1-A8	15.0 ± 1.6	
	G1-A9	12.5 ± 1.7	
	G1-A10	11.0 ± 1.6	
Potato	P1-T12	36.7 ± 0.7	35.6 ± 0.5

4.2 Earth Observation Data: CHRIS/PROBA Imagery

The European Space Agency (ESA) Project for On-Board Autonomy (PROBA) is a technology demonstration experiment to take advantage of autonomous pointing capabilities of a generic platform suitable for EO purposes. Among different sensors, the PROBA instrument payload includes the Compact High Resolution Imaging Spectrometer (CHRIS) [6]. The coupled CHRIS/PROBA system, launched on October 22, 2001, provides high spatial resolution hyperspectral and multiangular data, which constitutes a new generation of remote sensing information to be processed and exploited.

The platform acquires the images at times when the zenith angle of the platform with respect to the fly-by position is equal to a set of Fly-by Zenith Angles (FZA): 0, ±36, or ±55. Negative MZA (Minimum Zenith Angles) values correspond with target locations east of the ground track and negative FZAs with acquisition geometries when the satellite already passed over the target position. On the other hand, CHRIS measures over the visible/near-infrared (NIR) bands from 400 to 1050 nm, with a spectral sampling interval ranging between 1.25 (at 400 nm) and 11 nm (at 1000 nm). It can operate in different modes in order to find the best combination between spatial resolution and the number of spectral channels, considering the limits to onboard storage.

For this study, we worked with a multiangular hyperspectral CHRIS/PROBA image, as shown in Fig. 4, collected on July 14, 2003, at 11:32 GMT from five different view angles during a single orbital overpass. These images (namely A1, A2, A3, A4, A5) were acquired in Mode-1 with a spectral resolution of 62 bands over the visible/near-infrared bands from 400 to 1050 nm in a spatial resolution of 34 m. The acquisition geometry for the images is shown in Table 2. Radiometric calibration and atmospheric and geometric correction of CHRIS imagery was preliminarily performed; in particular, a dedicated atmospheric correction algorithm has been applied jointly with radiometric calibration of the data [21]. Examples of the BRDF features of two different crops are shown in Fig. 5a, b.

Table 2 CHRIS/PROBA acquisition geometry

	Minimum		satellite	zenith	angle
14/07/	A1	A2	A3	A4	A5
2003	57,3	42,4	27,6	42,5	57,4

5 Canopy Reflectance Model Inversion

5.1 An Inverse Ill-Posed Problem

By means of model inversion, information on surface canopy parameters can be obtained from observed EO data. According to the least-square error method, a

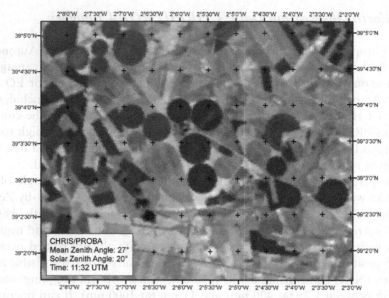

Fig. 4 CHRIS/PROBA imagery over Barrax, July 14, 2003

cost function is defined to provide a measure of "goodness of fit" so that the optimal values of the parameters x^* are found to minimize the error between the observed and the computed reflectance that depend on x as follows:

$$\min_x \left[F(x) = \frac{1}{2} \sum_{i=1}^{n} \omega_i \left[\rho_{\text{obs},i} - \rho_{\text{mod},i}(x) \right]^2 = \frac{1}{2} \| r(x) \|^2 \right] \tag{2}$$

Subject to upper and lower bound constraints (LB, UB), defined though *a priori* information,

$$LB \leq x \leq UB$$

where $\rho_{\text{obs},i}$ and $\rho_{\text{mod},i}$ are the observed canopy reflectance for n spectral bands and the corresponding modeled canopy reflectance, respectively. ω_i is the weight given to an observation. Each residual r_i should be weighted by the uncertainty associated with the observation i.

If the model accurately represents canopy reflectance, and there are no measurement errors, the value of the cost function should approach zero.

However, the inversion of CR models is by nature an ill-posed problem. Then, it may happen that the relation between canopy reflectance and the set of input canopy parameters is not unique (i.e., there is more than one minimum). That means different optimal model parameter combinations can produce almost identical spectra [2, 3, 5, 10, 12], for instance, demonstrated that the spectral reflectance of sparse canopy with mostly horizontal leaf orientation is similar to a dense canopy with mostly vertical leaf orientation. The average deviations

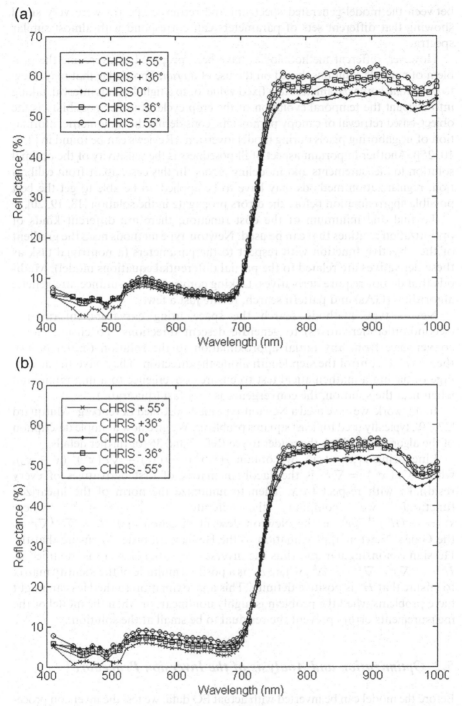

Fig. 5 (a) CHRIS/PROBA spectral and directional reflectance acquired over alfalfa (Barrax site on July 14, 2003, at 11:30 GMT). (b) CHRIS/PROBA spectral and directional reflectance acquired over corn (Barrax site on July 14, 2003, at 11:30 GMT)

between the model-generated spectrum and retrieved spectra were very small showing that different sets of parameters can correspond with almost similar spectra.

However, different methodologies have been proposed to overcome the problem of multiple solutions, based on the use of *a priori* knowledge that may allow to constraint some parameters to a fixed value, or to a tight bounds interval, taking into account the temporal evolution of the crop cycle. Another possibility is the object-based retrieval of canopy parameters, considering the radiometric information of neighboring pixels during model inversion. (Reviews can be found in [2, 3, 10, 25]). Another important aspect of ill-posedness is the sensitivity of the optimal solution to measurements and modeling errors. In this case, apart from calibration, regularization methods may have to be applied, to be able to get the best possible approximation before the errors propagate in the solution [18, 19, 20].

To find one minimum of the cost function, there are different kinds of optimization routines that can be used. Newton-type methods need the gradient of the objective function with respect to the parameters (a nontrivial task as these derivatives are related to the partial differential equations model). Methods that do not require derivatives, lacking convergence assurance, are genetic algorithms (GAs) and pattern search, to cite just a few.

Newton-type methods search the space using curvature information (second-order derivatives) to generate descent directions. To ensure global convergence from any initial approximation to the solution (initial point), they have to control the step length along the direction. They have the advantage of having a mathematical test to ensure convergence to a minimum, and when near the solution, the convergence is very fast (quadratic).

In this work, we have used a Newton-type method, called Levenberg–Marquard [27, 30], typically used for least squares problems. We give now a simple description of the algorithm and ask the reader to see Refs. 7 and 36 for further details.

Linearizing $r(x)$ at x^k, we obtain $\tilde{r}(x, x^k) = r(x^k) + \nabla r(x^k)^T(x - x^k)$, where $\nabla r(x^k) = \nabla r^k$ is the Jacobian matrix of first derivatives of every residual r with respect to x. Then to minimize the norm of the linearized function $\tilde{r}$, we compute the iterates $x^{k+1} = x^k + s^k$, where $s^k = -(H^k)^{-1} \nabla r^k$ is the Newton descent direction and $H^k = \nabla r^{kT} \nabla r^k$ is the Gauss–Newton approximation to the Hessian. In order to ensure that the Hessian is nonsingular and thus the inverse exists, the Hessian is modified as $\tilde{H}^k = \nabla r^{k\,T} \nabla r^k + \Delta^k$, where Δ^k is a positive multiple of the identity matrix to ensure that $\tilde{H}^k$ is positive definite. This approximation to the Hessian might have problems when the problem is highly nonlinear, or when the model or the measurements errors prevent the residual to be small at the solution.

5.2 Optimization and Analysis of the Inversion Procedure

Before the model can be inverted with actual EO data, we test the inversion procedure by using synthetically generated data (which are "error-free" by definition).

To test the inverse methodology, we have created three sets of synthetic data; for each one of them, the canopy BRDF has been calculated by using the model in forward mode for three different LAI values (0.5, 3.0, and 5.0). The other parameters were considered fixed, as described in Table 3. Also, the lower and upper bounds used of the inversion of these parameters are given here. Then, the minimum of the cost function value, the number of cost function evaluations, and the inversion time have been determined.

To test if there are several minima with good match to the data, we started the inversion process from different initial parameter values: first, by using as initial values the set of parameters used in the generation of the synthetic data (forward model simulation) except for the LAI parameter, which was set up equal to 3.0 (when the true values were LAI = 0.5 or LAI = 5) and to 1.5 (when it was LAI = 3.0); then, by using the lower and the upper bounds, LB and UB, respectively. Results for the estimation of LAI_{est} are presented in Table 4, considering initially 62 spectral bands and then successively only 4 (similar to the one of Landsat-TM satellite), and 5 view directions. In the table is also shown the number of function evaluations (f-count), the optimization time in seconds (Time), and the minimum of the cost function $f(x)$.

When the value of LAI is 0.5, the inversion results showed in general good accuracy for LAI estimation. When using 62 spectral bands and 4 spectral bands, the cost function values are of the order 10^{-5}, except when the inversion process is started from the upper bound for 62 bands. In this case, the optimization routine was not able to get an optimal solution. The cost function value was high, $f(x) =$ 0.4, despite the good accuracy in the estimated LAI = 0.49. To overcome this problem, a RE-START was performed to the L-M routine, using as initial values the set of parameters estimated from the previous optimization. The retrieved LAI was equal to 0.49 with a new minimum of the cost function of 7×10^{-6}.

For LAI = 3.0 and LAI $\doteq$ 5.0, the results are only accurate when the full spectral information is exploited (62 bands) in the inversion of the PSH model. This might be caused by a lower sensitivity of the model to higher LAI values.

The relevance of using 62 or 4 bands and the effect of searching for a low or high LAI is shown in Fig. 6a, b. The error surface for LAI and ALA parameters is plotted considering a cost function based on 62 or 4 bands, respectively.

The error surface is relatively sharp and well-behaved for low LAI (LAI <2) and when more spectral information is provided. It becomes flat for high LAI values and when using only four bands.

Table 3 Set of PSH model parameters used in forward simulation to generate the synthetic data set and the lower and upper bounds used in the inversion

Model Parameters	N	Chl_{a+b}	C_w	C_m	LAI	ALA	Hot	α_{soil}
Synthetic data	1.8	50	0.011	0.0055	var.	57	0.0057	1.0
Lower bound (LB)	1.3	30	0.011	0.001	0.01	30	0.001	0.8
Upper bound (UB)	2.0	70	0.1	0.01	6.00	80	1.000	1.2

Table 4 LAI$_{est}$ estimates for three LAI LAI forward simulations (0.5, 3.0, and 5.0) by using 62 and 4 spectral bands. The optimization process starts from different initial parameter values

True Value	InitialValues	62 Bands: 5 View Directions				4 Bands: 5 View Directions			
		LAI$_{est}$	f-Count	Time	Min $f(x)$	LAI$_{est}$	f-Count	Time	Min$f(x)$
LAI = 0.5	LAI = 3.0	0.49	403	103	6×10^{-6}	0.50	327	66	4×10^{-4}
	LB	0.50	404	102	1×10^{-5}	0.50	533	108	4×10^{-6}
	UB	0.49	848	205	0.4	0.49	318	65	2×10^{-3}
LAI = 3.0	LAI = 1.5	3.00	347	80	6×10^{-5}	3.00	409	81	2×10^{-4}
	LB	2.99	363	92	5×10^{-4}	2.78	426	87	0.0027
	UB	3.00	558	141	1×10^{-6}	2.77	459	93	0.0033
LAI = 5.0	LAI = 3.0	5.00	518	134	2×10^{-6}	4.03	402	84	0.012
	LB	4.99	895	232	2×10^{-6}	4.76	447	89	8×10^{-4}
	UB	5.00	607	157	2×10^{-6}	5.60	517	106	7×10^{-4}

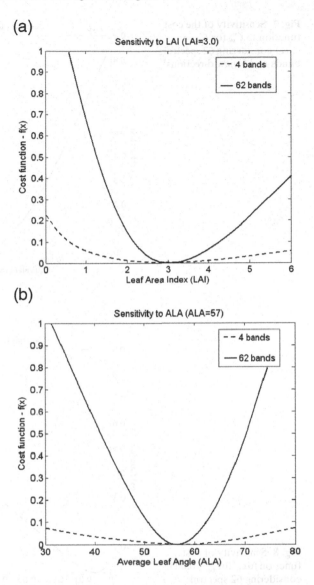

Fig. 6 (a) Sensitivity of the cost function to LAI for LAI = 3.0 considering 62 and 4 spectral bands and 5 view directions. (b) Sensitivity of the cost function to ALA for ALA = 57 and LAI = 3.0 considering 62 and 4 spectral bands and 5 view directions

The sensitivity of the cost function to the parameters C_w and C_m is plotted in Figs. 7 and 8, respectively. As a consequence of the low influence of the leaf water content on the spectral region considered for the analysis, the C_w parameter has a low sensitivity on the cost function, whereas greater influence was shown by C_m parameter (Fig. 8). This consideration can be used in model inversion parameterization and in fixing variables that denote a lower influence on the optimization.

Finally, the inversion was repeated by using 16 bands; the results are shown in Table 5 for LAI = 5.0, starting the optimization from the lower and the upper bounds. The estimated LAI, the cost function value, and the inversion time all

Fig. 7 Sensitivity of the cost function to C_w for $C_w =$ 0.011 considering 62 spectral bands and 5 view directions

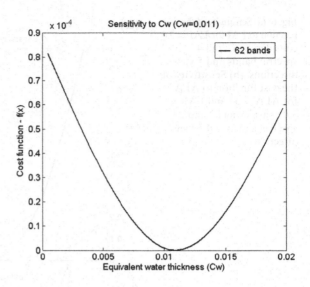

Fig. 8 Sensitivity of cost function to c_m for $_m = 0.0055$ considering 62 spectral bands and 5 view directions

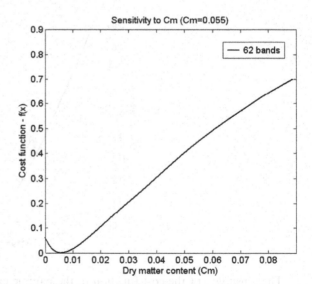

Table 5 LAI estimation accuracy for LAI = 5 by using 16 bands and 5 view directions

	16 Bands: 5 View Directions				Values of Other Parameters Estimated						
	LAI_{est}	f-Count	Time	Min$f(x)$	N	Chl_{a+b}	C_w	C_m	ALA	Hot	α_{soil}
LB	5.12	475	99	7×10^{-4}	1.78	50.29	0.011	0.0056	56.67	0.058	0.959
UB	5.00	758	161	2×10^{-6}	1.80	50.00	0.011	0.0055	56.98	0.057	0.998

Table 6 Estimation accuracy for LAI = 0.5, 3.0, and 5.0 by using 62 bands and 1 view direction

	62 Bands: 1 View Direction			
	LAI_{est}	f-Count	Time	Min $f(x)$
LAI = 0.5	0.38	78	48	0.0015
LAI = 3.0	2.54	73	35	0.043
LAI = 5.0	6.00	932	51	0.0683

indicate that some of the problems reported before by using only four spectral bands can be avoided adding spectral information, until at a certain point after that it starts to be redundant.

The directional information was finally analyzed. The inversions were performed again for the three LAI (0.5, 3.0, and 5.0) values by using only one view angle. The results are shown in Table 6. It was confirmed how in the inversion of model-generated data, the directional information plays an important role on model inversion regularization.

5.3 Inverting PSH Model with Real CHRIS/PROBA Data

As shown in the previous paragraph, the ability to determine canopy parameters correctly depends on the model inversion parameterization, the model accuracy, and on the use of sufficient spectral information. Finally, it depends on appropriate sampling of the surface BRDF.

CHRIS/PROBA data may provide the potential to overcome some of the difficulties associated with physical-based model inversion from satellite data, thanks to its simultaneous spectral and directional sampling. This latter characteristic is of great value to explore the anisotropic reflectance behavior of vegetated surfaces, thus contributing to decouple the counterbalancing effect between LAI and ALA on the spectral signal. In this sense, multidirectional information helps in circumventing the ill-posedness of numerical inversion of PSH (and similar) models. Thus, the only regularization taken into account in this paper has been a physical coherent bound on the parameter values.

For the inversion of the PSH model, we have considered several fields with different crops; in each field a small set of spectrally uniform and adjacent pixels has been selected in the multiangular CHRIS/PROBA image. The corresponding average values of reflectance in each band have been calculated to be used as input data in the inversion process.

Once the reflectance data for each field have been extracted from the CHRIS/PROBA images, the bounds and the initial values of the PSH model parameters have been defined. The N and Hot parameter value bounds have been left as broad as possible as no field measurements techniques are available for them. Chl_{a+b} has been bound in the range 30–70, C_w between 0.015 and 0.1, and C_m between 0.001 and 0.01. The parameters to be retrieved by model

Table 7 Input parameters, units, initial values, and bounds

Parameters	Units	Initial Values	Lower Bound	Upper Bound
N	–	1.3	1.3	2.0
Chl_{a+b}	g/cm^{-2}	30.0	30.0	70.0
C_w	g/cm^{-2}	0.015	0.015	0.100
C_m	g/cm^{-2}	0.001	0.001	0.010
LAI/LAI	–	0.1	0.1	6.5
Hot	–	0.0	0.0	1.0
ALA	Deg.	30	30	80
α_{soil}	–	0.80	0.80	1.20

inversion, LAI and ALA, have been allowed to vary in the range 0.1–6.5 and 30°–80° (starting from the values 0.1° to 30°), respectively. In Table 7, all input parameters with corresponding bounds are listed.

The input soil reflectance [22] is calculated by averaging soil spectra measured in selected field locations during the campaign. A wavelength-independent scaling factor, α_{soil}, has been introduced to consider the variability of soil reflectance due to soil moisture; this parameter has been left variable within a range of ±20% from the mean.

To understand the contribution of directional information in LAI estimation, the PSH model was inverted by using first one (A3), then three (A2, A3, A4), and finally five (A1, A2, A3, A4, A5) view angles.

To reduce redundancy and to understand the contribution of the spectral information on the directionality, the experiment has been first performed with the full spectral 62 CHRIS bands. Successively, the process has been repeated with a selection of optimal spectral bands in the visible (441, 542, 563, 583, 605, 664, 674, and 694 nm), in the red edge (706, 718, 731, 745, and 758 nm), and in the infrared (773, 780, 831, and 889 nm) parts of the spectrum (according to the results of previous works found in literature [39]) and finally by using 4 bands similar to Landsat-TM spectral configuration.

In Figs. 9 and 10, the scatterplot of the measured versus estimated LAI is reported for one and five view angles, respectively.

In case of alfalfa and even more of corn, the LAI estimation accuracy improves for each fixed spectral configuration when we add directional information. Diversely, for each fixed directional configuration, the addition of spectral information does not improve LAI estimation accuracy for alfalfa, but it does for corn; here we have a remarkable increase in estimation accuracy going from 4 to 17 spectral bands; however, the increase is less evident going from 17 to 62 spectral bands.

Considering these results, the contribution of directional information seems to be more important than the spectral dimensionality for the estimation of LAI. Concerning the LAI accuracy analysis for potato crops, the results indicate the impossibility to achieve reasonable values. Referring to LAI-2000 measurements, the LAI was systematically underestimated by using model

Fig. 9 Results of the PSH model inversion by using five view angles. Measured versus estimated LAI for three different canopies: alfalfa, corn, and potato

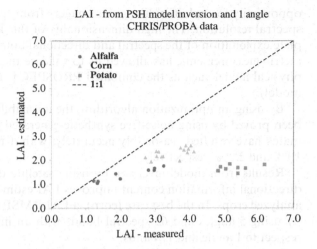

Fig. 10 Results of the PSH model inversion by using five view angles. Measured versus estimated LAI for three different canopies:alfalfa, corn, and potato

inversion. This may be due to two main reasons. Looking at field book notes and photos, it may be related to the agronomic practices of growing potato: during the satellite overpass, the potato field revealed deep grooves, partly filled with water. Perhaps additional restrictions on the soil reflectance should be considered in the model inversion parameterization for this crop.

6 Conclusions

This work aimed to perform LAI estimation from EO data, in particular using data obtained from innovative experimental satellites.

In this context, the CHRIS/PROBA technology demonstration mission and the ESA SPARC campaign offered the scientific community the unique opportunity to exploit multiangular imagery from space with high spatial and spectral resolution. The high dimensionality of this kind of data, with a complete exploitation of the spectral and directional domains of the canopy radiometric measurements, has allowed us to validate the inversion of the complex physical model such as the combined PROSPECT and SAILH models (PSH model).

By using an optimization algorithm, the invertibility of the PSH model has been proved by using noise-free synthetic-generated data. LAI and ALA estimates have resulted reasonably accurately, with a mean percentage error of 10% and 2.2%, respectively.

Results from model inversion by using satellite data have shown that the directional information content improves LAI estimation for two of the three analyzed crops. In the best case (corn), a LAI RMSE of 0.41 has been achieved by using 5 angles and 62 spectral bands with an improvement of 65% with respect to 1 angle and 17 bands.

It seems also that the directional information is predominant on the spectral one, thus suggesting the design of space-borne instruments in the future with better capabilities to sample the surface reflectance anisotropy. By using multiangular EO data, we can improve the estimation accuracies of land surface parameters on a large scale, which are more and more required by the community in several scientific and operative applications.

Acknowledgments S.G. has carried out this work in the framework of GNCS Visiting Professor Program 2007 and of Italy–Mexico Bi-lateral Project 2007–2008 on "Problemi inversi in idrologia: metodologie numeriche innovative."

References

1. Andrieu, B., Baret, F., Jacquemoud, S., Malthus T., Steven, M. (1997). Evaluation of an improved version of SAIL model to simulate bi-directional reflectance of sugar beet canopies. Remote Sensing of Environment, 60, 247–257.
2. Atzberger, C. (2002). Object-based retrieval of structural and biochemical canopy characteristics using SAIL + PROSPECT canopy reflectance model: A numerical experiment. In J. Sobrino (Ed.), Recent Advances in Quantitative Remote Sensing. Universitat de Valencia, Spain 129–138.
3. Atzberger, C. (2004). Object-based retrieval of biophysical canopy variables using neural nets and radiative transfer models. Remote Sensing of Environment, 93, 53–67.
4. Baret, F., Guyot, G., Major, D. (1989). TSAVI: a vegetation index which minimizes soil brightness effects on LAI and APAR estimation. In 12th Canadian Symposium on Remote Sensing and IGARSS'90, Vancouver, Canada, 10–14 July 1989, 4 pp.
5. Baret, F., Guyot, G. (1991). Potentials and limits of vegetation indices for LAI and APAR assessment. Remote Sensing of Environment, 35, 161–173.
6. Barnsley, M.J., Settle, J.J., Cutter, M., Lobb, D., Teston, F. (2004). The PROBA/CHRIS mission: A low-cost smallsat for hyperspectral, multiangle, observations of the earth surface and atmosphere. IEEE Transactions on Geoscience and Remote Sensing, 42(7), 1512–1520.
7. Bertzekas, D.P. (1999). Nonlinear Programming. Athena Scientific. USA.

8. Casa, R. (2003). Multiangular remote sensing of crop canopy structure for plant stress monitoring. Ph.D. thesis. University of Dundee.
9. Clevers, J. (1989). The application of a weighted infrared-red vegetation index for estimating leaf area index by correcting for soil moisture. Remote Sensing of Environment, 29, 25–37.
10. Combal, B., Baret, F., Weiss, M., Trubuil, A., Macé, D., Pragnère, A., et al. (2002). Retrieval of canopy biophysical variables from bidirectional reflectance using prior information to solve the ill-posed inverse problem. Remote Sensing of Environment, 84, 1–15.
11. D'Urso G., Santini, A. (1996). A remote sensing and modeling integrated approach for the management of irrigation distribution systems. In Evapotranspiration and Irrigation Scheduling; Proc. Am. Soc. Agric. Engin. (ASAE), Intern. Workshop, San Antonio, TX, 1996; 435–441.
12. D'Urso, G., Dini, L., Vuolo, F., Guanter, L. (2004). Preliminary analysis and modelling of BRDF on CHRIS/PROBA data from SPARC 2003. Rivista Italiana di Telerilevamento, vol. II, 1015–1020.
13. Food Agriculture Organisation (FAO). (1998). Crop evapotraspiration. Guidelines for computing crop water requirements. Irrigation and Drainage Paper, Rome, Italy, 56.
14. Fernández G., Moreno, J., Gandía, S., Martínez, B., Vuolo, F., Morales, F. (2005). Statistical variability of field measurements of biophysical parameters in SPARC-2003 and SPARC-2004 data campaigns. Proc. of the SPARC Final Workshop, ESA Proceedings WPP-250, Nordwijk, The Netherlands.
15. Gandia, S., Fernández, G., Moreno, J. (2005). Chlorophyll content measurements in the SPARC campaigns. Proc. of the SPARC Final Workshop, ESA Proceedings WPP-250, Nordwijk, The Netherlands.
16. Goel, N.S. (1988). Models of vegetation canopy reflectance, their use in estimation of biophysical parameters from reflectance data. Remote Sensing Reviews, 4, 1–212.
17. Goel, N.S. (1991). Inversion of canopy reflectance models for estimation of biophysical parameters from reflectance data. In: G. Asrar (Ed.), Theory and Applications of Optical Remote Sensing. New York: Wiley Interscience, 205–251.
18. Gómez, S., Gosselin, O., Barker, J. (2001). Gradient-based history—Matching with a global optimization method. Journal of Petroleum Science and Engineering, 200–208.
19. Gómez, S., Pérez A., Dilla, F., Alvarez, R.M. (2002). On the automatic calibration of a confined aquifer. Computational Methods in Water Resources XIV. Dordrecht, The Netherlands: Kluwer Academic Publishers.
20. Gomez, S., Ono, M., Gamio, C., Fraguela, A. (2003). Reconstruction of capacitance tomography images of simulated two-phase flow regimes. Applied Numerical Mathematics, 46, 197–208.
21. Guanter, L., Alonso, L., Moreno, J. (2005). A method for the surface reflectance retrieval from PROBA/CHRIS data over land: application to ESA SPARC campaigns. IEEE Transactions on Geoscience and Remote Sensing, 43(12), 2908–2917.
22. Huete, A.R., Jackson, R.D., Post, D.F. (1985). Spectral response of a plant canopy with different soil background. Remote Sensing of Environment, 17, 37–53.
23. Huete, A.R. (1987). Soil and sun angle interactions on partial canopy spectra. International Journal of Remote Sensing, 8, 1307–1317.
24. Jacquemoud, S., Baret, F. (1990). PROSPECT: a model of leaf optical properties spectra. Remote Sensing of Environment, 34, 75–91.
25. Kimes, D.S., Knyazikhin, Y., Privette, J.L., Abuelgasim, A.A., Gao, F. (2000). Inversion methods for physically-based models. Remote Sensing of Environment, 18, 381–439.
26. Lewis, P., Saich, P. (2006). From Web courses material at the Department of Geography, University College London.
27. Levenberg, K. (1944). A method for the solution of certain problems in least squares. Quarterly of Applied Mathematics, 2, 164–168.

28. LI-COR. (1992). LAI-2000 plant canopy analyzer instruction manual. Lincoln, NE: LI-COR.
29. Major, D., Schaalje, G., Wiegand, C., Blad, B. (1992). Accuracy and sensitivity analysis of SAIL model-predicted reflectance of maize. Remote Sensing of Environment, 41, 61–70.
30. Marquardt, D. (1963). An algorithm for least-squares estimation of nonlinear parameters. SIAM Journal on Applied Mathematics, 11, 431–441.
31. Moreno, J.F., Melix, J., Sobrino, J.A., Martinez-Lozano, J.A., Calpe-Maravilla, J., Calera Belmonte A. (2004). The SPECTRA Barrax Campaign (SPARC): an overview and first results from CHRIS data. Proceedings of the 2nd SPARC Workshop, ESA Proceedings, Nordwijk, The Netherlands.
32. Monteith, J.L. Unsworth, M.H. (1990). Principles of Environmental Physics, 2nd ed. London: Edward Arnold.
33. Monsi, M., Saeki, T. (1954). Uber den Lichtfaktor in den Pflanzengessellschafte und seine Bedeuting fur die Stoffproduktion. Journal of Japanese Botany, 14, 22–52.
34. Myneni, R.B., Ross, J., Asrar, G. (1989). A review on the theory of photon transport in leaf canopies. Agricultural and Forest Meteorology, 45, 1–153.
35. Nicodemus, F.E., Richmond, J.C., Hsia, J.J., Ginsberg, I.W., Limperis, T. (1977). Geometrical Considerations and Nomenclature for Reflectance. NBS Monograph, No. 160. National Bureau of Standards, U.S. Department of Commerce, 52.
36. Nocedal, J., Wright, S.J. (1999). Numerical Optimization. Springer Series in Operations Research. New York: Springer Verlag.
37. Pinty, B., Gobron, N., Widlowski, J.L., Gerstl, S.A.W., Verstraete, M.M., Antunes, M., Bacour, C., Gascon, F., Gastellu, J.-P., Goel, N., Jacquemoud, S., North, P., Qin, W., Thompson, R. (2000). The RAdiation transfer Model Intercomparison (RAMI) exercise. Journal of Geophysical Research. 106, D11: 11937–11956.
38. Sellers, P.J. (1985). Canopy reflectance, photosynthesis and transpiration. International Journal of Remote Sensing 6, 1335–1372.
39. Thenkabail, P.S., Enclona, M.S., Ashton, B., Van Der Meer, F.D. (2004). Accuracy assessments of hyperspectral waveband performance for vegetation analysis applications. Remote Sensing of Environment, 91, 354–376.
40. Tucker, C.J. (1978). A comparison of satellite sensor bands for monitoring vegetation. Photogrammetric Engineering & Remote Sensing 44, 1369–1380.
41. Verhoef, W. (1984). Light scattering by leaf layers with application to canopy reflectance modelling: the SAIL model. Remote Sensing of Environment, 16, 125–141.
42. Verhoef, W. (1998). Theory of radiative transfer models applied in optical remote sensing of vegetation canopies. Ph.D. thesis. National Aerospace Lab., Amsterdam, The Netherlands.
43. Vuolo, F. (2006). Physically based approaches for monitoring vegetation from space. New generation sensors and operational perspectives. PhD Thesis, University of Naples "Federico II."

A Stochastic Dynamic Programming Model for Valuing a Eucalyptus Investment

M. Ricardo Cunha and Dalila B.M.M. Fontes

Abstract This work proposes an exercise-dependent real options model for the valuation and optimal harvest timing of a forestry investment in eucalyptus. Investment in eucalyptus is complex, as trees allow for two cuts without replantation and have a specific time and growth window in which they are suitable for industrial processing into paper pulp. Thus, path dependency in the cutting options is observed, as the moment of exercise of the first option determines the time interval in which the second option may be exercised. Therefore, the value of the second option depends on the history of the state variables rather than on its final value. In addition, the options to abandon the project or convert land to another use, are also considered. The option value is estimated by solving a stochastic dynamic programming model. Results are reported for a case study in the Portuguese eucalyptus forest, which show that price uncertainty postpones the optimal cutting decisions. Moreover, optimal harvesting policies deviate from current practice of forest managers and allow for considerable gains.

1 Introduction

Traditionally, forestry investment decisions were analyzed by using discounted cash flow (DCF) techniques. Given their inability to account for flexibility, these approaches tend to systematically undervalue investments. DCF techniques are based on the assumption that future cash flows follow a constant pattern and can be accurately predicted. The project uncertainty, which may arise from the uncertainty about costs, selling prices, weather and legal conditions, to name but a few, and the flexibility, given to managers to react to changing conditions, are dealt with only superficially [20]. Another disadvantage of DCF is that it is linear and static in nature. It also assumes the project to be either irreversible or, if reversible, only decisions of now-or-never are allowed [10]. In particular, net present value (NPV) takes the project risk into

D.B.M.M. Fontes (✉)
Faculdade de Economia da Universidade do Porto Rua Dr. Roberto Frias, 4200–464 Porto, Portugal
e-mail: Fontes@fep.up.pt

P.J. Papajorgji, P.M. Pardalos (eds.), *Advances in Modeling Agricultural Systems*, DOI 10.1007/978-0-387-75181-8_16, © Springer Science+Business Media, LLC 2009

consideration by discounting expected cash flows to the present moment. This discount considers risk either by using certainty-equivalent cash flows [24] or by discounting cash flows using a risk-adjusted discount rate that can be obtained either through the use of the capital asset pricing model [25] or from a comparison with the market return rates of "similar or risk-equivalent" investments. This last approach is the most generalized among practitioners. Using this decision process, all projects with positive NPV are undertaken, as they provide an "effective" growth in the wealth of the investor (or market value of the firm). Forest investors, namely paper pulp production companies, use the NPV to decide whether to enter or not a specific investment project. The NPV technique, by not accounting for managerial flexibility, provides an underestimate of the project value, relative to real options. Therefore, when there is an option element in an investment, traditional DCF techniques may result in wrong project valuation and hence inadequate decisions (see, e.g., [21] and [10]).

Although the writings of Aristotle in ancient Greece already mention the existence of such features in business and investment, only since the 1970s have applications to natural resources management have known [14, 3]. Many other researchers have treated the question of replacing a forest stand with the same or other land use, the value of both being stochastic (e.g., [1, 23, 30]).

In a real options model of forestry, and particularly in eucalyptus, the basis starting point is that the owner of the forest holds a call option[1] (i.e., an option to acquire timber at an exercise price given by the cost of cutting the timber). This forestry option is similar to what in the finance literature is called an American option, as it can be exercised at any moment within the time interval during which the wood is suitable for pulp and paper production. Traditionally, studies addressing real options in a forest context have applied a single-option approach [18]. Such studies are, however, not suitable for the valuation of multiple harvesting forestry investments as investments in eucalyptus. Our study goes consequently beyond previous studies by considering two cutting options with potential extensions. Therefore, we have to decide when to harvest having in mind that a second harvest is possible and also that the latter must occur within a time window dependent on when the previous harvest has taken place. Unlike the single-rotation problem, the multirotation case represents a path-dependent option, which is far more complex (see [28]). In this case, the option value depends also on the history of the underlying state variables rather than only on their final value. Therefore, the project value today depends on the quantity of timber, which in turn depends on the timing of the previous harvest. In addition, we also consider the options to convert land usage or to abandon project.

Results are provided for a case study involving the investment in eucalyptus pulp production for the Portuguese paper industry, one of the most developed

---

[1] An option is the right, but not the obligation, to take some action in the future under specified terms. A call option gives the holder the right to buy a stock at a specified future date (maturity) by a specified price (exercise or strike price). This option will be exercised (used) if the stock price on that date exceeds the exercise price.

in the world. In this case study, two scenarios are considered: one where eucalyptus wood is sold to pulp and paper companies (base problem) and another where vertical integration is considered, that is, wood is processed into white paper pulpwood, which is then sold (extended problem).

After determining the maximum expected value of the investment, we retrieve the policy associated with such a value. Then, to test the quality of the cutting strategy and its robustness, we apply the strategies obtained to randomly generated data and compare the values obtained with the current practice of forest managers.

2 Literature Review

In 1973 for the first time, Black and Scholes [5] and Merton [19] provided a closed form solution for the equilibrium price of a call option, leading to a new paradigm in asset valuation. The model they developed gave origin to numerous papers and empirical research in contingent claims analysis and to applications to many types of contingent claims. Among these, valuation models for real investments have appeared, giving birth to the real options theory of investment decisions.

To date, not many studies have addressed forestry investments using a real options perspective and, as far as the authors are aware, no previous work has been developed in what concerns investment in eucalyptus forests. Morck et al. [20] developed a contingent claims analysis of a long-term investment in renewable resource investments. They have extended the general model of valuation of natural resources developed for a gold mining application [6]. Their model values a forestry lease and determines the optimal timing to harvest the timber. The value of the lease is considered to be the value of an option to cut down the trees at the best possible timing. In their model, the timber selling price and the inventory of timber are stochastic processes that follow geometric brownian motions. Zinkhan [30] proposes a Black–Scholes type of approach for the valuation of the land use conversion option when valuing timberlands. The conversion option represents the ability of the timberland owner to convert the land use from timber to some other alternative use. Thomson [26] develops this model further in an attempt to increase its consistency and simplicity. In both works, it is assumed that the land owner holds a European option rather than an American one. Bailey [4] proposes a model to value an agricultural producer considering optimal shutting and reopening and volatile output and demand. However, in what concerns forestry, the model is built only for productive trees with periodical (e.g., annual) harvest. This characteristic of the model makes it useless for our purpose of valuation of forestry investment in eucalyptus for paper pulp production, where after two cuts the asset is "worthless" for the paper industry. Abildtrup and Strange [1] analyze the decision to convert a natural or seminatural forest into Christmas tree production, when groundwater contamination is irreversible and future returns on noncontaminated groundwater resources and

Christmas tree production are uncertain. These authors concluded that conventional expected NPV analysis may not lead to an optimal decision rule and show that the option to postpone conversion and acquire new information should be included. Yin [29] discusses harvest timing, land acquisition, and entry decisions combining forest-level analysis and options valuation approach. Other authors have studied the impact of assuming that stochastic prices follow a mean reverting process rather than geometric Brownian motion (see, e.g., [16, 13]).

Most of the studies addressing real options in the forest investment context have applied a single-option approach. However, exceptions can be found in the literature. Some authors consider additional problem features, whereas others consider more than a single option. For example, Malchow-Mollera et al. [18] include temporal and spatial arrangement of harvests. These authors consider that constraints upon harvesting options on adjacent forest stands are imposed and examine the optimal harvest rules under adjacency restrictions and uncertainty. They conclude that the costs of adjacency constraints tend to increase with uncertainty and that the optimal harvesting strategies become rather complex due to the interaction of the involved stochastic variables. Duku-Kaakyire and Nanang [11] developed a forestry investment analysis where four management options are considered: delay of reforestation, capacity expansion, investment abandonment, and a compound option obtained considering simultaneously all three previous options. The models developed are valued on a standard binomial lattice. The authors solve an example with which they show that DCF techniques value the investment as unprofitable, consequently rejecting it. Real options, however, show the investment to be highly valuable. Insley and Rollins [16, 17] model the harvesting problem assuming infinite rotations as a linear complementarity problem that was solved numerically. The accuracy of the solution, obviously, depends on the grid used. Given the computational experiments performed, the authors were able to conclude that the value of the harvesting option can be very sensitive to the number of discrete levels of the stochastic variable.

In this work, the main objective is to develop a methodology that allows for the valuation of the investment in eucalyptus forest and also to find an optimal harvesting policy, considering the managerial flexibility inherent to the process. Although a similar problem to that of Insley and Rollins [16, 17] is addressed, their methodology cannot be applied to eucalyptus forestry investments, as they consider an infinite time horizon framework. Furthermore, this study goes beyond the usual valuation approach as the harvesting policies found are tested by being reapplied to specific data, both real and random.

3 Problem Description

The investment decision under consideration includes three main options: the option to cut down the trees; the option to postpone or defer harvesting of the trees until the price or quantity is favorable; and the option to convert the land to another use or to abandon the project. We model and solve the problem by considering a compound option obtained by considering all the aforementioned options simultaneously. It should be noticed that the decision of when to harvest the trees is associated with both harvesting options (cut and postponement). Furthermore, these decisions and, thus, options must be taken twice as the *Eucalyptus globulus* plantations allow for two rotations without replantation.

Many variables guide the profitability of the investment in eucalyptus. Nature, for example, has a major role in the growth of trees. Therefore, the inventory of timber is a very important decision factor. In this work, timber inventory is considered to follow a deterministic process with a known growth pattern depending on the region under consideration, the time since last cut (or plantation), and rotation. Also, an obviously very important factor for the decision making is the paper pulpwood selling price, as it determines the main positive cash flow. This price is stochastic and it is assumed to follow a geometric Brownian motion. This is further discussed in Section 4.1.

3.1 The Investment Decisions

During the investment process in forestry, and particularly in eucalyptus, when the tree is mature, the landowner possesses the option to cut down the trees and receive the resulting cash flow. On the other hand, such a decision may be postponed if price and/or wood inventory conditions are not favorable. This option is similar to a financial option on a stock, where the underlying asset is the amount of wood in the land and the strike price is the cost of cutting the trees and logging the wood. However, as previously explained, the investment process in eucalyptus is normally composed of two rotations (production cycles). The trees planted at the beginning of the process are suitable for two cuts – after the first cut trees regrow, permitting a second cut without replantation. Regarding the timing of cut, the characteristics of the wood that could be extracted from trees, which are up to 7 years old, do not fit the requirements of the industrial producers of pulpwood. The same happens after 16 years of growth, when the diameter of the trees becomes too large for its industrial processing and the fiber in the wood is not of the required standards. Thus, the window time frame is 9 years for both rotations. However, as the time of exercise of the first option determines the time window during which the second option may be exercised, its temporal location depends on the timing of the first cut. Furthermore, the total expected payoff to be received is given by the sum of the first and second harvesting expected cash flows. Given the path dependency observed, the cash flow received from the second harvesting depends upon the timing of the first harvesting. This dependency leads to a multiplicity of possible paths and strategies, as can be seen in the state transition diagram given in Fig. 1. In this diagram we have represented the time t and the elapsed time since last cut or plantation x_t, with a 2-year step interval.

Fig. 1 State transition
diagram: 2-year step interval

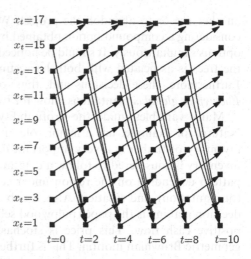

The cash flow to be received from harvesting the trees depends on the cutting period – as the quantity of wood, although known, evolves through time – and on wood market price and future value expectations. Moreover, the decision of cutting the trees in the first rotation must consider not only the immediate cash flow but also the expectation of the cash flow from the second rotation.

In our analysis, the eucalyptus forest investor is assumed to hold the land. Therefore, at any moment in time the investor holds the option to abandon the project and put it aside, or receive an estimated value from converting the land into an alternative use. Activities like tourism and hunting have increased the demand for forest land and, hence, its value. This alternative value (or capacity to abandon without a loss greater than the initial land acquisition investment) must then be considered in the valuation of eucalyptus forest investment.

4 Methodology

The approach developed consists of a dynamic programming model that is evaluated on a discrete-valued lattice. The uncertainty of the underlying risky asset, wood or paper pulp selling price, is modeled through the use of a standard binomial lattice. This approach is described in the following section. To develop the dynamic programming model, we first must define and explain the decision variables as well as the state variables. The decision variables are associated with the possible strategic decisions that investors and managers may undertake during the investment process, whereas the state variables are related to time (see Section 4.2). In Section 4.3, we present the dynamic programming model and explain how to solve it through backward induction. The optimal value of the investment project and the corresponding optimal harvesting strategy are given by the solutions to the proposed model.

4.1 The Binomial Lattice

The analysis performed in this work makes use of the multiplicative binomial model of Cox, Ross, and Rubinstein [8], the standard tool for option-pricing in discrete time. In this approach, the stochastic variable (the selling price in our case) is assumed to be governed by a geometric diffusion, which implies that at each period there is only one constant growth/decay rate. If this is assumed, a natural way of obtaining a valued-lattice for the stochastic variable is to discretize it through a standard binomial lattice (Fig. 2).

A node of price value P_t^i can lead to two nodes with their values being given by $P_{t+1}^{i+1} = uP_t^i$ and $P_{t+1}^{i+1} = dP_t^i$ with probability p and $q = 1 - p$, respectively. The probability of reaching each of these nodes is the usual equivalent Martingale measure used in the binomial option pricing model of Cox et al. (1979):

$$p = \frac{(1 + r_f) - d}{u - d} \quad \text{and} \quad q = 1 - p \tag{1}$$

where r_f is the risk-free interest rate over the interval Δ_t, $u = \exp(\sigma\sqrt{\Delta t})$, $d = \exp(-\sigma\sqrt{\Delta t})$ and σ is the standard deviation. It should be noticed that, as $\Delta t \to 0$, the parameters of the multiplicative binomial process converge to the geometric Brownian motion.

The advantages of this approach are that (i) it can be used to price options other than European, like American and path-dependent options; (ii) it does not depend on the investor subjective probabilities of an upward/downward price movement, as it uses risk-neutral probabilities; and (iii) it is intuitive and simple to understand and implement.

The binomial model breaks down the investment horizon into n time intervals or steps. The lattice is then developed in a forward movement by finding the value of selling prices. At each step, the price moves up or down by an amount obtained from its volatility, which is assumed to be constant and known, and

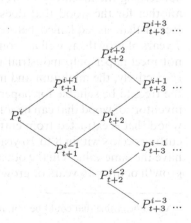

Fig. 2 A lattice discretizing selling price

the time step length. Thus, the lattice provides a representation of possible selling prices throughout the whole project life.

4.2 Decision and State Variables

The selling price variable P_t^i is modeled as a stochastic variable, which follows a binomial process, as in the previous section.

In terms of decisions, as previously described, we consider basically three possibilities: to cut the trees, delay trees cutting, and abandon project or convert land to another usage. Hence, we define a decision variable D_t^i that may assume three values according to the decision taken at time t and state (price index expectation) i:

$$\begin{cases} 1, & \text{if trees are cut,} \\ 2, & \text{if cutting is postponed and the investment maintained,} \\ 3, & \text{if the project is abandoned or the land converted to another use.} \end{cases}$$

The inventory of wood is considered to be a deterministic process, following a known growth pattern according to the considered region, the time since last cut (or plantation), and the rotation. The evolution of the time since the first cut (or plantation) x_t is modeled as:

$$x_{t+1} = \begin{cases} 2 & \text{if } D_t^i = 1, \\ x_t + 1 & \text{if } D_t = 2, \end{cases}$$

and initialized as $x_1 = 1$. In the second rotation, x_{t+1} is initialized as 2, rather than 1, as the trees are already there and thus do not need to be planted. If the decision is to abandon the project or convert land to another use, then the project ends.

Regarding the inventory of wood, we need two variables to represent it: one concerning the inventory of wood that is good for paper pulp production and another for the wood that does not meet paper pulp industrial requirements. Recall that, as explained before, the wood obtained from trees having up to 7 years of growth, as well as from trees with more than 16 years of growth, does not meet paper pulp industrial requirements. Let $tc_{min} = 8$ and $tc_{max} = 16$ be, respectively, the minimum and maximum time from plantation (or first cut) for the trees to be suitable for paper production. Let $SQ(x_t, i)$ and $CQ(x_t, i)$ be the inventory of wood that can be sold to paper pulp production and the inventory of wood that is extracted from cutting the trees, respectively. On one hand, if tree cutting occurs within 8 to 16 years of growth, then these two figures, obviously, have the same value. On the other hand, if trees are cut before reaching 8 years of growth or after 16 years of growth, then SQ is zero[2] and CQ is positive

---

[2] Any other value that could be obtained for the wood can be used in our model.

$$SQ(x_t, t) = \begin{cases} 0, & \text{if } x_t < tc_{\min} \text{ or } x_t > tc_{\max}, \\ Q_1(x_1), & \text{if } x_t = t, \\ Q_2(x_t), & \text{if } x_t < t, \end{cases}$$

$$CQ(x_t, t) = \begin{cases} Q_1(x_1), & \text{if } x_t = t, \\ Q_2(x_t), & \text{if } x_t < t, \end{cases}$$

where $Q_1(x_t)$ and $Q_2(x_t)$ are the quantities of wood extracted from cutting the trees that have been growing for x_t years at rotation 1 and rotation 2, respectively.

4.3 Dynamic Programming Model

In each period, the forest investor must decide whether the forestry investment is going to be kept or not. As both for project abandonment and land use conversion options only a lump sum is received, no subdivision is considered regarding these options. However, for the former case, the forest investor still has to make a decision regarding the timing of tree harvesting. The investor will choose either to harvest at the current period or to leave the trees there for the next period, whichever yields better expected cash flows.

Harvesting the trees yields the owner revenue from selling the wood but also involves costs K incurred with cutting, peeling, and transporting the wood. Therefore, at period t, given the elapsed time since last cut or plantation x_t and the price expectation index i, the net revenue π obtained if the cutting option is exercised is given by

$$\pi(x_t, i, t) = P_t^i \times SQ(x_t, t) - K \times CQ(x_t, t) \tag{2}$$

As described before, in order to allow for earlier exercise, the valuation procedure begins at the last stage and works backwards to the initial moment. At the final lattice nodes (i.e., at the end of the project life $t = T$), all possible selling price values and elapsed time since last cut values are computed. For each of these terminal nodes, the project value is then given by the largest of (i) the final revenue plus residual value if a harvesting decision is still possible and exercised or (ii) the residual value. Here, by residual value R we mean the maximum between abandoned land value and land conversion value. The land conversion value is assumed to have no real appreciation or depreciation. Evidence shows that agricultural and forest land values have been stationary. However, in the empirical sections, we perform some model sensitivity analysis to the value of R. It should be noticed that, after second harvest or at the end of the period of analysis if no second harvest takes place, the project value is given by the residual value R. Furthermore, this value is the same both for the base and the extended problems, which is in accordance with the fact that no additional investment costs have been considered for latter problem

$$V(x_T, i, T) = \max \begin{cases} \pi(x_T, i, T) + R \\ R \end{cases} \tag{3}$$

The project value at each intermediate lattice node is computed by performing a backward induction process. The project value at intermediate steps is used to compute the project value at previous steps by using risk-neutral probabilities. The decision made at any period (and state variable) has implications not only on the cash flow of the current period but also on the expected cash flow of future periods. Therefore, the optimal project value is obtained by maximizing the sum of the current period's net revenue with the optimal continuation value considering all possible decisions. The optimal project value at period t, given the elapsed time since last cut or plantation x_t and the price expectation index i, is then given by

$$V(x_t, i, t) = \max_{D_t^i} \begin{cases} \pi(x_t, i, t) + \frac{pV(x_{t+1}, i+1, t+1) + (1-p)V(x_{t+1}, i-1, t+1)}{1+r_f} & \text{if } D_t = 1, \\ \frac{pV(x_{t+1}, i+1, t+1) + (1-p)V(x_{t+1}, i-1, t+1)}{1+r_f} & \text{if } D_t = 2, \\ R & \text{if } D_t = 2, \end{cases} \tag{4}$$

where $\pi(x_t, i, t)$ is the net revenue from the selling of the wood, r_f is the real risk-free interest rate, p is the risk-neutral probability of an upward movement in the binomial price, and R is the largest of the abandonment value or conversion to another use value.

The estimated project net value is then given by $V(1, 1, 1,)/(1 + r_f)$ minus the initial investment. [3] Furthermore, the solution to this model also provides an optimal decision strategy for the forest manager.

In the following section, the dynamic programming model given by equations (3) and (4) is empirically tested through the application to a case study of investment in eucalyptus for the paper pulp industry in Portugal.

5 Case Study

In this section, we start by giving a brief characterization of the Portuguese forest sector. Second, we describe the case study to which our model is to be applied and end by presenting the results obtained.

---

[3] The initial investment is considered to be given by the land acquisition costs, the plantation costs, and the cost of a maintenance contract for the full length of the project. However, this contract does not include cutting, peeling, and transportation of the harvested wood, which account for the exercise price of the cutting option.

5.1 Brief Characterization of the Portuguese Forest Sector

Portugal, due to its natural characteristics, possesses unique natural and eco-logical conditions for forestry production, which only recently has started to be methodologically and professionally managed. This fact is particularly true for high-productivity forest species such as *Eucalyptus globulus* for the paper industry.

According to forest inventories [9], forest and wooded land represent 37.7% of the Portuguese continental regions. Nowadays, the main species existent in Portuguese forests are pinus pinaster (29.1%), cork oak (21.3%), and eucalyptus (20.1%), the latter being mainly represented by the *Eucalyptus globulus* species.

Eucalyptus globulus, the main subspecies in our country and the one that possesses the best characteristics for paper production, is a fast-growing tree. Originally biologically adapted to the poor soils of the Australian continent, in southern Europe, and Portugal in particular, this species grows very rapidly. Normally, the eucalyptus are cut with an age of 12 years and used to produce pulpwood, providing cellulose fibers that have remarkable qualities for the production of high-quality paper. In what concerns ownership, the pulp indus-try manages approximately 30% of the eucalyptus area in Portugal. The increase in the paper and pulpwood production translated into an increase in real terms of about 50% of the gross value added of the forestry sector, which reveals a much faster growth than the one observed for the rest of the Portu-guese economy [7].

Forest and forest-related industries are a key sector in the Portuguese econ-omy, generating wealth and employment. The modernization of this sector can be a source of competitive advantage for the country, given the favorable ecological and natural conditions. Due to the development of the paper pro-duction industry and the natural characteristics of the species, *Eucalyptus globulus* has shown itself to be a key player in the sector, with a huge wave of investment in the species in the past decades.

5.2 Data and Parameters

5.2.1 The Initial Investment: Plantation and Maintenance Costs

We assume that the investor contracts all expected operations to a specialized firm at the start of the investment and pays the contract in advance. The operations and their present value costs add up to 3212 euros per hectare, from which 1712 euros correspond with the plantation and the 2 rotations maintenance contract (which can last from 15 to 31 years, with 23 years expected duration), and 1500 euros correspond with the acquisition cost of 1 hectare of land. The details of how these values were obtained are given in Appendix A. Recall that in this study, two scenarios are considered. The base problem considers that the eucalyptus forest is owned by a forest investor and that its wood is sold to pulp and paper companies. The extended problem considers that the eucalyptus forest is owned by the paper industry and that the wood is processed into white pulp, which is then sold. The two problems are included in order to distinguish outputs for different types of agents. Hence, the pulpwood productive capacity is considered to be present and its investment is disregarded and treated as a sunk cost, not relevant for undergoing investment projects.

5.2.2 Wood and White Paper Pulpwood Prices

The price of wood in the base problem and the price of white paper pulpwood in the extended problem are assumed to follow a binomial stochastic process, as previously explained. We use 2002 prices: 45 euros per cubic meter of peeled eucalyptus wood; and 500 euros per cubic meter of white paper pulp [2]. The prices volatility was extracted from the time series of prices [7, 12], considering constant prices of 2002 and using the consumer price index as deflator [15]. The extracted volatility corresponds with the volatility of returns of the 3-year moving average of prices. This average was calculated in order to smooth the price series, as jumps in prices were periodical (3-year intervals between jumps). The price volatility measured by the standard deviation has been computed to be 0.0717 for wood and 0.1069 for white pulpwood.

5.2.3 Wood and White Paper Pulpwood Quantities

To compute the quantities of available wood per hectare of eucalyptus forest, we have used the inventory model Globulus by Tomé, Ribeiro, and Soares [27]. We have considered wood quantities for the three main Portuguese regions in what concerns eucalyptus investment: north central coast (region 1), central coast (region 2), and river Tejo valley (region 3). For each region and rotation, we consider different tree growth patterns. The average growth curve for the amount of wood in a hectare of eucalyptus forest land in each of the regions can be observed in Appendix B. In the extended problem, which uses pulpwood valuation, it has been considered that about 3.07 cubic meters of wood are required to produce 1 ton of white pulp, which corresponds with approximately 1.1 cubic meters of white pulp [27].

5.2.4 The Exercise Price for the Cutting Option K

The exercise price of the cutting option is given by the costs incurred in cutting the trees, peeling the wood, and transporting it to the factory. Considering constant prices of 2002, the total cost is estimated to be 22.5 euros per cubic meter of wood [2]. When considering the extended problem, which takes into account transformation into white pulp, the cost of this transformation must also be taken into consideration. The industrial processing cost is approximately 320 euros per ton of white pulpwood[4] (200 euros of variable costs and 120 euros of fixed costs). The cost of cutting, peeling, and transporting the wood necessary for a ton of white paper pulp is 69.14 euros. Therefore, a ton of white paper pulp costs 389.14 euros to transform. As 1 ton of white paper pulp corresponds with 1.1 cubic meters, the cost of processing 1 cubic meter of white paper pulp is then 353.76 euros [22].

5.2.5 The Risk-Free Interest Rate r_f

As we are working with real prices from the year 2002, we have considered a real risk-free interest rate of 3%. This value was approximated through the observation of the euro yield curve (average nominal risk free interest rate of 5% for long-term investments such as the eucalyptus) and the 2% expected inflation target in the euro area (Eurostat).

5.2.6 The Abandonment and Conversion to Another Land Use Value R

In each time step, the investor possesses the option to abandon the investment project, putting the land aside and/or receiving the market price for agricultural and forest land.

In other situations, the investor may convert the land use into other activities (e.g., tourism, hunting, or real estate development). In this work, we consider both situations. We assume that demand may exist for the land as it is, or for the land to be converted to another use. Several possible values for the conversion value R are considered, from the situation where no value is given to the land due to lack of demand, to situations of high conversion values that may arise, for example, due to real estate speculation.

6 Results

In this section, we specify the computational experiments conducted as well as the results obtained. All experiments have been performed using three regions in continental Portugal: 1, north central coast; 2, central coast; and 3, river Tejo valley, which present different productivity, as shown in Appendix B.

The results are divided into two sections: one regarding the dynamic programming model discussed in the previous section and another regarding the application of the harvesting strategies obtained to randomly generated data sets. In the first section, we report on the results obtained by applying the aforementioned dynamic programming model to the base and extended problems. In the second section, we present results regarding the implementation of the optimal strategies provided by the dynamic programming model to both the

---

[4] This value was obtained through the analysis of the operating costs of the paper pulpwood companies and their installed capacity, and also through inquiries to experts.

base and the extended problem. To do so, we randomly generate sets of price data, to which we apply these optimal strategies. The application of these strategies to the randomly generated data only considers the fact that price movements are upwards or downwards, disregarding the magnitude of the movement.

6.1 Results for the Base and Extended Problems

To start with, and in order to have a standard project value, we compute the value of the project considering that cuts are performed at years 12 and 23 respectively, the industry common practice. The price values used are the expected value of the price at years 12 and 23. Several possible values are considered for the land residual value. This valuation is in nature a net present value approach. The cut timing strategy considered has been observed to be the most common among eucalyptus forest managers.

The results, reported in Table 1, are computed, for the three above mentioned regions, as

$$\sum_{k=1}^{12} p_{k,12} \frac{\pi(12, k, 12)}{(1 + r_f)^{12}} + \sum_{k=1}^{23} p_{k,23} \frac{\pi(12, k, 23)}{(1 + r_f)^{23}} + \frac{R}{(1 + r_f)^{23}}, \tag{5}$$

where $p_{k,12}$ and $p_{k,23}$ are the probabilities of having a price index k at the 12th and 23rd years, respectively, and are computed as

$$V(x_t, i, t) = \max_{D_t^i} \begin{cases} p^{t-1} & \text{if } k = 1, \\ (1 - p)^{t-1} & \text{if } k = t, \\ (1 - p) \cdot p_{k+1,t-1} + p \cdot p_{k-1,t-1} & \text{otherwise.} \end{cases} \tag{6}$$

Regarding the values obtained by our model, and in order to show that the postpone and abandonment/conversion options interact, we value the project considering (i) that only the harvesting and postponement options exist and

Table 1 Common practice project value for varying residual values (in euros per hectare)

Residual Value	Base Problem			Extended Problem		
	Region 1	Region 2	Region 3	Region 1	Region 2	Region 3
0	11695	9705	6137	34605	28785	17958
1000	12201	10212	6643	35111	29291	18464
2000	12708	10719	7150	35618	29798	18971
3000	13215	11225	7657	36125	30305	19478
4000	13721	11732	8164	36632	30811	19985
5000	14228	12239	8670	37138	31318	20491

Table 2 Real options project value for varying residual values (in euros per hectare)

Residual Value	Base Problem			Extended Problem		
	Region 1	Region 2	Region 3	Region 1	Region 2	Region 3
0	15959	13084	8186	51295	42116	26030
1000	16371	13496	8599	52114	42935	26849
2000	16784	13909	9012	52934	43755	27667
3000	17196	14322	9426	53753	44574	28488
4000	17609	14735	9842	54573	45396	29308
5000	18022	15149	10261	55393	46219	30128

(ii) that all options exist. Furthermore, the project is valued considering several land residual values R.

For the results presented in Table 2, where only the harvesting and postponement options exist, the residual value R is only recovered at the end of the project. This happens at time T or either whenever the two possible cuts have already been performed or no further cutting is allowed.

When considering all options, the residual/conversion land value is obtained in the same conditions as above or whenever it is optimal to give up further harvesting. In the experiments performed, we have considered that the value to be recovered is the residual value R, as a better comparison with the results obtained when considering only the harvesting and postponement options is possible. It should be noticed that by doing so, we are considering an abandonment option. In here, we only report one set of results, as the results obtained for cases (i) and (ii) were the same. Differences in the project value when considering all options began to appear only when there is an abnormal demand for the land, which appreciates its value to levels several times the considered acquisition value.

The differences in valuation between the standard NPV approach, typically used by the industry, and the options approaches here presented are clear. The real options approach, due to the consideration of the flexibility in the management of the forest, allows for much higher values. It provides a revenue increase that can go up to 36.46% for the base problem and up to 49.15% for the extended problem. It should be noticed that the increase decreases with the residual value, as the higher the residual value is, the smaller is the impact of the harvesting strategy on project value.

Another main conclusion that can be drawn is that the different sources of flexibility clearly interact. However, the difference in valuation considering the presence of all options and only the main cutting options is only relevant if considering very high residual values. This does not come as a surprise, as we consider that the residual value is a fixed value in time. Given that we are discounting future cash flows, then the residual present value decreases with project time horizon. Furthermore, we are not considering a real and specific land conversion value, which can often be much larger than the abandonment value.

Table 3 Project value differences between base and extended problems for varying residual values (in euros per hectare)

	Harvest and Postpone Options			All Options		
Residual Value	Region 1	Region 2	Region 3	Region 1	Region 2	Region 3
0	35339	29035	17845	35339	29035	17844
1000	35344	29040	17850	35345	29040	17851
2000	35344	29040	17850	35345	29040	17851
3000	35355	29051	17861	35359	29056	17866
4000	35344	29040	17850	35345	29040	17851
5000	35366	29062	17872	35374	29070	17881

Before comparing the results obtained for the base problem and the extended problem, let us recall what these problems represent. The base problem considers the situation where the forest is considered to be privately owned, and the eucalyptus wood is sold to paper industry companies. The extended problem, however, considers vertical integration of wood and paper production. The value added by the ownership of forest land being in the hands of the pulp industries (or the value added through vertical integration) can be estimated through the difference between the project values for the base and the extended problems, as given in Table 3. Recall that, for the extended problem, no additional investment costs are being considered, thus to the value added through vertical integration we still have to deduct the aforementioned investment costs.

As it can be observed, the difference in project value is small and basically constant across the different residual values considered, that is, it almost does not depend on the residual value. Nevertheless, the value added through vertical integration seems to be extremely high and justifies the observed direct and indirect investment of paper industry companies, out of their core business of paper pulp production, into producing the inputs themselves. Furthermore, as expected, the larger differences are observed for the most productive regions. However, care should be taken in drawing conclusions, as non-neglectful investments may be required for the vertical integration to take place. As these costs have not been accounted for in the results, the benefits of vertical integration are overestimated. This overestimation is, however, not as large as it might seem, as the residual value – the value of the firm at the end of the project – is the same for both problems.

6.2 Applying the Optimal Strategies

To evaluate the quality of the optimal strategy produced by our dynamic programming model, we apply the strategy provided by the model when considering all options to randomly generated price data sets. The project value obtained in this way is then compared with the one that would be obtained by the common practice of harvesting after 12 years of growth. Five data sets have

Table 4 Characteristics of the randomly generated data sets

	Price Values		
Data Set	Minimum(%)	Maximum(%)	Average(%)
1	80	100	80
2	100	100	90
3	100	100	100
4	100	100	110
5	100	120	120

been randomly generated according to the characteristics given in Table 4. The characteristics reported are relative to the price values used in the case study.

We report the project value obtained by applying the optimal strategies devised both for the base and the extended problems. These results are provided for the five randomly generated data set types. We also report on the common practice project value (CPV) obtained by performing the harvest at years 12 and 23 (see Table 5, Table 6, Table 7, and Table 8 for different residual values).

As it can be seen, in the above situations the project has a larger value when computed using the real options approach. Region 1, the north central coast, is the most appropriate area in the country for eucalyptus and, thus, the most productive. Results for the other two regions also provide the same conclusions.

Table 5 Project value considering a residual of 0 for all regions when decision strategies are applied to specific data sets (in euros per hectare)

	Residual 0					
	Region 1		Region 2		Region 3	
Data Set	Value	CPV	Value	CPV	Value	CPV
1	33492	29619	27578	24271	16631	16476
2	13813	12920	11548	10943	6060	6112
3	48288	29479	39219	24500	26473	15363
4	34748	48615	28550	40416	17540	25296
5	46278	43898	37600	36559	25306	22645
Average	35324	32906	28899	27338	18402	17178

Table 6 Project value considering a residual of 1000 for all regions when decision strategies are applied to specific data sets (in euros per hectare)

	Residual 1000					
	Region 1		Region 2		Region 3	
Data Set	Value	CPV	Value	CPV	Value	CPV
1	33892	30126	27978	24778	17031	16983
2	14213	13426	11948	11450	6460	6619
3	48688	29986	39619	25006	26873	15869
4	35148	49121	28950	40922	17940	25803
5	46678	44404	38000	37066	25706	23151
Average	35724	33413	29299	27844	18802	17685

Table 7 Project value considering a residual of 3000 for all regions when decision strategies are applied to specific data sets (in euros per hectare)

| | Residual 3000 | | | | | |
| | Region 1 | | Region 2 | | Region 3 | |
Data Set	Value	CPV	Value	CPV	Value	CPV
1	34692	31139	28778	25792	17831	17996
2	15013	14440	12748	12463	7260	7632
3	49488	30999	40419	26020	27673	16883
4	35948	50135	29750	41936	18740	26816
5	47478	45418	38800	38079	26506	24165
Average	36524	34426	30099	28858	19602	18698

Table 8 Project value considering a residual of 5000 for all regions when decision strategies are applied to specific data sets (in euros per hectare)

| | Residual 5000 | | | | | |
| | Region 1 | | Region 2 | | Region 3 | |
Data Set	Value	CPV	Value	CPV	Value	CPV
1	35492	32153	29578	26805	18631	19009
2	15813	15453	13548	13476	8060	8645
3	50288	32013	41219	27033	28473	17896
4	36748	51148	30550	42949	19540	27830
5	48278	46431	39600	39093	27306	25178
Average	37324	35440	30899	29871	20402	19712

It should be noticed that the harvesting decisions typically take place after 16 years of growth for both rotations. This corresponds with the maximum time growth allowed. Thus, it can be concluded that the cutting option is exercised almost always at maturity, which allows for the investor to take advantage of the full tree growth. This fact is inconsistent with the common practice of most forest producers. The urge to cash in the value of their eucalyptus wood, due to high risk aversion, typically made at the 12th year seems to be not optimal. A possible explanation may be that once the second cut is performed, then the project can be repeated and, given the human life expectancy, it may not make sense to wait longer than the minimum required time.

7 Conclusions

In this work, we address the valuation of an investment in eucalyptus forest land for the paper pulp industry by using real options theory. The modeling of the eucalyptus forest investment decisions using real options theory allows for valuation results more consistent with the normal positive investment outcomes than with the ones obtained using traditional valuation techniques. The interpretation of the two possible tree cuts as two exercise interdependent call options on the wood allows for the introduction of managerial flexibility in the forest management process. The inclusion of the abandonment and conversion to another use options increases the closeness to the real decision process.

The evidence in this chapter, through the application of the developed stochastic dynamic programming model to the eucalyptus investments in three Portuguese regions and for randomly generated situations, is consistent with the experts' warnings of the undervaluation performed by traditional discounted cash flow techniques. A rationale for the trend of vertical integration of paper pulp and wood production is also provided by the value difference between the two problems considered.

It is also shown that typical cutting time decisions by forest owners are not consistent with cash flow maximization. According to our model, cutting should be almost in all situations performed at the end of the time interval when wood is suitable for paper pulp production.

Acknowledgments The authors would like to thank FCT for the financial support through scholarship BD/12610/03 and project POCI/MAT/61842/04.

Appendix A

Table 9 Initial investment: Operation costs and present values*

		Operation Costs (Euros)	
Year	Maintenance Operations	2002	Present Value
1	Soil preparation and fertilization		
	Plantation		
	Infrastructure – Paths	1100	1100
1	Replantation (10–15% of planted trees)	60	60
2	Soil fertilization	62.5	60.6
3	Soil cleaning (milling cutter)	90	84.8
4	Soil fertilization		
	Infrastructures cleaning	100	91.5
7	Soil cleaning (disk harrow)		
	Infrastructure cleaning	100	83.7
12	Infrastructure cleaning	10	7.2
14	Soil fertilization	55	37.4
15	Road selection	90	59.5
16	Soil fertilization		
	Infrastructure cleaning	100	64.1
19	Soil cleaning (disk harrow)		
	Infrastructure cleaning	100	58.7
24	Infrastructure cleaning	10	5
Value of plantation and maintenance contract			1712

*The data was provided by Aliança Florestal [2].

Appendix B

Table 10 Quantities of eucalyptus wood (in m³ per ha)

	Rotation 1			Rotation 2		
Year	Region 1	Region 2	Region 3	Region 1	Region 2	Region 3
3	7.22	6.84	4.05	15.78	17.69	8.38
3	18.09	16.06	10.29	15.78	17.69	8.38
3	32.84	28.14	18.91	15.78	17.69	8.38
3	50.16	42.06	29.14	15.78	17.69	8.38
3	69.02	57.05	40.4	15.78	17.69	8.38
3	88.67	72.59	52.23	15.78	17.69	8.38
3	108.63	88.31	64.32	15.78	17.69	8.38
3	128.54	103.97	76.46	15.78	17.69	8.38
3	148.2	119.43	88.51	15.78	17.69	8.38
3	167.47	134.59	100.38	15.78	17.69	8.38
3	186.25	149.37	112	15.78	17.69	8.38
3	204.49	163.76	123.33	15.78	17.69	8.38
3	222.18	177.72	134.36	15.78	17.69	8.38

References

1. Abildtrup, J. & Strange, N. (1999), 'The option value of non-contaminated forest watersheds,' Forest Policy and Economics 1, 115–125.
2. Aliança Florestal (2002), Technical Internal Report: www.alflorestal.pt.
3. Arrow, K. & Fisher, A. C. (1974), 'Environmental preservation, uncertainty and irreversibility,' Quarterly Journal of Economics 88, 312–319.
4. Bailey, W. (1991), 'Valuing agricultural firms: An examination of the contingent claims approach to pricing real assets,' Journal of Economic Dynamics and Control 15, 771–791.
5. Black, F. & Scholes, M. (1973), 'The pricing of options and corporate liabilities,' Journal of Political Economy 27, 637–654.
6. Brennan, M. & Schwartz, E. (1985), 'Evaluating natural resource investments,' Journal of Business 58, 135–157.
7. CELPA (Associação da Indústria Papeleira) : www.celpa.pt.
8. Cox, J., Ross, S. & Rubinstein, M. (1979), 'Option pricing: a simplified approach,' Journal of Financial Economics 12, 229–263.
9. DGRF – Database (2001), Inventário Florestal Nacional by Direcção Geral dos Recursos Florestais: www.dgf.min-agricultura.pt.
10. Dixit, A. & Pindyck, R. S. (1994), Investment Under Uncertainty, Princeton University Press, Princeton, NJ.
11. Duku-Kaakyire, A. & Nanang, D. M. (2004), 'Application of real options theory to forestry investment analysis,' Forest Policy and Economics 6, 539–552.
12. FAO (Food and Agricultural Organization of the United Nations): www.fao.org.
13. Gjolberg, O. & Guttornmsen, A. G. (2002), 'Real options in the forest: What if prices are mean reverting,' Forest Policy and Economics 4, 13–20.
14. Henry, C. (1974), 'Option values in the economics of irreplaceable assets,' Review of Economic Studies 41, 89–104.
15. INE – Infoline (2001), Instituto Nacional de Estatística: www.ine.pt.

16. Insley, M. (2002), 'A real options approach to the valuation of a forestry investment,' Journal of Environmental Economics and Management 44, 471–492.
17. Insley, M. & Rollins, K. (2005), 'On solving the multirotational timber harvesting problem with stochastic prices: a linear complementarity formulation,' American Journal of Agricultural Economics 87, 735–755.
18. Malchow-Mollera, N., Strangeb, N. & Thorsen, B. J. (2004), 'Real-options aspects of adjacency constraints,' Forest Policy and Economics 6, 261–270.
19. Merton, R. C. (1973), 'An intertemporal capital asset pricing model,' Econometrica 41, 867–887.
20. Morck, R., Schwartz, E. & Stangeland, D. (1989), 'The valuation of forestry resources under stochastic prices and inventories,' Journal of Financial and Quantitative Analysis 24(4), 473–487.
21. Pindyck, R. S. (1991), 'Irreversibility, uncertainty, and investment,' Journal of Economic Literature 29(3), 1110–1148.
22. Portucel Setúbal (2002), Technical Internal Report: www.portucelsoporcel.com.
23. Reed, W. (1993), 'The decision to conserve or harvest old-growth forest,' Ecological Economics 8, 45–49.
24. Robichek, A. A. & Myers, S. (1966), 'Conceptual problems in the use of risk-adjusted discount rates,' Journal of Finance 21, 727–730.
25. Sharpe, W. F. (1964), 'Capital asset prices: A theory of market equilibrium under conditions of risk,' Journal of Finance 19(3), 425–442.
26. Thomson, T. A. (1992), 'Option pricing and timberlands land-use conversion option: comment,' Land Economics 68, 462–469.
27. Tomé, M., Ribeiro, F. & Soares, P. (2001), O modelo glóbulus 2.1, Technical report, Grupo de Inventariação e Modelação de Recursos Naturais, Departamento de Engenharia Florestal, Instituto Superior de Agronomia, Universidade Técnica de Lisboa.
28. Wilmott, P. (1998), Derivatives, the Theory and Practice of Financial Engineering, Wiley, Chickester, UK.
29. Yin, R. (2001), 'Combining forest-level analysis with options valuation approach – a new framework for assessing forestry investment,' Forest Science 47(4), 475–483.
30. Zinkhan, F. (1991), 'Option pricing and timberlands land-use conversion option,' Land Economics 67, 317–325.

Modelling Water Flow and Solute Transport in Heterogeneous Unsaturated Porous Media

Gerardo Severino, Alessandro Santini, and Valeria Marina Monetti

Abstract New results concerning flow velocity and solute spreading in an unbounded three-dimensional partially saturated heterogeneous porous formation are derived. Assuming that the effective water content is a uniformly distributed constant, and dealing with the recent results of Severino and Santini (*Advances in Water Resources* 2005;28:964–974) on mean vertical steady flows, first-order approximation of the velocity covariance, and of the macrodispersion coefficients are calculated. Generally, the velocity covariance is expressed via two quadratures. These quadratures are further reduced after adopting specific (i.e., exponential) shape for the required (cross)correlation functions. Two particular formation structures that are relevant for the applications and lead to significant simplifications of the computational aspect are also considered.

It is shown that the rate at which the Fickian regime is approached is an intrinsic medium property, whereas the value of the macrodispersion coefficients is also influenced by the mean flow conditions as well as the (cross)-variances σ^2_γ of the input parameters. For a medium of given anisotropy structure, the velocity variances reduce as the medium becomes drier (in mean), and it increases with σ^2_γ. In order to emphasize the intrinsic nature of the velocity autocorrelation, good agreement is shown between our analytical results and the velocity autocorrelation as determined by Russo (*Water Resources Research* 1995;31:129–137) when accounting for groundwater flow normal to the formation bedding. In a similar manner, the intrinsic character of attainment of the Fickian regime is demonstrated by comparing the scaled longitudinal macrodispersion coefficients $\frac{D_{11}(t)}{D_{11}(\infty)}$ as well as the lateral displacement variance $\frac{X_{22}(t)}{X_{22}(\infty)} = \frac{X_{33}(t)}{X_{33}(\infty)}$ with the same quantities derived by Russo (*Water Resources Research* 1995;31:129–137) in the case of groundwater flow normal to the formation bedding.

G. Severino (✉)
Division of Water Resources Management, University of Naples Federico II, Naples, Italy
e-mail: severino@unina.it

P.J. Papajorgji, P.M. Pardalos (eds.), *Advances in Modeling Agricultural Systems*,   361
DOI 10.1007/978-0-387-75181-8_17, © Springer Science+Business Media, LLC 2009

1 Introduction

Water flow and solute transport in a heterogeneous unsaturated porous formation (hereafter also termed *vadose zone*) are determined by the inherent variability of the hydraulic formation properties (like the conductivity curve), flow variables (e.g., the flux), and boundary conditions imposed on the domain. Because the formation properties are highly variable, the resultant uncertainty of their spatial distribution is usually modelled within a stochastic framework that regards the formation property as a random space function (RSF). As a consequence, flow and transport variables become RSFs, as well.

For given statistics of the flow field, solute spreading can be quantified in terms of the first two moments of the probability density function of the solute particles displacement. This task has been accomplished by [4, 20] for transport by groundwater and unsaturated flow, respectively. However, in the case of unsaturated flow, it is much more difficult due to the nonlinearity of the governing equations. By using the stochastic theory developed in [20, 35, 36], Russo [20] derived the velocity covariance that is required to model transport in a vadose zone. Then, Russo [20] predicted the continuous transition from a convection-dominated transport applicable at small travel distances to a convection-dispersion transport valid when the solute body has covered a sufficiently large distance. Assuming that the mean flow is normal to the formation bedding, Russo [24] derived the velocity covariance tensor and the related macrodispersion coefficients showing that the rate of approaching the asymptotic (Fickian) regime is highly influenced by the statistics of the relevant formation properties. The assumption of mean flow normal to the formation bedding was relaxed in a subsequent study [25].

All the aforementioned studies were carried out by assuming that, for a given mean pressure head, the water content can be treated as a constant because the spatial variability of the retention curve is normally found very small compared with that of the conductivity curve [19, 26, 29]. More recently, Russo [27] has investigated the effect of the variability of the water content on flow and transport in a vadose zone. He has shown that the water content variability increases the velocity variance, and therefore it enhances solute spreading. Such results were subsequently refined by Harter and Zhang [10]. Further extensions of these studies were provided by Russo [28] who considered flow and transport in a bimodal vadose zone.

In this chapter, we derive new expressions for the velocity covariance. Unlike the previous approach mainly developed by [20, 23, 27] and relying on the stochastic theory of [35, 36], we make use of the recent results of [30]. Such a choice is motivated by the fact that we can obtain very simple results. Then, by adopting the Lagrangian approach developed in the past, we analyze solute spreading in a vadose zone with special emphasis on the impact of the spatial variability of the formation properties upon attainment the Fickian regime.

2 The General Framework

The study is carried out in two stages. In the first, we relate the statistical moments of the flux to those of the porous formation as well as mean flow conditions, and in the second, we use the previously derived statistics to analyze solute spreading .

2.1 *Derivation of the Flux Statistics*

We consider an unbounded domain of a vadose zone with statistically aniso-tropic structure in a Cartesian coordinate system x_i ($i = 1, 2, 3$) with x_1 down-ward oriented. In order to relate the statistical moments of the flux $\mathbf{q}$ to those of the porous formation, the following assumptions are employed: (i) the local steady-state flow obeys the unsaturated Darcy's law and continuity equation

$$\mathbf{q}(\mathbf{x}) = -K(\Psi, \mathbf{x})\nabla(\Psi + x_1) \quad \nabla \cdot \mathbf{q} = 0 \tag{1}$$

respectively, and (ii) the local relationship between the conductivity $K = K(\Psi, \mathbf{x})$ and the pressure head Ψ is nonhysteretic, isotropic, and given by the model of [8]

$$K(\Psi, \mathbf{x}) = K_s(\Psi, \mathbf{x}) \exp[\alpha(\mathbf{x})\psi] \tag{2}$$

where α is a soil pore-size distribution parameter, whereas $K_s(\mathbf{x}) = K_s(0, \mathbf{x})$ represents the saturation conductivity. The range of applicability of the con-ductivity model (2) has been discussed by [18], and its usefulness to analyze transport has been assessed [20].

Owing to the scarcity of data, and because of their large spatial variations, K_s and α are regarded as stationary RSFs defined through their log-transforms Y and ζ as follows:

$$K_s = K_G = \exp(Y') \quad \alpha = \alpha_G \exp(\zeta') \quad K_G = \exp(\langle Y \rangle) \quad \alpha_G = \exp(\langle \zeta \rangle) \tag{3}$$

in which $\langle \rangle$ denotes the expected value, and the prime symbol represents the fluctuation. In line with numerous field findings (e.g., [13, 17, 34, 26]), we assume that the various (cross)correlation functions $C_\gamma (\gamma = Y, \zeta, Y\zeta)$ have axisymmetric structure, that is,

$$C_\gamma(x) = \sigma_\gamma^2 \rho_\gamma(x) \quad x = |\mathbf{x}| = \sqrt{\left(\frac{x_1}{I_v}\right)^2 + \left(\frac{x_2}{I}\right)^2 + \left(\frac{x_3}{I}\right)^2} \tag{4}$$

where σ_γ^2 and $\rho_\gamma(x)$ represent the variance and the autocorrelation, respectively, whereas I_v and I are the vertical and horizontal integral scale of the γ-para-meters. Whereas there is a large body of literature devoted to the spatial variability of Y (for a wide review, see [5]), limited information on ζ is available. However, some studies ([16, 19, 26, 33, 34]) have shown that the γ-RSFs exhibit very similar integral scales. Thus, based on these grounds, we shall assume that $I'_v \equiv I_v$, $I^\lambda \equiv I$ and $\rho_\gamma(x) \equiv \rho(x)$. Furthermore, we shall consider hereafter a flow normal to the formation bedding that represents the most usual configuration encountered in the applications ([9, 37]).

By dealing with the same flow conditions (i.e., mean vertical steady flow in an unbounded vadose zone) considered by [30], the first-order fluctuation $\mathbf{q}'$ is given by (see second of Equation (8) in [30]):

$$\mathbf{q}'(\mathbf{x}) = -\exp(\langle\Psi\rangle)[Y'(\mathbf{x}) + \Psi_1(\mathbf{x}) + \langle\Psi\rangle\zeta'(\mathbf{x})]\mathbf{e}_1 + \nabla\Psi_1(\mathbf{x}) \qquad (5)$$

(where $\mathbf{e}_1$ is the vertical unit vector). Let us observe that the fluctuation (5) has been rewritten after scaling the lengths by α_G^{-1} and the velocities by K_G (although for simplicity we have retained the same notation). Then, by using the spectral representation of the quantities appearing on the right-hand side of (5) leads to

$$q_1'(\mathbf{x}) = -\exp(\langle\Psi\rangle) \int \frac{d\mathbf{k}}{(2\pi)^{3/2}} \exp(-j\mathbf{k}\cdot\mathbf{x})\left[\tilde{Y}'(\mathbf{k}) + \langle\Psi\rangle\tilde{\zeta}'(\mathbf{k})\right]\frac{k^2 - k_1^2}{k^2 + jk_1} \qquad (6a)$$

$$q_m'(\mathbf{x}) = -\exp(\langle\Psi\rangle) \int \frac{d\mathbf{k}}{(2\pi)^{3/2}} \exp(-j\mathbf{k}\cdot\mathbf{x})\left[\tilde{Y}'(\mathbf{k}) + \langle\Psi\rangle\tilde{\zeta}'(\mathbf{k})\right]\frac{k_1 km}{k^2 + jk_1} \qquad (6b)$$

with $\tilde{f}(\mathbf{k}) = \int \frac{d\mathbf{x}}{(2\pi)^{3/2}} \exp(j\mathbf{k}\cdot\mathbf{x})f(\mathbf{x})$ being the Fourier transform of $f(\mathbf{x})$ (hereafter $m = 2, 3$). The flux covariance C_{q_i} (we assume that the coordinate axes are principal) is obtained by averaging the product $q_i'(\mathbf{x})q_i'(\mathbf{y})$, that is,

$$C_{q_i}(\mathbf{r}) = \sigma^2 \exp(2\langle\Psi\rangle\chi_i(\mathbf{r})) \qquad (7)$$

with $\sigma^2 = \sigma_Y^2 + 2\langle\Psi\rangle\sigma_{Y\zeta} + (\langle\Psi\rangle\sigma_\zeta)^2$. Because of the linear dependence of the fluctuation $\mathbf{q}'$ upon the Fourier transform of Y and ζ [see Equations (6a)–(6b)], the covariance (7) depends only upon the separation distance $\mathbf{r} = \mathbf{x} - \mathbf{y}$ between two arbitrary points $\mathbf{x}$ and $\mathbf{y}$ belonging to the flow domain. The functions χ_i needed to evaluate (7) are defined as

$$\chi_i(\mathbf{r}) = \begin{cases} \int \frac{d\mathbf{k}}{(2\pi)^{3/2}} \exp(-j\mathbf{k}\cdot\mathbf{r})\tilde{\rho}(\mathbf{k})\frac{(k^2 - k_1^2)^2}{k^4 + k_1^2} & i = 1 \\[2mm] \int \frac{d\mathbf{k}}{(2\pi)^{3/2}} \exp(-j\mathbf{k}\cdot\mathbf{r})\tilde{\rho}(\mathbf{k})\frac{(k_1 ki)^2}{k^4 + k_1^2} & i = m \end{cases} \qquad (8)$$

and they quantify the impact of the medium heterogeneity structure on the flux covariance . Before dealing with the most general case, we want to consider here two particular anisotropic heterogeneity structures that enable one to simplify the computational effort and are at the same time relevant for the applications.

The first pertains to highly vertically anisotropic formations. It applies to those soils where the parameters can be considered correlated over very short vertical distances. The assumption of high vertically anisotropic formations is quite reasonable for sedimentary soils for which it has been shown that the rate of decay of the hydraulic properties correlation in the vertical direction is very rapid, and the integral scale in that direction may be much smaller than the integral scale in the horizontal direction (e.g., [1, 13, 17]). From a mathematical point of view, this implies that $\gamma(x_1 + r_v)$ and $\gamma(x_1)$ can be considered uncorrelated no matter how small is the vertical separation distance r_v. This type of approximation was already adopted by [7, 11] to investigate one-dimensional flow in partially saturated bounded formations and well-type flow, respectively. Thus, by employing such an assumption, the χ_i functions (8) can be written as

$$
\chi_i(\mathbf{r}) = \begin{cases} \frac{S_0}{2\pi^2} \int d\mathbf{k}_h \exp(-j\mathbf{k}_h \cdot \mathbf{r}_h) \tilde{\rho}(k_h) k_h^4 \int_0^\infty \frac{du \cos(r_v u)}{(u^2 + k_h^2)^2 + u^2} & i = 1 \\[2mm] \frac{S_0}{2\pi^2} \int d\mathbf{k}_h \exp(-j\mathbf{k}_h \cdot \mathbf{r}_h) \tilde{\rho}(k_h) k_i^2 \int_0^\infty \frac{du u^2 \cos(r_v u)}{(u^2 + k_h^2)^2 + u^2} & i = m \end{cases}
\tag{9}
$$

where the parameter S_0 depends only upon the shape of $\rho(r)$ (see, e.g., [38]), whereas r_h is the horizontal separation distance.

The integrals over u are evaluated by the residual theory yielding

$$
\int_0^\infty \frac{du \cos(r_v u)}{(u^2 + k_h^2)^2 + u^2} = \frac{\pi}{2} \left[\Theta\left(\frac{\rho_-}{2}\right) - \Theta\left(\frac{\rho_+}{2}\right) \right] \left(1 + 4k_h^2\right)^{-1/2}
\tag{10a}
$$

$$
\int_0^\infty \frac{du u^2 \cos(r_v u)}{(u^2 + k_h^2)^2 + u^2} = \frac{\pi}{2} \left[\frac{\rho_+}{2} \Theta\left(\frac{\rho_+}{2}\right) - \frac{\rho_-}{2} \Theta\left(\frac{\rho_-}{2}\right) \right] \left(1 + 4k_h^2\right)^{-1/2}
\tag{10b}
$$

$$
\Theta(a) = \frac{\exp(-|r_v|\sqrt{a})}{\sqrt{a}} \qquad\qquad \rho_\pm = 1 + 2k_h^2 \pm \left(1 + 4k_h^2\right)^{1/2}
\tag{10c}
$$

Hence, substitution into Equation (9) permits one to express χ_i via a single quadrature as (after introducing polar coordinates and integrating over the angle):

$$
\chi_i(\mathbf{r}) = \begin{cases} \frac{S_0}{2} \int_0^\infty \frac{dk_h \tilde{\rho}(k_h) k_h^5}{\sqrt{1 + (2k_h)^2}} \left[\Theta\left(\frac{\rho_-}{2}\right) - \Theta\left(\frac{\rho_+}{2}\right) \right] J_0(r_h k_h) & i = 1 \\[2mm] \frac{S_0}{4} \int_0^\infty \frac{dk_h \tilde{\rho}(k_h) k_h^3}{\sqrt{1 + (2k_h)^2}} \left[\rho_+ \Theta\left(\frac{\rho_+}{2}\right) - \rho_- \Theta\left(\frac{\rho_-}{2}\right) \right] \left[J_0(r_h k_h) - \frac{J_1(r_h k_h)}{r_h k_h} \right] & i = m \end{cases}
\tag{11}
$$

(J_n denotes the n-order Bessel's function).

The second case is that of a medium with fully vertically correlated parameters. Such a model is appropriate to analyze flow and transport at regional scale [3]. Because in this case the Fourier transform $\tilde{\rho}$ is

$$\tilde{\rho}(k) = \sqrt{2\pi}\tilde{\rho}_h(k_h)\delta(k_1) \tag{12}$$

(ρ_h being the horizontal autocorrelation), it is easily proved from Equation (8) that

$$\chi_i(r_h) = \begin{cases} \lambda^{-1}\rho_h(r_h) & i = 1 \\ \\ 0 & i = m \end{cases} \tag{13}$$

In line with past studies (e.g., [3]), Equation (13) clearly shows that assuming fully vertically correlated parameters implies that the formation may be sought as a bundle of noninteracting (i.e., $\chi_m = 0$) soil columns. Furthermore, it is also seen that the vertical flux autocorrelation decays like λ^{-1}.

Scaling k_1 by the anisotropy ratio $\lambda = \frac{I_v}{I}$ (for simplicity, we retain the same notation), the general expression (8) can be rewritten as

$$\chi_i(\mathbf{r}) = \begin{cases} \frac{\lambda^3}{(2\pi)^{3/2}} \int \frac{d\mathbf{k}\,\exp(-j\mathbf{k}\cdot\mathbf{r})\tilde{\rho}(k)\left(k^2-k_1^2\right)^2}{\left[k_1^2+\lambda^2\left(k^2-k_1^2\right)\right]^2+(\lambda k_1)^2} & i = 1 \\ \\ \frac{\lambda}{(2\pi)^{3/2}} \int \frac{d\mathbf{k}\,\exp(-j\mathbf{k}\cdot\mathbf{r})\tilde{\rho}(k)(k_1 k_i)^2}{\left[k_1^2+\lambda^2\left(k^2-k_1^2\right)\right]^2+(\lambda k_1)^2} & i = m \end{cases} \tag{14}$$

The advantage of such a representation stems from the fact that ρ becomes isotropic, and therefore its Fourier transform will depend upon the modulus of $\mathbf{k}$, solely. As a consequence, we can further reduce the quadratures appearing into Equation (14) by adopting spherical coordinates and integrating over the azimuth angle, the result being:

$$\chi_i(r) = \begin{cases} \sqrt{\frac{2}{\pi}} \int_0^\infty \int_0^{1/\lambda} \frac{dk\,du\,k^4\tilde{\rho}(k)\cos(\lambda rku)}{u^2+k^2[1+(1-\lambda^2)u^2]^2}\left[1-(\lambda u)^2\right]^2 & i = 1 \\ \\ \frac{1}{\sqrt{2\pi}} \int_0^\infty \int_0^{1/\lambda} \frac{dk\,du\,k^4\tilde{\rho}(k)\cos(\lambda rku)}{u^2+k^2[1+(1-\lambda^2)u^2]^2}\left[1-(\lambda u)^2\right]u^2 & i = m \end{cases} \tag{15}$$

The relationship (15) together with (7) provides a σ_γ^2-order representation of the flux covariance valid for arbitrary autocorrelation ρ.

In order to carry out the computation of (11), (13), and (15), we shall consider two types of models for ρ, namely the exponential and the Gaussian one. While the exponential model has been widely adopted (see, for instance, [20, 35, 36, 37]), there are very few studies (e.g., [30]) dealing with the Gaussian ρ. Nevertheless, many field findings (e.g., [1, 13, 17]) have shown that the

Gaussian model is also applicable. Before going on, it is worth emphasizing here that the general results (11), (13), and (15) are limited to the case of a flow normal to the formation bedding. The more general case of a flow arbitrary inclined with respect to the bedding was tackled by [25]. Finally, let us observe that the flux variance $\sigma_{q_i}^2$ is obtained from Equation (7) as

$$\sigma_{q_i}^2 = \sigma^2 \exp(2\langle\Psi\rangle)\chi_i(0) \qquad i = 1, 2, 3. \tag{16}$$

2.2 Results

We want to derive here explicit expressions for χ_i. Thus, assuming exponential ρ permits us to carry out the quadrature over k into Equation (15) yielding:

$$\chi_1(r) = \int_0^{1/\lambda} du \bar{Y}(u,r) \left[\frac{1 - u\zeta(u)}{1 - \zeta^2(u)}\right]^2 \qquad \chi_m(r) = \int_0^{1/\lambda} du \bar{Y}(u,r) \frac{u\zeta(u)[1 - u\zeta(u)]}{2[1 - \zeta^2(u)]^2} \tag{17a}$$

$$\bar{Y}(u,r) = 2\zeta^3(u) \exp[\lambda ru\zeta(u)] + \lfloor 1 - \lambda ru + (\lambda ru - 3)\zeta^2(u)\rfloor \exp(-\lambda ru) \tag{17b}$$

where we have set $\zeta(t) = \frac{t}{1+(1-\lambda^2)t^2}$. For an isotropic formation (i.e., $\lambda = 1$), we can express Equation (17a) in analytical form:

$$\chi_1(r) = \lfloor 1 - (r + 1) \exp(-r)r^{-2}\rfloor \tag{18a}$$

$$\chi_m(r) = \frac{1}{2} \left\{ [E_i(-r) - E_i(-2r)] \exp(r) + (1 - \ln 2 + r^{-2}) \exp(-r) + (r - 1)r^{-2} \right\}. \tag{18b}$$

Instead, in the case of Gaussian ρ, the quadratures appearing into Equation (15) have to be numerically evaluated.

For $r = 0$, and dealing with small λ, we obtain

$$\chi_1(0) = \sqrt{\frac{\pi}{2}} \int_0^\infty \frac{dk\tilde{\rho}(k)k^3}{\sqrt{1 + (2k)^2}} \qquad \chi_m(0) = \frac{\chi_1(0)}{2}. \tag{19}$$

In particular, assuming exponential structure for ρ yields $\chi_1(0) = \frac{1}{3}\left(1 + \frac{2\pi}{9}\sqrt{3}\right)$, and for the Gaussian model one has $\chi_1(0) = \frac{1}{4}\left[1 + \left(\pi - \frac{1}{2}\right) \exp\left(\frac{1}{4\pi}\right) \text{erfc}\left(\frac{1}{2\sqrt{\pi}}\right)\right]$.

In the case of an isotropic formation, it yields

$$\chi_1(0) = \frac{1}{2} \qquad \chi_m(0) = \frac{3}{4} - \ln 2 \tag{20}$$

for exponential ρ, whereas for the Gaussian model one quadrature is required, the result being

$$\chi_1(0) = 0.504 \qquad\qquad \chi_m(0) = 0.060. \qquad\qquad (21)$$

In the general case, $\chi_i(0)$ are analytically computed for exponential autocorrelation:

$$\chi_1(0) = \frac{(2\lambda^2 - 1)(2\lambda - 1)}{(4\lambda^2 - 3)(\lambda^2 - 1)} + \frac{\ln(1 + \lambda)}{(\lambda^2 - 1)^2} + \frac{sg(1 - \lambda)(6\lambda^4 - 6\lambda^2 + 1)}{(4\lambda^2 - 3)\sqrt{|4\lambda^2 - 3|}(\lambda^2 - 1)^2} \quad (22a)$$

$$\chi_m(0) = -\frac{1}{2}\left[\frac{2\lambda - 1}{(4\lambda^2 - 3)(\lambda^2 - 1)} + \frac{\ln(1 + \lambda)}{(\lambda^2 - 1)^2} + \frac{sg(1 - \lambda)(4\lambda^4 - 2\lambda^2 - 1)}{(4\lambda^2 - 3)\sqrt{|4\lambda^2 - 3|}(\lambda^2 - 1)^2}\right] \quad (22b)$$

with

$$\alpha(\lambda) = \begin{cases} 2\left\{\tan^{-1}\left[\frac{P_2(\lambda)}{\lambda\sqrt{3 - 4\lambda^2}}\right] + \tan^{-1}\left(\frac{1}{\sqrt{3 - 4\lambda^2}}\right)\right\} & 3 \geq 4\lambda^2 \\[2mm] \ln\left\{\left[\frac{P_2(\lambda) + sg(\lambda - 1)\lambda\sqrt{4\lambda^2 - 3}}{P_2(\lambda) - sg(\lambda - 1)\lambda\sqrt{4\lambda^2 - 3}}\right]\left[\frac{1 + sg(\lambda - 1)\sqrt{4\lambda^2 - 3}}{1 - sg(\lambda - 1)\sqrt{4\lambda^2 - 3}}\right]\right\} & 3 \leq 4\lambda^2 \end{cases}$$

being $P_2(\lambda) = 2\lambda^2 - \lambda - 2$, whereas $sg(x)$ is the signum function.
For Gaussian ρ, we have (after integrating over k):

$$\chi_1(0) = 2\int_0^{1/\lambda} du\,[1 - u\zeta(u)]^2\bar{\Phi}[\zeta(u)], \qquad \chi_m(0) = 2\int_0^{1/\lambda} du\,u\zeta(u)[1 - u\zeta(u)]\bar{\Phi}[\zeta(u)] \quad (23a)$$

$$\bar{\Phi}(a) = \frac{1}{2} + \frac{a^2}{\pi}\left[a\exp\left(\frac{a^2}{\pi}\right)\mathrm{erfc}\left(\frac{a}{\sqrt{\pi}}\right) - 1\right]. \qquad\qquad (23b)$$

Observe that in the case of Equations (22a)–(22b), we can easily derive the following asymptotics:

$$\chi_1(0) = \frac{1}{3}\left(1 + \frac{2\pi}{9}\sqrt{3}\right) + O(\lambda^2) \qquad \chi_m(0) = \frac{\chi_1(0)}{2} + O(\lambda^2) \qquad \lambda << 1 \quad (24a)$$

$$\chi_1(0) = O(\lambda^{-1}) \qquad \chi_m(0) = O(\lambda^{-3}) \qquad \lambda >> 1. \quad (24b)$$

3 Macrodispersion Modelling

We want to use the above results concerning water flow to model macrodispersion. To do this, we briefly recall the Lagrangian description of a solute particle motion by random flows [31]. Thus, the displacement covariance $\chi_{ii}(t)$ of a moving solute particle has been obtained by Dagan [4] as follows:

$$\chi_{ii}(t) = 2 \int_0^t d\tau (t - \tau) u_{ii}(\tau), \tag{25}$$

where $u_{ii} = u_{ii}(t)$ represents the covariance of the pore velocity, this latter being defined as the ratio between the flux q and the effective water content ϑ_e (i.e., the volumetric fraction of mobile water). The Fickian dispersion coefficients can be derived as asymptotic limit of the time-dependent dispersion coefficients D_{ii} (hereafter also termed *macrodispersion coefficients*) defined by

$$D_{ii}(t) = \frac{1}{2}\frac{d}{dt}X_{ii}(t) \tag{26}$$

[5]. To evaluate the covariance u_{ii}, in principle one should employ a second-order analysis similar to that which has led to Equation (7). However, field studies (e.g., [19, 26, 29]) as well as numerical simulations [21, 22] suggest that in uniform mean flows, the spatial variability of the effective water content may be neglected compared with the heterogeneity of the conductivity K. Thus, based on these grounds and similar to previous studies (e.g., [20, 24, 25]), we shall regard ϑ_e as a constant uniformly distributed in the space, so that the pore velocity covariance is calculated as

$$u_{ii}(r) = \sigma^2 \exp(2\langle\Psi\rangle)\chi_i(r). \tag{27}$$

Inserting (27) into (25) provides the displacements covariance tensor

$$\frac{X_{ii}(t)}{I^2} = 2\sigma^2 \exp(2\langle\Psi\rangle) \int_0^t d\tau (t - \tau)\chi_i(\tau), \tag{28}$$

and concurrently the macrodispersion coefficients

$$\frac{D_{ii}(t)}{UI} = \sigma^2 \exp(2\langle\Psi\rangle) \int_0^t d\tau \chi_i(\tau). \tag{29}$$

Let's observe that the time has been scaled by $\frac{I}{U}$ with $U = \frac{K_G}{\vartheta_e}$ (even if we have kept the same notation). As it will be clearer later on, one important aspect related to Equation (29) is that the time dependency of D_{ii} is all encapsulated in the integral of the term χ_i, which implies that (in the spirit of the assumptions underlying the current chapter) the rate of attainment of the Fickian regime is an intrinsic formation property. Furthermore, given the formation structure and for fixed mean pressure head, the enhanced solute spreading is due to an increase in the variance σ_γ^2 in agreement with previous results (see [20, 21, 24]).

At the limit $t << 1$, the behaviour of $\chi_i(t)$ and $D_{ii}(t)$ is the general one resulting from the theory of [31], and it is

$$\frac{X_{ii}(t)}{I^2} = \sigma^2 \exp(2\langle\Psi\rangle)\chi_i(0)t^2 \qquad \frac{D_{ii}(t)}{UI} = \sigma^2 \exp(2\langle\Psi\rangle)\chi_i(0)t. \qquad (30)$$

The large time behaviour will be considered in the following.

4 Discussion

We want to quantify the impact of the relevant formation properties as well as mean pressure head upon the velocity spatial distribution and solute spreading . Before going on, we wish to identify here some practical values of the input parameters σ_γ^2.

Following the field studies of [19, 26, 29], we shall assume that Y and ζ are independent RSFs, so that their cross-correlation can be ignored. The assumption of lack of correlation between Y and ζ may be explained by the fact that Y is controlled by structural (macro) voids, whereas ζ (or any other parameter related to the pore size distribution) is controlled by the soil texture [12].

The variance of ζ can be either larger or smaller than that of Y. For example, [32] reported the ζ-variance in the range 0.045–0.112 and the variance of Y in the range 0.391–0.960, whereas [26] found $\sigma_\gamma^2 = 0.425$ compared with $\sigma_\gamma^2 = 1.242$. Instead, [34], and [19] found the variances of ζ and Y to be of similar order, whereas [14, 15] observed the variance of ζ to exceed that of Y. Thus, in what follows we shall allow $\frac{\sigma_\gamma^2}{\sigma_Y^2}$ to vary between 10^{-2} and 10 in order to cover a wide range of practical situations.

4.1 Velocity Analysis

In Fig. 1, the scaled velocity variances $\frac{u_{ii}(0)}{(U\sigma)^2}\exp(-2\langle\Psi\rangle) = \chi_i(0)$ have been depicted versus the anisotropy ratio λ. Generally, the curves $\chi_i(0)$ are monotonic decreasing approaching to zero for large λ. In particular, $\chi_m(0)$ vanish faster than $\chi_1(0)$ in accordance with Equation (13). At small λ, the curves $\chi_i(0)$ assume the constant values (20)–(21). From Fig. 1, it is also seen that $\chi_m(0) < \chi_1(0)$. The physical interpretation of this is that fluid particles can circumvent inclusions of low conductivity laterally in two directions, causing them to depart from the mean trajectory to a lesser extent than in the longitudinal direction. Moreover, assuming different shapes of ρ does not significantly impact $\chi_i(0)$. Finally, it is worth observing that the $\chi_i(0)$ curves calculated for the exponential model lead to the same expression of the flux variances obtained by [36].

To grasp the combined effect of the medium heterogeneity and the mean pressure head upon the velocity variance, we have depicted in Fig. 2 the contour lines of the ratio $\frac{u_{ii}(0)}{\chi_i(0)(U\sigma_Y)^2} = \Lambda$.

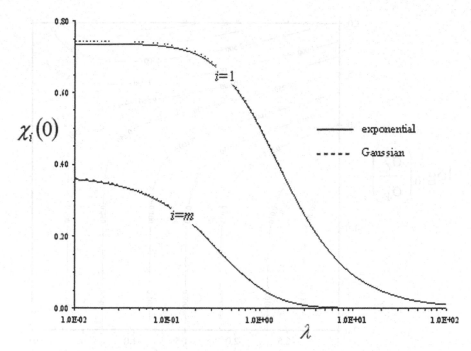

Fig. 1 Dependence of the scaled variances $\chi_i(0)$ for both the exponential (solid line) and the Gaussian model (dashed line) upon the anisotropy ratio λ

The Λ-function decreases when (i) the variability of the capillary forces reduces compared with that of the saturated conductivity (for fixed $<\Psi>$) and (ii) the porous formation becomes drier in mean (for given $\frac{\sigma_\zeta^2}{\sigma_Y^2}$). This last result is due to the fact that an increase of suction reduces *de facto* the fluid particles mobility. For $\sigma_\zeta^2 < \frac{\sigma_Y^2}{10}$, the isolines are practically vertical suggesting that in this case, the velocity variance is mainly influenced by the mean flow conditions. To the contrary, for $\sigma_\zeta^2 > \sigma_Y^2$, the isolines tend to become horizontal implying that the flow conditions have a negligible effect upon Λ.

The relationship (7) shows that the contributions of the mean flow combined with the variance of the input parameters and the domain structure are separated into a multiplicative form. As a consequence, the velocity autocorrelation

$$\rho_{u_{ii}}(r) = \frac{C_{u_{ii}}(r)}{C_{u_{ii}}(0)} = \frac{\chi_i(r)}{\chi_i(0)} \tag{31}$$

is a formation property, and therefore it is applicable to both saturated and unsaturated porous media. This is demonstrated by the good agreement (Fig. 3a, b) between (31) calculated for the exponential ρ (solid line) and the velocity autocorrelation (symbols) (obtained by [23]) for the same model of ρ,

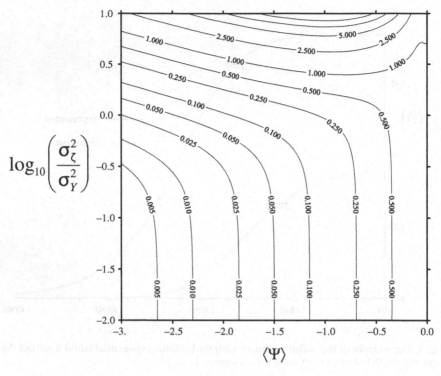

Fig. 2 Contour lines of Λ versus the scaled mean pressure head $\langle \Psi \rangle$, and the ratio $\sigma_\zeta^2 / \sigma_Y^2$

and dealing with a mean groundwater flow normal to the formation bedding (Equations (15a)–(15b) in [23]). Such a result was anticipated by Russo [24], who used the velocity covariance obtained for saturated flow conditions to analyze transport in a vadose zone.

For completeness, in Fig. 3a, b we have also reported the velocity autocorrelation (dashed line) corresponding with the Gaussian model of ρ. At a first glance, it is seen that the shape of ρ plays an important role in determining the behaviour of the velocity correlation. More precisely, close to the origin the autocorrelation ρ_{u_i} evaluated via the exponential ρ decays faster than that obtained by adopting the Gaussian model, whereas the opposite is observed at large r. This is due to the faster vanishing with distance of the Gaussian model compared with the exponential one. Similar results were obtained by other authors (for a wide review, see [38]).

The dependence of the velocity autocorrelation (Fig. 3a, b) upon the anisotropy ratio λ can be explained as follows: as water moves in the porous medium, a low conductivity inclusion (say a flow barrier) occurs that makes the velocity covariance decay with the separation distance. The possibility to circumvent a flow barrier will drastically decrease, and concurrently the distance over which the velocity is correlated will increase, as the typical horizontal size of the flow barrier is prevailing upon the vertical one, and vice versa. This justifies the

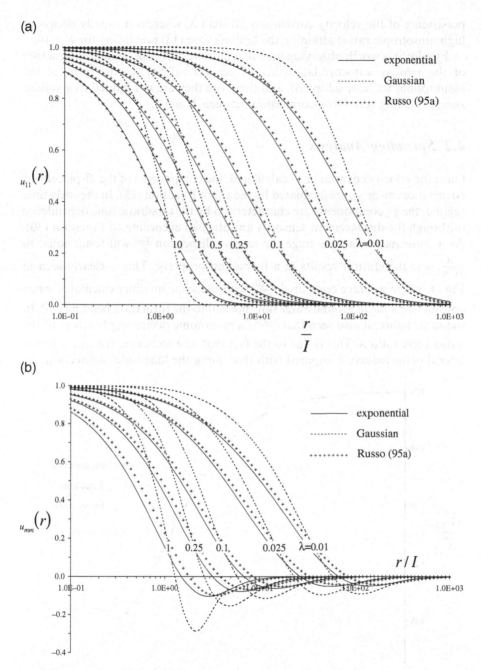

Fig. 3 (a) Longitudinal and (b) transversal components of the autocorrelation velocity versus the dimensionless distance (r/I) for various values of the anisotropy ratio λ

persistence of the velocity correlation at small λ, whereas it rapidly decays at high anisotropic ratios attaining the limiting case (13) practically for $\lambda \geq 10$.

Finally, it is worth observing here that basic results concerning the structure of the velocity autocorrelation are of general validity (in the spirit of the assumptions we have adopted), and therefore they are not limited to a vadose zone exhibiting zero cross-correlation between Y and ζ.

4.2 Spreading Analysis

Once the velocity covariance is calculated, the components of the displacement covariance tensor χ_{ii} are evaluated by the aid of Equation (28). In the early time regime, the χ_{ii} components are characterized by the quadratic time dependence (although the displacement tensor is anisotropic) according to Equation (30). As a consequence, at this stage the relative dispersion $\frac{\chi_{mm}}{\chi_{11}}$ will result equal to $\frac{\chi_m(0)}{\chi_1(0)}$, and therefore it results in a formation property. This is clearly seen in Fig. 4, where we have compared $\frac{\chi_m(0)}{\chi_1(0)}$ with the same quantity calculated when considering a mean groundwater flow normal to the formation bedding [23]. In the same figure, it also seen that $\frac{\chi_{mm}}{\chi_{11}}$ is a monotonic decreasing function of the anisotropic ratio λ. This is due to the fact that as λ increases, the tortuosity of lateral paths reduces compared with that along the longitudinal direction. As

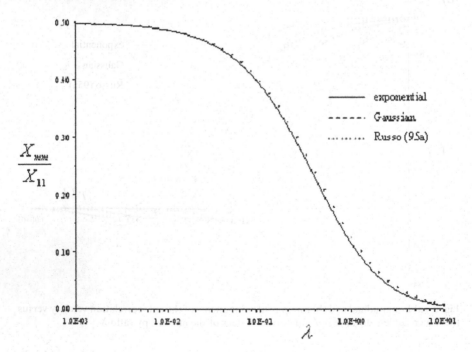

Fig. 4 Dependence of the early time relative dispersion X_{mm}/X_{11} upon the anisotropy ratio λ

expected from Equation (19), at small anisotropic ratios (say for $\lambda < 0.01$), $\frac{X_{mm}}{X_{11}}$ is equal to $\frac{1}{2}$ irrespective of ρ. At any rate, even at higher λ, the shape of ρ is found practically immaterial.

By considering exponential ρ, the computation of $\chi_{ii}(t)$ is reduced to one integral solely:

$$X_{11}(t) = 2\left(\frac{I\sigma}{\lambda}\right)^2 \exp(2\langle\Psi\rangle) \int_0^{1/\lambda} du \bar{\Gamma}(u,t)\left\{\frac{1 - u\zeta(u)}{u[1 - u\zeta^2(u)]}\right\}^2 \qquad (32a)$$

$$X_{mm}(t) = \left(\frac{I\sigma}{\lambda}\right)^2 \exp(2\langle\Psi\rangle) \int_0^{1/\lambda} du \bar{\Gamma}(u,t) \frac{\zeta(u)}{u} \frac{[1 - u\zeta(u)]}{[1 - u\zeta^2(u)]^2} \qquad (32b)$$

$$\bar{\Gamma}(u,t) = 2\zeta(u)\exp[-\lambda u\zeta(u)t] + \lfloor(\lambda ut - 1)\zeta^2(u) - (\lambda ut + 1)\rfloor \exp(-\lambda ut) + [\zeta(u) - 1]^2. \qquad (33)$$

Like groundwater flow [5], for an isotropic formation we can express (32a)–(32b) in analytical form, that is,

$$X_{11}(t) = 2(I\sigma)^2 \exp(2\langle\Psi\rangle)\{t - [\ln t - E_i(-t) + E]\} \qquad (34a)$$

$$X_{mm}(t) = (I\sigma)^2 \exp(2\langle\Psi\rangle)\{[E_i(-t) - E_i(-2t)]\exp(t) - (\ln 2 - 1)\exp(-t) + 2\ln 2 - 1\} \qquad (34b)$$

(E represents the constant of Euler–Mascheroni).

The dependence of the scaled longitudinal covariance $\frac{X_{11}(t)}{(I\sigma)^2}\exp(-2\langle\Psi\rangle) = \bar{X}_{11}(t)$ upon the time and for several values of the anisotropic ratio has been depicted in Fig. 5. As the solute body invades the flow system, $\bar{X}_{11}$ grows monotonically with the time. The early as well as large time regime are those predicted by the theory of Taylor [31], that is, a nonlinear time dependence (whose persistence increases as the anisotropic ratio reduces) on short times, whereas for t large enough, $\bar{X}_{11}$ grows linearly with time. Regarding the dependence of $\bar{X}_{11}(t)$ upon the anisotropy ratio, preliminary simulations have shown that for $\lambda \leq 0.01$, the longitudinal covariance components can be evaluated by adopting Equation (11) for $\chi_1(\tau)$. Instead, for $\lambda > 10$, we obtain from (13) and (28)–(29) the following results:

$$\frac{X_{11}(t)}{I^2} = \frac{2\sigma^2}{\lambda}\exp(2\langle\Psi\rangle)\Xi(t) \qquad \frac{D_{11}(t)}{UI} = \frac{\sigma^2}{\lambda}\exp(2\langle\Psi\rangle)\Delta(t) \qquad (35)$$

with

$$\Xi(t) = t + \rho_h(t) - 1 \qquad \Delta(t) = 1 - \rho_h(t) \qquad (36)$$

for exponential ρ_h, whereas

$$\Xi(t) = t\,\mathrm{erf}(\sqrt{\pi}t) + \frac{2}{\pi}[\rho_h(t) - 1] \qquad \Delta(t) = \mathrm{erf}(\sqrt{\pi}t) \qquad (37)$$

for Gaussian ρ_h.

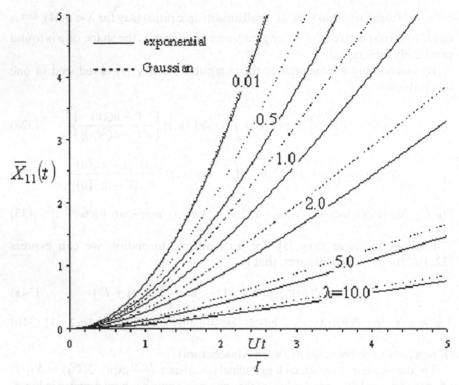

Fig. 5 Longitudinal component of the time-dependent scaled displacement covariance tensor $\bar{X}_{11}(t)$ for different values of the anisotropy ratio λ

The behaviour of the transverse components in the early time regime is similar to X_{11} in the sense that $\frac{X_{mm}(t)}{(I\sigma)^2}\exp(-2\langle\Psi\rangle) = \bar{X}_{mm}(t)$ increases with the travel time at a faster rate than linearly. However, unlike the longitudinal components, X_{mm} approach constant asymptotic values as the solute body moves (see also [20]). In particular, for exponential ρ, integration of Equation (32b) with respect to u is carried out after taking the limit $t \to \infty$ to yield

$$X_{mm}(\infty) = \frac{sg(4\lambda^2 - 3)}{\lambda^2|4\lambda^2 - 3|^{3/2}}\left[(1 - 2\lambda)\sqrt{|4\lambda^2 - 3|} + 2(2\lambda^2 - 1)\beta(\lambda)\right] \quad (38a)$$

$$\beta(\lambda) = \begin{cases} 2\arctan\left(\frac{\sqrt{3-4\lambda^2}}{1+2\lambda}\right) & 4\lambda^2 \leq 3 \\ \ln\left\{\left[\frac{P_2(\lambda)+\lambda\sqrt{4\lambda^2-3}}{P_2(\lambda)-\lambda\sqrt{4\lambda^2-3}}\right]\left(\frac{1+\sqrt{4\lambda^2-3}}{1-\sqrt{4\lambda^2-3}}\right)\right\} & 4\lambda^2 \geq 3. \end{cases} \quad (38b)$$

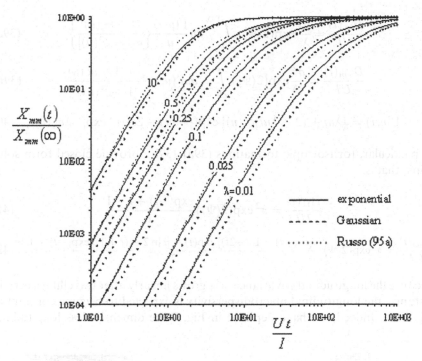

Fig. 6 Lateral component of the time-dependent scaled displacement covariance tensor $X_{mm}(t)/X_{mm}(\infty)$ for different values of the anisotropy ratio λ

To illustrate the impact of the formation anisotropy on the rate at which $X_{mm}(t)$ approach their asymptotic values, we have depicted in Fig. 6 the normalized transverse variance $\frac{X_{mm}(t)}{X_{mm}(\infty)}$ for both the exponential (solid line) and the Gaussian model (dashed line) versus the time, and for different values of λ.

At given time, as λ decreases, the ratio $\frac{X_{mm}(t)}{X_{mm}(\infty)}$ reduces, and vice versa. In fact, a large λ means that any low conductivity inclusion is less than an obstruction to the vertical flow, as streamlines can easily bypass it laterally, thus reducing the transverse spread. This enables the solute particles to approach the Fickian regime faster. Similarly to the previous section, a fundamental consequence of the separation structure of Equation (28) is that the distance needed to reach Fickian conditions is an intrinsic formation property. To show this, we have depicted in Fig. 6 the ratio (symbols) between the transverse displacement covariance and their asymptotic values as calculated from equations (21b) and (23) reported in [23], which are valid for groundwater flow normal to the formation bedding.

The macrodispersion coefficients are calculated in a straightforward manner by combining the general result (29) with the above-derived displacement covariance tensor. Thus, for $\rho = \exp(-r)$, we can express D_{ii} in terms of one quadrature solely:

$$\frac{D_{11}(t)}{UI} = \frac{\sigma^2}{\lambda}\exp(2\langle\Psi\rangle)\int_0^{1/\lambda} du\frac{\bar{\bar{\Gamma}}(u,t)}{u}\left\{\frac{1-u\zeta(u)}{u[1-\zeta^2(u)]}\right\}^2 \tag{39a}$$

$$\frac{D_{mm}(t)}{UI} = \frac{\sigma^2}{2\lambda}\exp(2\langle\Psi\rangle)\int_0^{1/\lambda} du\bar{\bar{\Gamma}}(u,t)\frac{\zeta(u)[1-u\zeta(u)]}{[1-\zeta^2(u)]^2} \tag{39b}$$

$$\bar{\bar{\Gamma}}(u,t) = [\lambda ut + (2-\lambda ut)\zeta^2(u)]\exp(-\lambda ut) - 2\zeta^2(u)\exp[-\lambda u\zeta(u)t] \tag{40}$$

In particular, for isotropic formations (39a)–(39b), provide closed form solutions, that is,

$$\frac{D_{11}(t)}{UI} = \sigma^2\exp(2\langle\Psi\rangle)\frac{\exp(-t)+t-1}{t} \tag{41}$$

$$\frac{D_{mm}(t)}{UI} = \sigma^2\exp(2\langle\Psi\rangle)\frac{t[E_i(-t)-E_i(-2t)]\exp(t)+[t\ln 2-(t+1)]\exp(-t)+1}{2t}. \tag{42}$$

Because the longitudinal covariance X_{11} grows linearly after travelling a certain distance, the longitudinal macrodispersivity is expected to become constant at this limit. Indeed, we have depicted in Fig. 7 the dimensionless longitudinal

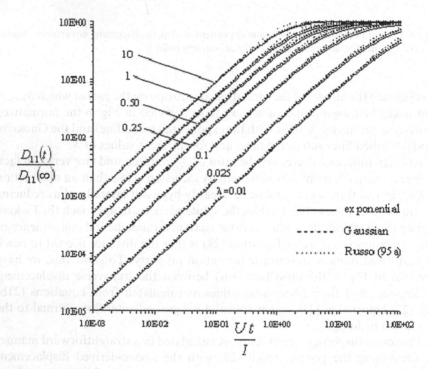

Fig. 7 Longitudinal scaled macrodispersion coefficient $D_{11}(t)/D_{11}(\infty)$ as function of the dimensionless time Ut/I and for different values of the anisotropy ratio λ

macrodispersion coefficient $\frac{D_{11}(t)}{D_{11}(\infty)}$ versus the time for few values of the aniso-tropy ratio. The curves are plotted by considering the exponential as well as Gaussian model for ρ. The lower λ is, the larger is the distance the solute body has to traverse in order to get its asymptotic regime. This is because low values of λ require the solute plume to travel a longer distance to fully sample the formation heterogeneity and therefore to reach the Fickian regime. In the same figure, we have also depicted the same ratio $\frac{D_{11}(t)}{D_{11}(\infty)}$ calculated for groundwater flow normal to the bedding (see [23]) to emphasize the intrinsic character of the distance needed to reach the asymptotic regime.

The transverse components of the macrodispersion tensor tend to zero at large t. This is a direct consequence of the fact that at this limit, X_{mm} tend to a finite value (see Fig. 6). Nevertheless, the magnitude of the transverse macro-dispersion coefficients is deeply influenced by the formation anisotropy. Indeed, in Fig. 8 we have reported the quantity $\frac{D_{mm}(t)}{U I\sigma^2}\exp(-2\langle\Psi\rangle)$ as a function of time. After an initial growth faster than linearly in t, D_{mm} drops to zero more slowly as λ decreases (in line with the physical interpretation given to explain the behaviour of X_{mm}). Finally, let's observe that the lateral dispersion is higher for Gaussian ρ at early times, whereas the opposite is seen asymptotically. This

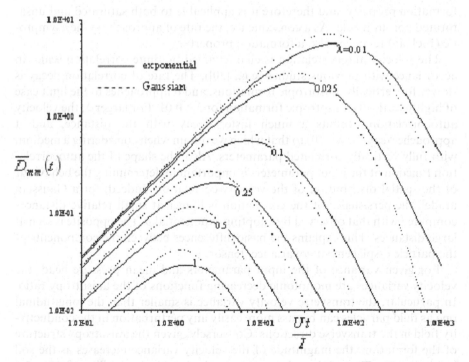

Fig. 8 Time dependence of the scaled lateral macrodispersion coefficients $\overline{D}_{mm}(t)$ for different values of the anisotropy ratio λ

is a consequence of the fact that at small arguments (see Fig. 3b), the Gaussian function causes a longer persistence of the correlation compared with the exponential model, whereas the opposite is observed at higher arguments.

5 Conclusions

The basic aim of the current chapter has been to analyze flow velocity and solute spreading in partially saturated heterogeneous porous formations by means of analytical tools. We have obtained simple results that permit one to grasp the main features of transport phenomena taking place in a vadose zone. This has been accomplished by combining the general Lagrangian formulation developed in the past (e.g., [5, 20]), and relating the spatial moments of a moving solute body to the velocity field, with the analytical results of [30] concerning steady unsaturated flow in an unbounded domain.

One of the main findings is the representation (27) of the velocity covariance . It is expressed by means of two quadratures, and it is valid for arbitrary (cross)correlation functions. In addition, simple results are derived by dealing with highly vertically anisotropic formations (i.e., $\lambda < < 1$), and with the case of $\lambda > > 1$. It is demonstrated that the velocity autocorrelation is an intrinsic formation property, and therefore it is applicable to both saturated and unsaturated porous media. As a consequence, the rate of approaching the asymptotic (Fickian) regime is also a formation property.

The velocity autocorrelation is characterized by finite correlation scales in accordance with previous studies (e.g., [20]). The rate of correlation decay is slower for vertically anisotropic formations, and it approaches to the limit case of highly vertically anisotropic formation for $\lambda < 0.01$. For larger λ, the velocity autocorrelation exhibits a much faster decay with the distance, and it approaches (say for $\lambda > 10$) to that which we obtain when considering a medium with fully vertically correlated parameters. Also, the shape of the autocorrelation function of the input parameters is important in determining the behaviour of the spatial distribution of the velocity correlation. Indeed, for a Gaussian model, the persistence of the correlation is longer at small relative distances compared with that observed by adopting exponential ρ. The opposite is seen at large distances. This explains the basic differences between the components of the particle displacement covariance tensor.

For given variance of the input parameters and mean pressure head, the velocity variances are monotonic decreasing functions of the anisotropy ratio. In particular, the transverse velocity variance is smaller than the longitudinal one as fluid particles can bypass more easily any perturbation in the conductivity field in the transverse directions. Conversely, given the anisotropy structure of the formation, the magnitude of the velocity variance increases as the soil becomes more saturated, and it is enhanced when the variability of the saturated conductivity is small compared with that of the capillary forces.

The dimensionless $\frac{\chi_{ii}(t)}{(I_\sigma)^2}\exp(-2\langle\Psi\rangle)$ components of the displacement covariance tensor are in agreement with the general Lagrangian results concerning flow in random fields (see [5, 23, 24]). An increase of suction reduces the fluid particles mobility, and therefore the magnitude of the displacements covariance. In the case of exponential model for ρ, analytical closed forms are derived for χ_{mm} in the asymptotic regime. The behaviour of the macrodispersion coefficients $D_{ii}(t)$ reflects that of the covariance displacements. Thus, $D_{11}(t)$ asymptotically tends to the constant (Fickian) value $D_{11}(\infty)$. The rate of approaching to $D_{11}(\infty)$ is an intrinsic formation property. It is also shown that the transverse macrodispersion coefficients $D_{mm}(t)$ asymptotically tend to zero with a rate that is faster for Gaussian ρ.

We would like to emphasize that the current analysis of water flow and solute transport in a vadose zone is based on a small-perturbation analysis following the linear theory of Dagan [4]. It is generally accepted [5] that under saturated conditions, the linear theory captures the foremost features of transport phenomena whenever the variability of the saturated hydraulic conductivity is small (i.e., $\sigma_Y^2 \ll 1$). However, numerical simulations [2] as well as theoretical studies [6] on transport under steady groundwater flow show a broader range of applicability (say $\sigma_Y^2 \approx 1$) because of the mutual cancellation effects of the errors induced by the linearization procedure. In the case of unsaturated flow, the nonlinear effects are expected to be larger because of the larger variability of the random input parameters [20]. However, because we have shown that the structure of the velocity autocorrelation is an intrinsic formation property, it is reasonable to expect that even for the unsaturated flow, the linear theory is applicable under a relatively wide range of practical situations. This has been tested numerically by Russo [18].

Finally, we want to underline that although the Fickian regime attainment is an intrinsic medium property, it is worth recalling that such a result relies on the assumptions underling the current study. Hence, it should not be employed in situations (like unsaturated flow close to the groundwater table or when accounting for the water content variability) that are not in compliance with such assumptions.

Acknowledgments The authors express sincere thanks to Dr. Gabriella Romagnoli for reviewing the manuscript. This study was supported by the grant PRIN (# 2004074597).

References

1. Byers, E., and Stephans, D., B., 1983, Statistical and stochastic analyses of hydraulic conductivity and particle-size in fluvial sand, Soil Sci. Soc. Am. J., 47, 1072–1081.
2. Chin, D., A. and Wang, T., 1992, An investigation of the validity of first-order stochastic dispersion theories in isotropic porous media, Water Resour. Res., 28, 1531–1542.
3. Dagan, G., and Bresler, E., 1979, Solute dispersion in unsaturated heterogeneous soil at field scale: theory, Soil Sci. Soc. Am. J., 43, 461–467.

4. Dagan, G., 1984, Solute transport in heterogeneous porous formations, J. Fluid Mech., 145, 151–177.
5. Dagan, G., 1989, Flow and Transport in Porous Formations, Springer.
6. Dagan, G., Fiori, A., and Janković, I., 2003, Flow and transport in highly heterogeneous formations: 1. Conceptual framework and validity of first-order approximations, Water Resour. Res., 39, 1268, doi: 10.1029/2002WR001717.
7. Fiori, A., Indelman, P. and Dagan, G., 1998, Correlation structure of flow variables for steady flow toward a well with application to highly anisotropic heterogeneous formations, Water Resour. Res., 34, 699–708.
8. Gardner, W., R., 1958, Some steady state solutions of unsaturated moinsture flow equations with application to evaporation from a water table, Soil Sci., 85, 228–232.
9. Gelhar, L., W., 1993, Stochastic Subsurface Hydrology, Prentice Hall.
10. Harter, Th., and Zhang, D., 1999, Water flow and solute spreading in heterogeneous soils with spatially variable water content, Water Resour. Res., 35, 415–426.
11. Indelman, P., Or, D. and Rubin, Y., 1993, Stochastic analysis of unsaturated steady state flow through bounded heterogeneous formations, Water Resour. Res., 29, 1141–1147.
12. Jury, W., A., Russo, D., and Sposito, G., 1987, The spatial variability of water and solute transport properties in unsaturated soil, II. Analysis of scaling theory, Hilgardia, 55, 33–57.
13. Nielsen, D., R., Biggar, J., W., and Erh, K., T., 1973, Spatial variability of field measured soil-water properties, Hilgardia, 42, 215–259.
14. Ragab, R., and Cooper, J., D., 1993a, Variability of unsaturated zone water transport parameters: implications for hydrological modelling, 1. In-situ measurements, J. Hydrol., 148, 109–131.
15. Ragab, R., and Cooper, J., D., 1993b, Variability of unsaturated zone water transport parameters: implications for hydrological modelling, 2. Predicted vs. in-situ measurements and evaluation methods, J. Hydrol., 148, 133–147.
16. Reynolds, W., D., and Elrick, D., E., 1985, In-situ measurement of field-saturated hydraulic conductivity, sorptivity, and the α-parameter using the Guelph permeameter, Soil Sci., 140, 292–302.
17. Russo, D., and Bresler, E., 1981, Soil hydraulic properties as stochastic processes, 1, Analysis of field spatial variability, Soil Sci. Soc. Am. J., 45, 682–687.
18. Russo, D., 1988, Determining soil hydraulic properties by parameter estimation: on the selection of a model for the hydraulic properties, Water Resour. Res., 24, 453–459.
19. Russo, D, and Bouton, M., 1992, Statistical analysis of spatial variability in unsaturated flow parameters, Water Resour. Res., 28, 1911–1925.
20. Russo, D., 1993, Stochastic modeling of macrodispersion for solute transport in a heterogeneous unsaturated porous formation, Water Resour. Res., 29, 383–397.
21. Russo, D., Zaidel, J., and Laufer, A., 1994a, Stochastic analysis of solute transport in partially saturated heterogeneous soils: I. Numerical experiments, Water Resour. Res., 30, 769–779.
22. Russo, D., Zaidel, J., and Laufer, A., 1994b, Stochastic analysis of solute transport in partially saturated heterogeneous soils: II. Prediction of solute spread and breakthrough, Water Resour. Res., 30, 781–790.
23. Russo, D., 1995a, On the velocity covariance and transport modeling in heterogeneous anisotropic porous formations 1. Saturated flow, Water Resour. Res., 31, 129–137.
24. Russo, D., 1995b, On the velocity covariance and transport modeling in heterogeneous anisotropic porous formations 2. Unsaturated flow, Water Resour. Res., 31, 139–145.
25. Russo, D., 1995c, Stochastic analysis of the velocity covariance and the displacement covariance tensors in partially saturated heterogeneous anisotropic porous formations, Water Resour. Res., 31, 1647–1658
26. Russo, D., Russo, I., and Laufer, A., 1997, On the spatial variability of parameters of the unsaturated hydraulic conductivity, Water Resour. Res., 33, 947–956.

27. Russo, D., 1998, Stochastic analysis of flow and transport in unsaturated heterogeneous porous formations: effects of variability in water saturation, Water Resour. Res., 34, 569–581.
28. Russo D., 2002, Stochastic analysis of macrodispersion in gravity-dominated flow through bimodal heterogeneous unsaturated formations, Water Resour. Res., 38, 1114, doi: 10.1029/2001WR000850.
29. Severino, G., Santini, A., and Sommella, A., 2003, Determining the soil hydraulic conductivity by means of a field scale internal drainage, J. Hydrol., 273, 234–248.
30. Severino, G., and Santini, A., 2005, On the effective hydraulic conductivity in mean vertical unsaturated steady flows, Adv. Water Resour., 28, 964–974.
31. Taylor, G., I., 1921, Diffusion by continuous movements, Proc. Lond. Math. Soc., A20, 196–211.
32. Ünlü, K., Nielsen, D., R., Biggar, J., W., and Morkoc, F., 1990, Statistical parameters characterizing the spatial variability of selected soil hydraulic properties, Soil Sci. Soc. Am. J., 54, 1537–1547.
33. White, I., and Sully, M. J., 1987, Macroscopic and microscopic capillary length and time scales from field infiltration, Water Resour. Res., 23, 1514–1522.
34. White, I., and Sully, M., J., 1992, On the variability and use of the hydraulic conductivity alpha parameter in stochastic treatment of unsaturated flow, Water Resour. Res., 28, 209–213.
35. Yeh, T-C, J., Gelhar, L., W., and Gutjahr, A., 1985a, Stochastic analysis of unsaturated flow in heterogenous soils 1. Statistically isotropic media, Water Resour. Res., 21, 447–456.
36. Yeh, T-C, J., Gelhar, L., W., and Gutjahr, A., 1985b, Stochastic analysis of unsaturated flow in heterogenous soils 2. Statistically anisotropic media with variable α, Water Resour. Res., 21, 457–464.
37. Zhang, D., Wallstrom, T., C., and Winter, C., L., 1998, Stochastic analysis of steady-state unsaturated flow in heterogeneous media: comparison of the Brooks-Corey and Gardner-Russo models, Water Resour. Res., 34, 1437–1449.
38. Zhang, D., 2002, Stochastic Methods for Flow in Porous Media, Academic Press.

Genome Analysis of Species of Agricultural Interest

Maria Luisa Chiusano, Nunzio D'Agostino, Amalia Barone, Domenico Carputo, and Luigi Frusciante

Abstract In recent years, the role of bioinformatics in supporting structural and functional genomics and the analysis of the molecules that are expressed in a cell has become fundamental for data management, interpretation, and modeling. This interdisciplinary research area provides methods that aim not only to detect and to extract information from a massive quantity of data but also to predict the structure and function of biomolecules and to model biological systems of small and medium complexity. Although bioinformatics provides a major support for experimental practice, it mainly plays a complementary role in scientific research. Indeed, bioinformatics methods are typically appropriate for large-scale analyses and cannot be replaced with experimental approaches. Specialized databases, semiautomated analyses, and data mining methods are powerful tools in performing large-scale analyses aiming to (i) obtain comprehensive collections; (ii) manage, classify, and explore the data as a whole; and (iii) derive novel features, properties, and relationships. Such methods are thus suitable for providing novel views and supporting in-depth understanding of biological system behavior and designing reliable models.

The success of bioinformatics approaches is directly dependent on the efficiency of data integration and on the value-added information that it produces. This is, in turn, determined by the diversity of data sources and by the quality of the annotation they are endowed with. To fulfill these requirements, we designed the computational platform ISOLA, in the framework of the International Solanaceae Genomics Project. ISOLA is an Italian genomics resource dedicated to the Solanaceae family and was conceived to collect data produced by 'omics' technologies. Its main features and tools are presented and discussed as an example of how to convert experimental data into biological information that in turn is the basis for modeling biological systems.

M.L. Chiusano
Department of Soil, Plant, Environmental and Animal Production Sciences,
University of Naples Federico II, Naples, Italy
e-mail: chiusano@unina.it

P.J. Papajorgji, P.M. Pardalos (eds.), *Advances in Modeling Agricultural Systems*, 385
DOI 10.1007/978-0-387-75181-8_18, © Springer Science+Business Media, LLC 2009

1 Introduction

Bioinformatics is a discipline in which biology, computer science, statistics, applied mathematics, and information technology merge. The use of computers to study biological systems encompasses the design of methods and the implementation of algorithmic tools to facilitate the collection, organization, and analysis of large amounts of data. Most molecular data are in the form of a succession of letters that summarizes a basic level of the structural organization of a DNA, mRNA, or a protein molecule, commonly defined *primary structure*. This symbolic linear depiction is referred to as a *biomolecular sequence*. In this mold, molecular data can be easily edited, manipulated, compared, and analyzed. Computational methods proved early on to be especially suitable for many of these basic approaches. The growing availability of 'sequence data' combined with the 'high throughput' technologies makes bioinformatics essential in supporting the analysis of the structure and function of biological molecules. To this end, bioinformatics plays a key role in data mining and has broad applications in the molecular characterization of an organism's gene and protein space (i.e., genomics and proteomics), in the genome-wide study of mRNA expression (transcriptomics), the systematic study of the chemical fingerprints that specific cellular processes leave behind (metabolomics), drug discovery, and in the identification of biomarkers as biological indicators of disease, toxicity of pathogens, or effectiveness of healing. Nevertheless, the substantial improvement in technologies and data processing tools also drives the generation of 'digital data' and metadata that can be investigated by large-scale analytical strategies.

Despite the advances in experimental strategies and computational approaches that support the characterization of genomes and their products, only 10% of genome organization and functionality is today understood. This means that we are still far from achieving the ambitious goal of the *in silico* simulation of complex living systems. Indeed, this would require deeper knowledge of the structure and functions of an organism's genome, transcriptome, and proteome and of their links with cellular physiology and pathophysiology. The real challenge of bioinformatics is to design suitable computational methods for revealing the information that biological data still hide and to integrate the large amount of the so-called 'omics' data to approach a systems biology view. The long-term goal is the creation of models to simulate biological system behaviors and their exploitation for applications in medicine, biotechnologies, and agricultural sciences.

This chapter presents an overview of the importance of 'omics' approaches in agriculture and plant science. Bioinformatics strategies for exploiting 'omics' data follows. The discussion is mainly focused on the state of the art of genome data analysis and on the methods employed for gene discovery and for comprehension of gene functionalities. The purpose is to highlight the strong connection between bioinformatics and the need for models that describe genome functionalities of living organisms.

Finally, the chapter presents our efforts to organize and integrate 'omics' data to model genetic expression maps and study regulatory networks and mechanisms of genetic control. As partners of the consortium established by the International Solanaceae (SOL) Genomics Network (http://www.sgn.cornell.edu/solanaceae-project/), the authors designed and currently maintain an Italian Solanaceae genomics resource (ISOLA), which was conceived as a multilevel computational environment based on effective data integration using comparative and evolutionary approaches. The computational platform described as a sample application, together with the multiple aspects of data modeling it covers, may represent a reference for analysis and modeling of molecular mechanisms in species of agricultural interest. The strategy of *moving from the 'omics' to systems biology* is one of the possible pathways for tackling the incipient challenges in bioinformatics.

2 Genome Analysis and Applications in Agriculture

Genome analysis involves a wide variety of studies and is aimed at understanding the structure, function, and evolution of all the hereditary information of a given organism. Knowledge of the genetic make-up of living organisms could dramatically and effectively change our approach to agriculture. Species with large amounts of DNA make genetic studies and manipulations difficult. By contrast, species with a small DNA content can be more easily analyzed. Unsurprisingly, the latter species were the first candidates for genome sequencing projects. Because of major differences among living organisms in terms of fundamental characteristics, their genes must encode different enzymes that are necessary for their metabolic processes. Therefore, a wide range of variation among living organisms is found in terms of DNA quantity, karyotype (i.e., number, type, shape, etc., of chromosomes), and DNA sequence. As reported by Cullis [1], genomes show amazing diversity in terms of shapes (chromosomes) and sizes. Nuclear DNA content has been generally referred to as the amount of DNA contained in the nucleus of a gamete (the 1C value), without any regard to the actual ploidy of the gamete-producing individuals. The largest range of nuclear DNA content occurs in the plant kingdom, even between closely related species. In plants, one of the smallest genomes belongs to *Arabidopsis thaliana* (0.125 picogram), which is about 200 times smaller than other plant genomes (www.rbgkew.org.uk/cval/homepage.html). The largest genome reported to date belongs to *Fritillaria assyriaca*, with roughly 127 pg. This represents a 1000-fold difference between the largest and smallest genomes characterized to date in the plant kingdom. Living organisms show high variability also in terms of somatic (2n) chromosome numbers (e.g., among animals, *Homo sapiens* 2n = 46, *Felix domesticus* 2n = 38, *Drosophila melanogaster* 2n = 8; among plants, *Solanum lycopersicum* 2n = 24, *Arabidopsis thaliana* 2n = 10, *Prunus persica* 2n = 16). A relatively common phenomenon in the plant kingdom is the occurrence of polyploids. These have more than two sets of the basic chromosome number x, and species in which x is a multiple of 6 or 7 are frequent [2]. Additional differences among living organisms are given by their karyotypes. Most of the species possess a single centromere for each chromosome. However, some organisms have polycentric or holocentric chromosomes, with multiple or diffused centromeres.

At the molecular level, the large-scale organization of the chromosomes relies on repetitive DNA sequences. Early studies on the kinetics of the DNA reassociation [3] already indicated that the genome of higher eukaryotes consisted of large amounts of highly repeated DNA, whereas a small portion of the genomic sequence was present as unique or very low copy. Nowadays, most genes found in species with much larger genomes are also present in species with smaller genomes. The smallness of some species genomes is due to the lower abundance of highly repeated sequences compared with that of other species [2]. The proportion of the genome taken up by repetitive DNA varies widely and may represent more than 90% of the whole genome [4]. The sequences in a genome are generally classified with respect to the number of times they are represented. Three main classes can be identified: unique sequences (which probably represent the genes), moderately repetitive sequences, and highly repetitive sequences.

Technologies for acquiring data on genomes have progressed greatly in the past few years, and a large amount of cytological and molecular information has been generated. Particularly attractive is the availability of the complete DNA sequence of hundreds of viruses and bacteria and a few eukaryote organisms, including mammals and flowering plants. Also important are advances that are being made in understanding gene and protein structure, expression, and function. These developments are strongly accelerating the speed at which scientists can exploit the knowledge gained in understanding the biochemical and molecular basis of metabolic, physiological, developmental, and reproductive traits of interest. This can result in several practical applications to challenging problems in agriculture [5]. The generation of high-quality molecular maps, for example, can allow better characterization and manipulation of genes underlying complex noteworthy traits, such as those associated with stress tolerance and yield. In addition, it will make marker-assisted selection possible to a greater extent both for animal and plant breeding efforts [6–9]. Advances in knowledge may also provide novel strategies to stabilize agricultural yields in concert with the environment (e.g., through the development of plants that tolerate drought and pathogens). They will also contribute to the introduction of improved and novel crops/animals that may create new economies based on agricultural products and may increase the diversity of product packaging available to consumers. On this point, the creation of animals that produce biomedically useful proteins in their blood or milk is an attractive perspective. Of equal interest is the development of plants containing more essential macro- and micronutrients and the exploitation of the thousands of secondary metabolites that higher plants synthesize.

In addition, solutions to challenges related to environmental management and energy can be met through the exploitation of genomic knowledge and the application of novel technologies. Reduction in greenhouse gases achievable by the production of plant biofuels and rehabilitation of chemically contaminated sites by phytoremediation are two examples of these synergic effects.

As a whole, the major challenges facing mankind in the 21st century are the need for increased food and fiber production, a cleaner environment, and renewable chemical and energy resources. Plant-based technologies and a deeper understanding of numerous fundamental aspects of plant biology can play a major role in meeting each of these challenges, allowing plants to be used for biomass, chemical feedstocks, and production of biodegradable materials.

Indeed, information gained from genomics (mapping, sequencing, and understanding gene function) is being used to improve traits through genetic engineering and new breeding strategies. Genomics approaches are revolutionizing biology as they affect our ability to answer questions concerning the whole genome, which cannot be answered using a gene-by-gene approach. This could be the case of complex traits such as fruit development in agronomically important species like tomato and pepper, or tuber development in potato, which are controlled by many genes, each exhibiting a function not easily determined. In higher plants, sugars are known to be produced through photosynthesis in leaves and are then transported to other organs (fruits, tubers, seeds, roots) where they are metabolized or stored. This source-sink balance is of basic importance for human food and nutrition. Therefore, the understanding of its genetic determination is fundamental for crop yield and quality.

Besides the comprehension of basic biological processes, the genomics revolution allowed marker-assisted breeding to evolve into genomics-assisted breeding for crop improvement [10]. Indeed, the increasing amount of sequencing information today available for many agriculturally important species and the research that successfully unraveled metabolic pathways have allowed a huge number of new molecular markers for agronomic traits to be discovered and used by breeders. Genomic characterization and germplasm phenotyping are becoming fundamental tools for increasing crop genetic diversity, thus allowing a broad range of genes and genotypes for important traits to be identified. An extremely important advantage deriving from available molecular markers clearly distinguishing genotypes is the realization of genetic traceability of food products. Public questions over food safety are creating demand for food tracking and traceability systems to ensure it, where traceability is defined as the ability to follow and document the origin and history of a food or feed product. DNA-based traceability using Single Nucleotide Polymorphism (SNP) markers is applicable to every organism for which genetic variation is known, thus leading DNA to become nature's bar code to trace products from the consumer to the farm of origin. Moreover, genomics tools could increase the ability to guarantee and protect all agricultural products and their derivatives (milk, meat, vegetable, wine, cheese) that can be better distinguishable and characterized by PDO (Protected Designation of Origin), PGI (Protected Geographical Indication),

TSG (Traditional Speciality Guaranteed) certifications. A wide discussion is still open on the development of breeding and molecular strategies that can efficiently exploit the available genomic resources and genomics research for crop improvement [11], which should combine high-throughput approaches with automation and enhanced bioinformatics techniques.

3 Biological Data Banks and Data Integration

Biological data are collected in molecular databases that may be accessed using the Internet (see Database issue 35 of *Nucleic Acids Research* [2007] for a description of the currently available biological databases).

Biological knowledge is today distributed among many different data banks, which are increasingly indispensable and important tools in assisting scientists to understand and explain biological phenomena despite the trouble to ensure the consistency of information delivered. The spread of these collections was mainly due to the need to organize and distribute the amount of molecular data from 'omics' approaches. Different types of biological databases are available today: large-scale public repositories, meta-databases, genome browser databases, community-specific and project-specific databases. Large-scale public repositories are places for long-term storage and represent the current knowledge on the nucleotide or protein primary sequences of all organisms as well as the three-dimensional structure of biological molecules. They interchange the deposited information and are the source for many other specialized databases. Examples include GenBank for nucleotide sequences [12] and UniProt [13] for protein sequence information and the Protein Data Bank [14] for structure information. A meta-database can be considered a database of databases that collects information from different sources and usually makes them available in new, more convenient and user-friendly forms. Examples include Entrez [15] and GeneCards [16]. Another category of databases are the genome browsers, which enable scientists to visualize and browse entire genomes with annotated data that come usually from multiple diverse sources. Examples include the Ensembl Genome Browser [17], the GMOD Gbrowse Project [18], and the UCSC Genome Browser [19]. There are a number of community-specific databases that address the needs of a particular community of researchers. Examples of community-specific databases are those focused on studying particular organisms [20] or on specific types of data [21]. Project-specific databases are often short-lived databases that are developed for project data management during the funding period. Usually these databases and the corresponding Web resources are not updated beyond the funding period of the project. Since 1996, the journal *Nucleic Acids Research* has provided an annual compilation of the most important databases, noting their growing relevance to working scientists.

Because of the relevance of the data collected for the whole scientific community, data sharing has been the most interesting achievement in science in

recent years. Under the policy of the Bermuda principle (http://www.gene.
ucl.ac.uk/hugo/bermuda.htm) and its extension to large-scale results [22],
data are required to be publicly available and submitted to public repositories
(e.g., sequence data to GenBank). However, there are no standards established
for many data types (e.g., proteomics data, metabolomics data, protein localization, *in situ* hybridization, phenotype description, protein function information). Standards, specifications, and requirements for publication of data in
repositories should be established in general agreement and made accessible to
researchers early on in their data-generation and research activity processes.

Data integration is one of the major challenges in bioinformatics. By means
of the analytical methods available, data with the same content type, but which
originate from different experimental approaches, must be organized, integrated, and analyzed with distinct yet related cognitive goals. Furthermore,
bioinformatics approaches must be successfully applied to integrate different
data types (such as those from genomics, proteomics, and metabolomics
approaches as well as experimental or clinical measurements) to provide an
overview of the biological system under investigation. The long-term goal of
this level of integration is to drive the research community closer to understanding the biological system physiology at a more holistic level. This strategy
is useful for deeper comprehension of the response to stimulation mechanisms.
It can also be exploited for the development of novel diagnostic approaches and
therapies and for the identification of effective biomarkers.

4 Analysis of Biological Sequences: Sequence Comparison and Gene Discovery

Improvement of automated sequencing technology and proliferation of large-
scale sequencing projects have supported the building of bioinformatics infra-
structures for the management, processing, and analysis of genome sequence
data. Sequence comparison, which focuses on finding all significant regions of
similarities among two or more sequences, is the rationale that underpins
understanding of the function and evolutionary processes that act on genomes.
It is a widely used approach in bioinformatics whereby both similarities and
differences in molecular sequences, genomes, RNAs, and proteins of different
organisms are investigated to infer how selection has acted upon these elements.
Furthermore, sequence comparison is fundamental in molecule annotation,
that is, in defining the structural and functional role of molecules, exploiting
the paradigm *structure determines function*. However, whereas this paradigm
holds biologically true generally, caution must be exercised while assigning
function to orphan sequences [23]. Therefore, a suitable method to avoid
wrong or misleading (false positive) deductions is to cross-link structural and
functional information so as to obtain predictions that are as reliable as
possible.

As the value of genome sequences lies in the definition of their structural and functional role, the tasks of genome annotation is the first to be undertaken when facing a genome sequencing project. The primary task of genome annotation involves the identification of gene locations and definition of gene structures into raw genomic sequences. Several *in silico* methods can be used for gene annotation in a genome [24–26]. These methods can be based on *ab initio* predictions or on similarity based methods that identify sequences sharing high level of similarity in other genomes and/or in public sequence databases using specific algorithms (e.g., Smith-Waterman and BLAST [27]). The *ab initio* gene finder tools [28] attempt to recognize coding regions in genomic sequences, exploiting compositional properties in order to discriminate between coding and noncoding segments of the sequence and to detect signals involved in gene specification such as initiation and termination signals, exon and intron boundaries, and so forth. *Ab initio* methods need to be trained on a set of known genes, assuming that all coding regions within a particular genome will share similar statistical properties. Therefore, gene predictors require real models of genes that encode for the different possible classes of molecules such as mRNAs or other RNA types (tRNA, rRNAs, small RNAs) that a genome usually expresses.

Identification of a gene and definition of its structure (exon–intron regions) is the first step toward characterizing genes and hence genome functionality. However, recognizing genes in the DNA sequences remains a problem in genome analysis because the features that define gene structures are not clearly and univocally defined. Although much software for gene finding has been developed to detect both plant and animal genes [29, 30], it is still impossible to overcome limits concerning gene discovery and gene structure prediction because (i) the definition of the exact gene boundaries in a genomic region, such as the detection of the transcription start sites (TSS), is still not accurate; (ii) the prediction of small genes (300 nucleotides or less) or genes without introns is missed by many methods; (iii) partial knowledge about non–protein-coding RNAs (ncRNAs) such as microRNA or small nucleolar RNA (snoRNA); and (iv) the ambiguities concerning the prediction of alternative gene structures, such as alternative transcripts from the same gene locus or alternative splicing from the same gene product. All the issues mentioned above are research topics of the same area as bioinformatics, which is dedicated to designing methods for gene discovery. However, persistent efforts to improve the quality of gene predictions have not yet solved problems related to false-positive and false-negative predictions.

Currently, the most direct and valuable method for protein-coding gene identification and for gene structure definition relies on full-length sequencing cDNAs (i.e., more stable DNA molecules synthesized from mRNA templates), with subsequent alignments of the cDNA sequences to the genomic DNA [31, 32]. However, experimental determination of full-length cDNA sequences is an expensive and time-consuming approach when compared with high-throughput Expressed Sequence Tag (EST) sequencing. An EST represents a tiny portion of an entire mRNA and thus provides a 'tag level' association with an expressed gene sequence. Thus, despite EST intrinsic shortcomings due to limited sequence quality, these data represent a valuable source of information to accomplish the task of gene identification and gene model building [33, 34, 35].

A further aspect to consider while annotating a genome is the analysis of inter-genic regions (i.e., stretches of DNA sequences located between the genes) of 'repetitive DNAs' [36] (which are especially copious in plant genomes [37]) and of the so-called 'junk DNA' (i.e., DNA that has no apparent gene function). Such studies hold great promise for additional insights into the evolution and organization of genomes [38, 39].

5 Transcriptome Analysis

Gene expression is the process by which the information in a gene is made manifest as a biologically functional gene product, such as an RNA molecule or a protein. A genome expresses at any moment only a tiny portion of its genes, and none of the available modeling methods are yet able to suggest which genes are expressed under specific conditions.

Because among the RNA molecules expressed by a genome the mRNA is commonly accepted to be the one that most determines the specific functionality of a cell, the sequencing of mRNA molecules is the goal of many transcriptome projects. However, full-length mRNA sequencing is a more expensive and time-consuming approach than is the high-throughput sequencing of ESTs.

An EST is produced by one-shot sequencing of a cDNA. The resulting sequence is a relatively low-quality fragment whose length is limited by current technology to approximately 500 to 800 nucleotides. ESTs represent the first truly high-throughput technology to have populated the databases and have made the rapid growth of advanced computational studies in biology inevitable. Expressed sequence tags are generated and deposited in the public repository as redundant and unannotated sequences, with negligible biological information content. The weak signal associated with an individual raw EST increases when many ESTs are analyzed together so as to provide a snapshot of the transcriptome of a species. EST sequences are a versatile data source and have multiple applications: due to exponential growth in genome sequencing projects, they are widely used for gene location discovery and for gene structure prediction. As already discussed in the previous section, predictions are usually based on spliced-alignment of source-native ESTs onto the genomic sequences [33]. EST sequences can provide useful information on the putative functionality of the tissue that the mRNA they represent were collected from. They may provide digital gene expression profiles (digital Northern) to infer the expression levels of different genes. The strategy is based on the fact that the number of ESTs is reported to be proportional to the abundance of cognate transcripts in the tissue or cell type used to make the cDNA library [40]. Large-scale computer analyses of EST sequences can be used in the identification and analysis of coexpressed genes [41, 42]. It is important to find genes with similar expression patterns (coexpressed genes) because there is evidence that many functionally related genes are coexpressed and because this coexpression may reveal much about the genes' regulatory systems. Similar analyses can be carried out in order to detect genes exhibiting tissue- or stimuli-specific expression [43]. For this purpose, it is necessary to (i) reconstruct the full-length mRNA from EST fragments exploiting a clustering/assembling procedure [44, 45, 46]; (ii) assign a putative function to the mRNA [47, 48]; (iii) implement classification tools and data-mining techniques for reconstructing expression patterns. Transcript levels for many genes at once can be measured with DNA microarray technology, providing a systematic study of gene expression levels. Microarray experiments offer an efficient way to rapidly analyze the activity of thousands of genes simultaneously aiding gene functional assignment. Such experiments produce large amounts of data that are analyzed using data mining techniques. The key impasse is the interpretation of the results and the selection of those transcripts on which to focus attention.

The discussion on the *minimum* information required to interpret unambiguously and potentially reproduce and verify an array-based gene expression experiment is still open, as is the debate on how results from different gene expression experiments have to be interpreted for the definition of genes showing the same trend (i.e., expression profiles) or genes showing the same regulation pattern [49].

6 Systems Biology: The Major Challenge

This section describes emerging areas in bioinformatics. Indeed, *systems biology* may provide a relevant contribution to modeling and the comprehension of living organism's functionalities in the 'omics' era.

As discussed above, bioinformatics is still facing both computational and biological challenges due to the need to collect and organize large biological data collections and to manage the biological complexity revealed by increasing scientific knowledge. This is only the starting point for data analysis and, all the more so, for modeling biological processes. Computational tools and methods for data analysis are available, and the demand for more sophisticated bioinformatics solutions is expanding. Data need to be investigated to understand the structure, organization, and function of molecules for deriving peculiarities and similarities from specific cellular systems under specific conditions. Data are analyzed also to provide information on the building blocks of living organisms also in light of their evolution. This is the key to enhance our knowledge of what has determined the organization and physiology of living organisms. However, the study of biological systems cannot be limited to simply provide an exhaustive list of their components (chemical components, reactions, proteins, genes, cells, etc.) but has to point out how and when such components are assembled and how they interact with one another. In other words, structural and dynamic aspects must be considered simultaneously in order to understand the inner make-up and workings of the system to provide information on the mechanisms that control the expression of a genome in time (i.e., during development) and space (i.e., in the single compartments of a multicellular life form) from the cellular locations to the tissue and to more complex anatomic organizations.

Interest in employing methods of knowledge discovery and text and data mining is strong and consistent with pursuing this goal and generating models of biological systems. As scientific publications are the major tool for the exchange of new scientific facts, automatic methods for extracting interesting and non-trivial information from unstructured text become very useful. Indeed, the understanding and modeling of biological systems rely on the availability of numerical values concerning physical and chemical properties of biological macromolecules, and their behavior in a cell (e.g., kinetic parameters) is not yet widely reported in public standardized repositories like molecular databases. The gathering and convergence of these data, supported by automated processes and summarized by integrative approaches, represent the founding elements in systems biology. This can be considered the new challenge in biological research. As Denis Noble wrote: *"Systems biology . . . is about putting together rather than taking apart, integration rather than reduction"* [50]. Systems biology strategies can therefore be viewed as a combination of 'omics' approaches, data integration, and modeling [51]. They require not only high-throughput biological results such as those from DNA sequencing, DNA arrays, genotyping, proteomics, and so forth, but also computational power and space for model generation and for integration of different levels of biological information. This provides a global view on the data collected and yields models that describe system behavior as a whole.

The need of a 'whole-istic' biological view [52] is by now an accepted fact. This is manifested by coordinated efforts from a wide variety of scientists worldwide (computational biologists, statisticians, mathematicians, computer scientists, engineers, and physicists) in modeling metabolic networks, cell-signaling pathways, regulatory networks, and in developing syntactically and semantically sound ways to describe these models so as to construct virtual cells on computers [http://www.e-cell.org/;. http://www.nrcam.uchc.edu/]. The results of these efforts are still far from exhaustive. Nonetheless, the interdisciplinary field of systems biology is continuing to evolve. Therefore, the determination of standards and ontologies within systems biology [53], improvements in sampling methods for molecular simulation [54], and the expansion of computational toolboxes give cause for confidence in the impact of future applications.

7 An Italian Resource for Solanaceae Genomics

We present here our effort as partners of the consortium established by the International Solanaceae (SOL) Genomics Network. The long-term goal of the consortium is to build a network of resources and information dedicated to the biology of the Solanaceae family, which includes many species of major agricultural interest such as tomato and potato. The Solanaceae comprises about 95 genera and at least 2400 species. Many of these species have considerable economic importance as food (tomato, potato, eggplant, garden pepper), ornamental (petunia), and drug plants (tobacco). Solanaceae species show a wide morphologic variability and occupy various ecological niches though they share high genome conservation. The need to enhance our knowledge of the genetic mechanisms that determine Solanaceae diversification and adaptation has led scientific efforts to be gathered under the International Solanaceae (SOL) Genome Project. The cultivated tomato, *Solanum lycopersicum*, has been chosen by the SOL initiative for BAC-based [55] genome sequencing. The long-term goal is to exploit the information generated from the tomato genome sequencing project to analyze the genome organization, its functionality, and the evolution of the entire Solanaceae family. To address key questions raised by the rationale of the SOL project, large amounts of data from different 'omics' approaches are being generated. The raw data are of little use as they stand and need to be converted into biologically meaningful information. Therefore, bioinformatics approaches become preeminent, though their results may be far from exhaustive and complete.

We present here the design and organization of ISOLA (http://biosrv.cab. unina.it/isola/), an Italian **Solan**aceae genomics resource. It was conceived to collect, integrate, and converge results generated from different 'omics' approaches within the consortium and to manage the overwhelming amounts of data already available and under production for investigating multiple aspects of the Solanaceae genomics. ISOLA is a multilevel computational environment and consists of different data sources and tools that are necessary to enhance data quality, to extract their information content, and to exploit their integration efficiently. The multilevel structure of ISOLA summarizes the semantics of biological data entities (Fig. 1). ISOLA currently consists of two main levels: the genome and the expression level. The cornerstone of the genome level is represented by the *Solanum lycopersicum* genome draft sequences produced by the International Tomato Genome Sequencing Consortium and by the data generated during the genome annotation procedure. Instead, the basic element of the expression level is the transcriptome information from different Solanaceae species, in the form of species-specific comprehensive collections of ESTs.

Cross-talk between the genome and expression levels ensures the sharing of data sources. This is achieved by tools that extract information content from the levels' subparts and produce value-added biological knowledge. The multilevel

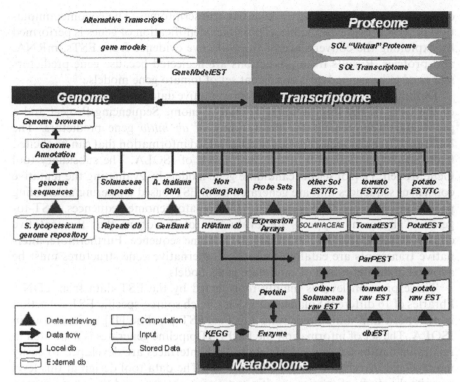

Fig. 1 ISOLA multilevel environment. Data collection and tools of the platform: The subparts of the Genome and the Transcriptome levels are included in the light area and the dark gray area, respectively. The data that are shared are located in the grid area. Subsidiary tools lay on the interface of the two levels (GeneModelEST). Value-added data gathered from the platform are listed and enclosed in the level to which they contribute more. Entry points for proteome and metabolome approaches are indicated

environment includes (i) 'basic' tools for enhancing data quality and increasing data information content and (ii) 'subsidiary' tools, which lie over the existing multilevel environment, exploiting the synergy between the levels. Each level can be independently accessed: the genome browser gateway was created for exploring the annotation of the draft tomato genome sequences, whereas the EST database gateway was created for browsing the EST-based transcriptome resources. Both the access points allow user-driven data investigation and are cross-linked to support Web-based navigation.

The genome level is enriched with reliable annotations of all the gene classes (mRNA and other non-protein-coding RNAs) and of the repeats (simple and complex ones). As already discussed, integration with the rich collections of preannotated Solanaceae ESTs represents a valuable resource for annotating tomato draft genomic sequences efficiently and effectively. In addition, annotation based on data from different species can be exploited for genome-based comparative analyses of the Solanaceae transcriptomes. The genome annotation process is made as reliable as possible: identification of genes is performed by exploiting the convergence of the collective evidence from ESTs, mRNA, and proteins. Results from predictions are neglected because gene predictors still need to be trained on a consistent set of tomato gene models.

Definition of good quality and a representative data set of gene models is one of the tasks of the International Tomato Genome Sequencing Project and a preliminary requirement for the training of *ab initio* gene predictors. The building of gene models represents value-added information that can be elicited from effective integration of the main levels of ISOLA. The subsidiary tool GeneModelEST [35] selects candidate gene models by evaluating the tentative consensus sequences that are generated from EST-based clustering/assembling procedures and are then aligned to the tomato genome sequences. EST-to-genome alignment supports the evaluation of the exons that the tentative consensus sequences describe along the genome sequence. Furthermore, alternative transcripts are catalogued because alternative gene structures must be avoided in the definition of candidate gene models.

The expression level is mainly represented by the EST data from cDNA libraries of 16 different Solanaceae species. Each source-specific EST collection is processed and annotated using the ParPEST pipeline [48], a *basic* tool in ISOLA. The set of information the ParPEST pipeline generates is deposited in relational databases, and user-friendly Web interfaces are developed to easily manage and investigate the EST collections. The data/tool design at this level permits the study of species-specific expression patterns and their time course, in normal or pathologic conditions and/or under specific biotic or abiotic stimuli. The protein-based functional annotation of EST sequences and the detection of the putative protein-coding region (Open Reading Frames) can undoubtedly enrich our knowledge of Solanaceae proteomes. This gives the opportunity to plug into ISOLA data generated from proteome efforts. In addition, the association of ESTs to metabolic pathways as described in the KEGG database [56] provides an entry point to integrate into ISOLA metabolome data, which can further support the definition of gene expression patterns. On the other hand, the association between EST sequences and the oligonucleotide probe-set from tomato expression arrays [57] [GeneChip Tomato Genome Array; http://www.affymetrix.com/products/arrays/specific/tomato.affx] leaves data from expression profiling arrays to be part and parcel of ISOLA.

Organization, classification, and annotation of source-specific EST data provide an estimation of the transcriptome space for each Solanaceae species. They also permit gene expression maps, regulatory networks, and metabolic processes to be modeled according to the paradigm 'from the omics to systems biology.' Annotation of the tomato draft genome sequences, on the other hand, provides assessment of the tomato gene space and shows how fundamental it is to achieve in-depth knowledge of a reference genome within a flowering plant family of such considerable economic and agronomic importance. All the information generated in annotating the tomato genome can be useful for the comprehension of the genome organization, its functionality, and the evolution of the entire Solanaceae family.

ISOLA is not a static 'platform.' It is under continuous evolution in that it considers the ongoing growth of data sources as well as the production of novel computational and experimental methods. We are well aware that ISOLA is still far from fulfilling the challenge of systems biology approaches. However, it lays the foundations for providing reliable molecular information on what is acting in a biological system (cell, tissue, organism), as well as where and when. Of course, methods must be improved to efficiently investigate the basis of comparative approaches, common features that could highlight how biological phenomena take place. We also believe that studying the structure, its function, and the evolution of the Solanaceae genomes is a suitable test-bench to challenge and expand this effort.

8 Conclusions

There are many bioinformatics applications in support of high-throughput experimental technologies, and high-level computational requirements are necessary. *Ad hoc* methods for data storage, data warehousing, data integration, data visualization, and data modeling are fundamental. The main target is to appropriately model structures and functions of molecules and the biological phenomena they give rise to.

The demand for bioinformatics tools and data banks able to fulfill such requirements also shows that bioinformatics must be based on multidisciplinary competences covering many different scientific aspects and challenging many different fields of research in the context of rapidly evolving scientific research and technologies. An additional requirement is therefore the need to train researchers for the integrated research environment, which also includes competences from those who plan and conduct experimental analyses and have in-depth knowledge of the biological systems to be modeled.

Acknowledgments We wish to thank Prof. Gerardo Toraldo for useful discussions and constant support. This is the contribution DISSPAPA book 3. Part of the presented work is supported by the Agronanotech Project (Ministry of Agriculture, Italy) and by the PRIN 2006 (Ministry of Scientific Research, Italy) and is in the frame of the EU-SOL Project (European Community).

References

1. Cullis, C.A.: Plant genomics and proteomics. Hoboken, NJ: John Wiley and Sons, pp. 214 (2004).
2. Heslop-Harrison, J.S., Murata, M., Ogura, Y., Schwarzacher, T., Motoyoshi, F.: Polymorphisms and genomic organization of repetitive DNA from centromeric regions of *Arabidopsis thaliana* chromosomes. Plant Cell 11: 31–42 (2000).

 3. Britten, R.J., Kohne, D.E.: Repeated sequences in DNA. Science 161: 529–540 (1968).
 4. Zwick, M.S., Hanson, R.E., McKnight, T.D., Islam-Faridi, M.N., Stelly, D.M., et al.: A rapid procedure for the isolation of Cot1 DNA from plants. Genome 40: 138–142 (1997).
 5. EPSO: European plant science: A field of opportunities. Journal of Experimental Botany 56: 1699–1709 (2005).
 6. Iovene, M., Barone, A.., Frusciante, L.., Monti, L.., Carputo, D.: Selection for aneuploid *Solanum commersonii-S. tuberosum* hybrids combining low wild genome content and resistance traits. Theoretical and Applied Genetics 119: 1139–1146 (2004).
 7. Barone, A., Frusciante L.: Molecular marker-assisted selection for resistance to pathogens in tomato. In: Marker-assisted selection: Current status and future perspectives in crops, livestock, forestry and fish. E.P. Guimaraes, J. Ruane, B.D. Scherf, A. Sonnino, J.D. Dargie (eds), FAO, Rome, Italy pp. 151–164 (2007).
 8. Barone, A.: Molecular marker-assisted selection for potato breeding. American Journal of Potato Research 81: 111–117 (2004).
 9. Ruane, J., Sonnino, A.: Marker-assisted selection as a tool for genetic improvement of crops, livestock, forestry and fish in developing countries: On overview of the issues. In: Marker-assisted selection: Current status and future perspectives in crops, livestock, forestry and fish. E.P. Guimaraes, J. Ruane, B.D. Scherf, A. Sonnino, J.D. Dargie (eds), FAO, Rome, Italy pp. 4–13 (2007).
10. Varsheney, R.K, Graner, A., Sorrells, M.E: Genomics-assisted breeding for crop improvement. Trends in Plant Science 10(12): 621–630 (2005).
11. Peleman, J.D., Rouppe van der Voort, J.: Breeding by design. Trends in Plant Science 8: 330–334 (2003).
12. Wheeler, D.L., Smith-White, B., Chetvernin, V., Resenchuk, S., Dombrowski, S.M., et al.: Plant genome resources at the national center for biotechnology information. Plant Physiology 138: 1280–1288 (2005).
13. Schneider, M., Bairoch, A., Wu, C.H., Apweiler, R.: Plant protein annotation in the UniProt Knowledgebase. Plant Physiology 138: 59–66 (2005).
14. Deshpande, N., Addess, K.J., Bluhm, W.F., Merino-Ott, J.C., Townsend-Merino, W., et al.: The RCSB Protein Data Bank: A redesigned query system and relational database based on the mmCIF schema. Nucleic Acids Research 33: D233–D237 (2005).
15. Ostell, J.: The NCBI Handbook: The Entrez Search and Retrieval System. National Library of Medicine (NLM), Washington, DC, USA Part 3, section 15, Rome, Italy (2003).
16. Safran, M., Chalifa-Caspi, V., Shmueli, O., Olender, T., Lapidot, M., et al.: Human Gene-Centric Databases at the Weizmann Institute of Science: GeneCards, UDB, CroW 21 and HORDE . Nucleic Acids Research 31: 142–146 (2003).
17. Hubbard, T. J. P., Aken, B.L., Beal, K., Ballester, B., Caccamo, M., et al.: Ensembl 2007. Nucleic Acids Research 35: D610–D617 (2007).
18. Stein, L.D., Mungall, C., Shu, S., Caudy, M., Mangone, M., et al.: The generic genome browser: A building block for a model organism system database. Genome Research 12(10): 1599–1610 (2002).
19. Kuhn, R.M., Karolchik, D., Zweig, A.S., Trumbower, H., Thomas, D.J., et al.: The UCSC Genome Browser database: Update 2007 Nucleic Acids Research. 35: D668–D673 (2007).
20. Yamazaki, Y., Jaiswal, P.: Biological ontologies in rice databases. An introduction to the activities in Gramene and Oryzabase. Plant Cell Physiology 46: 63–68 (2005).
21. D'Agostino, N., Aversano, M., Frusciante, L., Chiusano, M.L.: TomatEST database: In silico exploitation of EST data to explore expression patterns in tomato species. Nucleic Acids Research 35: D901–D905 (2007).
22. The Wellcome Trust. Sharing Data from Large-Scale Biological Research Projects: A System of Tripartite Responsibility. Fort Lauderdale, FL: Wellcome Trust (2003).

23. Noel, J.P., Austin, M.B., Bomati, E.K.: Structure-function relationships in plant phenyl-propanoid biosynthesis. Current Opinion in Plant Biology 8: 249–253 (2005).
24. Claverie, J.M.: Computational methods for the identification of genes in vertebrate genomic sequences. Human Molecular Genetics 6: 1735–1744 (1997).
25. Stormo, G.D.: Gene-finding approaches for eukaryotes. Genome Research 10(4): 394–397 (2000).
26. Davuluri, R.V., Zhang, M.Q.: Computer software to find genes in plant genomic DNA. Methods in Molecular Biology 236: 87–108 (2003).
27. Altschul, S.F., Gish, W., Miller, W., Myers, E.W., Lipman, D.J.: Basic local alignment search tool. Journal of Molecular Biology 215(3): 403–410 (1990).
28. Yao, H., Guo, L., Fu, Y., Borsuk, L.A., Wen, T.J., et al.: Evaluation of five ab initio gene prediction programs for the discovery of maize genes. Plant Molecular Biology 57(3): 445–460 (2005).
29. Zhang, M.Q.: Computational prediction of eukaryotic protein-coding genes. Nature Reviews Genetics. 3: 698–709 (2002).
30. Schlueter, S.D., Dong, Q., Brendel, V.: GeneSeqer@PlantGDB: Gene structure prediction in plant genomes. Nucleic Acids Research. 31: 3597–3600 (2003).
31. Seki, M., Naruska, M., Kamiya, A., Ishida, J., Satou, M., et al.: Functional annotation of a full-length *Arabidopsis* cDNA collection. Science 296: 141–145 (2002).
32. Alexandrov, N.N., Troukhan, M.E., Brover, V.V., Tatarinova, T., Flavell, R.B.,et al.: Features of *Arabidopsis* genes and genome discovered using full-length cDNAs. Plant Molecular Biology 60(1): 69–85 (2006).
33. Adams M.D., Kelley, J.M., Gocayne, J.D., Dubnick, M., Polymeropoulos, M.H., et al.: Complementary DNA sequencing: Expressed sequence tags and human genome project. Science 252: 1651–1656 (1991).
34. Zhu, W., Schlueter, S.D., Brendel, V.: Refined annotation of the *Arabidopsis* genome by complete expressed sequence tag mapping. Plant Physiology. 132: 469–484 (2003).
35. D 'Agostino, N., Traini, A., Frusciante, L., Chiusano, M.L.: Gene models from ESTs (GeneModelEST): An application on the *Solanum lycopersicum* genome. BMC Bioinformatics 8(Suppl 1): S9 (2007).
36. Lewin, B.: Genes VIII. Upper Saddle River, NJ: Prentice Hall (2003).
37. Jiang, N., Bao, Z., Zhang, X., Eddy, S.R., Wessler, S.R.: Pack-MULE transposable elements mediate gene evolution in plants. Nature 431: 569–573 (2004).
38. Morgante, M.: Plant genome organisation and diversity: The year of the junk! Current Opinion in Biotechnology 17(2): 168–173 (2006).
39. Morgante, M., De Paoli, E., Radovic, S.: Transposable elements and the plant pan-genomes. Current Opinion in Plant Biology 10(2): 149–155 (2007).
40. Audic, S., Claverie, J.M.: The significance of digital gene expression profiles. Genome Research 7(10): 986–995 (1997).
41. Ewing, R.M., Ben Kahla, A., Poirot, O., Lopez, F., Audic, S., et al.: Large-scale statistical analyses of rice ESTs reveal correlated patterns of gene expression. Genome Research 9: 950–959 (1999).
42. Wu, X., Walker, M.G., Luo, J., Wei, L.: GBA server: EST-based digital gene expression profiling. Nucleic Acids Research 33 (Web Server issue): W673–W676 (2005).
43. M égy, K., Audic, S., Claverie, J.M.: Heart-specific genes revealed by expressed sequence tag (EST) sampling. Genome Biology 3(12): RESEARCH0074, Rome, Italy, 11 (2003).
44. Burke, J., Davison, D., Hide, W.: d2_cluster: A validated method for clustering EST and full-length cDNA sequences. Genome Research 9: 1135–1142 (1999).
45. Pertea, G., Huang, X., Liang, F., Antonescu, V., Sultana, R., et al.: TIGR Gene Indices clustering tools (TGICL): A software system for fast clustering of large EST datasets. Bioinformatics 19: 651–652 (2003).
46. Kalyanaraman, A., Aluru, S., Kothari, S., Brendel, V.: Efficient clustering of large EST data sets on parallel computers. Nucleic Acids Research 31: 2963–2974 (2003).

47. Hotz-Wagenblatt, A., Hankeln, T., Ernst, P., Glatting, K.H., Schmidt, E.R., et al.: ESTAnnotator: A tool for high throughput EST annotation. Nucleic Acids Research 31: 3716–3719 (2003).
48. D 'Agostino, N., Aversano, M., Chiusano, M.L.: ParPEST: A pipeline for EST data analysis based on parallel computing. BMC Bioinformatics 6 (Suppl 4): S9 (2005).
49. Van Helden, J.: Regulatory sequence analysis tools. Nucleic Acids Research 31: 3593–3596 (2003).
50. Noble, D.: The music of life. Oxford: Oxford University Press (2006).
51. Ge, H., Walhout, A.J., Vidal, M.: Integrating 'omic' information: A bridge between genomics and systems biology. Trends in Genetics. 19(10): 551–560 (2003).
52. Chong, L., Ray, L.B.: Whole-istic Biology. Science 295(1): 1661 (2002).
53. Strömbäck, L., Hall, D., Lambrix, P.: A review of standards for data exchange within systems biology. Proteomics 7(6): 857–867 (2007).
54. Lei, H., Duan, Y.: Improved sampling methods for molecular simulation. Current Opinion in Structural Biology 17(2): 187–191 (2007).
55. Mueller, L.A., Tanksley, S.D., Giovannoni, J.J., van Eck, J., Stack, S., et al.: The Tomato Sequencing Project, the first cornerstone of the International Solanaceae Project (SOL). Comparative and Functional Genomics 6: 153–158 (2005).
56. Kanehisa, M., Goto, S., Hattori, M., Aoki-Kinoshita, K.F., Itoh, M., et al.: From genomics to chemical genomics: New developments in KEGG. Nucleic Acids Research 34: D354–D357 (2006).
57. Fei, Z., Tang, X., Alba R., Giovannoni, J.: Tomato Expression Database (TED): A suite of data presentation and analysis tools. Nucleic Acids Research 34: D766–D770 (2006).

Modeling and Solving Real-Life Global Optimization Problems with Meta-heuristic Methods

Antonio Mucherino and Onur Seref

Abstract Many real-life problems can be modeled as global optimization problems. There are many examples that come from agriculture, chemistry, biology, and other fields. Meta-heuristic methods for global optimization are flexible and easy to implement and they can provide high-quality solutions. In this chapter, we give a brief review of the frequently used heuristic methods for global optimization. We also provide examples of real-life problems modeled as global optimization problems and solved by meta-heuristic methods, with the aim of analyzing the heuristic approach that is implemented.

1 Introduction

A model is the representation of something. It consists in a set of mathematical functions and expressions describing an entity or a process. Real-life problems are modeled in order to solve them by mathematical or computational techniques, and many of them can be modeled as global optimization problems. An optimization problem is usually composed of an objective function and constraints, which are easy to interpret. However, some optimization problems emerging from real-life applications are very hard to solve. In a global optimization problem, the objective is to find the global minimum or maximum value of a given function $f : x \in A \to y \in B$, which is called the objective function. The solution to a minimization problem is the point $x^* \in A$ that corresponds with the global minimum value in B:

$$x^* \in A : f(x) \geq f(x^*) \, \forall x \in A.$$

There may be subsets of the domain A, where a solution is better than all of the other solutions within a small neighborhood. Such solutions are referred to as local optima. The word *global* is used to refer to the best local optimum solution

A. Mucherino (✉)
Center for Applied Optimization, University of Florida, Gainesville, FL, USA
e-mail: amucherino@ufl.edu

.

P.J. Papajorgji, P.M. Pardalos (eds.), *Advances in Modeling Agricultural Systems*,
DOI 10.1007/978-0-387-75181-8_19, © Springer Science+Business Media, LLC 2009

x^*, which is also the optimal value of function f in its whole domain A. The typical formulation of a global optimization problem is given as follows:

$$\min f(x)$$

subject to

$$c_i(x) = 0 \quad \forall i$$
$$c_j(x) \leq 0 \quad \forall j.$$

Functions c_i and c_j represent the equality and inequality constraints, which define domain A. A maximization problem has the exact same formulation, where the objective function is $h(x) = -f(x)$.

Examples of real-life problems that have been formulated as global optimization problems range from agriculture and biology to chemistry and many other applied fields. Examples of agricultural problems that can be formulated as global optimization problems are determining the economically optimal scheduling of fertilization and irrigation that maximizes the farmer's profits [12], finding the optimal compromise between inventory precision and price for forest inventories [1], or optimizing the water management of the irrigation systems in agriculture, which represent the major users of water in the world [20]. Examples from biology include the prediction or the simulation of protein molecules. These problems can be formulated as global optimization problems [14], where a certain energy function has to be minimized. In chemistry, the energetically stable clusters of atoms subject to different energy models, such as the Lennard-Jones energy [25] and the Morse energy [30], can be located by optimizing these potential energies.

Many optimization methods and algorithms have been developed, and researchers have the possibility to choose the one that better fits their needs. Methods may require that some hypothesis on the objective function or on the constraints must be satisfied, whereas other methods may be more flexible and applicable to any kind of optimization problem. Moreover, methods may be easy to implement and fast to converge, whereas other methods might be more complex and require more efforts for developing a software procedure. Generally, the easier methods can only provide approximation of the problem solution, whereas the more complex ones are also able to generate high-quality solutions. The researcher has the task to choose the best method for his personal purposes and find the compromise between simple implementations and a good solution quality.

The chapter is organized as follows. In Section 2, we will discuss the modeling issues arising when dealing with real-life problems. The focus will be on real-life problems modeled as global optimization problems and solved by meta-heuristic methods. A brief description of the most popular meta-heuristic methods for global optimization is provided in Section 3. In Section 4, some real-life problems from agriculture, chemistry, and biology will be presented, with the aim of analyzing the heuristic method implemented. Section 5 provides some conclusions.

2 Modeling Real-Life Problems

A model is a simplified description of a complex process or entity that uses a mathematical language for describing its features and the relations among these features. In real life, entities and processes may be quite complex, and developing models for them is not trivial. We can express these real problems mathematically by numbers and functions in an abstract way. For instance, the gravitational force between two planets can be expressed by the square of the inverse of the distance between them and by a physical law that combines the masses of the planets and a gravitational constant in a mathematical expression. If we want to solve our problem in a computational environment, then it must be written in a language based on mathematics that computers can understand. Moreover, the information on the computational environment, the methods available for solving the considered problem, and the expected quality of solution are important factors. In general, the main steps for solving a real-life problem on computers are as follows: (i) building a mathematical model that translates the problem into mathematical expressions, (ii) building a computational model, which takes all the implementation issues into consideration, and (iii) finally, developing robust software based on the computational model. There is strong feedback between these steps. For instance, if the developed software is too slow, then it can be modified by changing the computational model, whose changes may also affect the underlying mathematical model.

There are many examples of real-life problems that are formulated as optimization problems, where some features of the modeled real entity in these problems are to be optimized. We focus in this chapter on three such real-life problems arising from agriculture, chemistry, and biology fields. In the agriculture problem, forests are modeled and their features are expressed to optimize the cost and precision of forest inventories. The problem of finding stable clusters of molecules is a global optimization problem with the objective of minimizing the energy in clusters of molecules. Finally, a geometric model for protein folding, which provides a description of the protein conformations based only on geometric features, is optimized in order to simulate protein molecules.

Once a global optimization problem has been formulated, the usual approach is to attempt to solve it by using one of the many methods for global optimization [34]. The choice of the method that fits the structure of the problem is very important. An analysis of the complexity of the model is required, and the expected quality of the solution needs to be determined. The complexity of the problem can be derived from the data structures used and from the mathematical expression of the objective function and the constraints. If the objective function is linear, or convex quadratic, and the problem has box, linear, or convex quadratic constraints, then the optimization problem can be solved efficiently by particular methods, which are tailored to the objective function [6, 13, 18]. If the objective function and the constraints are nonlinear without any restriction, then more general approaches must be used. For differentiable functions, whose gradient vector can be computed, deterministic methods may be used subject to some hypotheses. Functions that are twice differentiable with a computable Hessian matrix can be locally approximated by a quadratic function. Typical examples are the Trust Region algorithms [9].

Meta-heuristic methods for global optimization are the focus of this chapter. In general, meta-heuristics are more flexible and faster and easier to implement than the deterministic methods. However, these methods are based on probabilistic mechanisms and they cannot guarantee that the solution will be found under certain hypothesis. Meta-heuristic methods are widely applied in many research fields. Because of their simplicity and flexibility, meta-heuristic methods are the choice of many researches who are not experts in computer science and numerical analysis. Among the developed heuristic methods for optimization, the implementation issues render a wide range of difficulty. The easier implementations generally provide approximations of the solution, whereas more complex implementations may produce higher-quality solutions. The researchers first seek to find out the method that is the best fit for their problem. This decision may result in trading off the quality of the solution with speed or ease of implementation. For high-quality solutions, modeling issues may usually become more complex requiring additional programming skills and powerful computational environments.

3 Meta-heuristic Methods

Many meta-heuristic searches have been developed to date for solving hard global optimization problems. Most of them took inspiration from animal behavior or natural phenomena and try to reproduce such processes on computers. The basic idea behind some of the most used methods is provided in the following.

3.1 Simulated Annealing Algorithm

In 1983, Kirkpatrick [23] proposed the simulated annealing (SA) algorithm. The inspiration from the physical annealing process was as follows. To obtain a crystalline structure, which is a stable low-energy configuration, the temperature of a given system can be slowly decreased. In SA, the mobility of a system is simulated by applying random perturbations to the current system state, which represents an approximation of the solution of the considered optimization problem. Such perturbations can be accepted or rejected by using a random mechanism controlled by the temperature parameter. At high temperatures, all system states are allowed. As the temperature decreases, the chance of accepting higher-energy state systems also decreases. In the SA algorithm, it is very important to set a cooling schedule: if the temperature is decreased too fast, the system can get stuck at a local minimum. This reflects the behavior of the physical annealing, in which a fast temperature decrease leads to a polycrystalline or amorphous state. On the other hand, if the temperature decreases too slowly, the simulation can become computational demanding. Therefore, the determining success factor in SA is the cooling schedule. Generally, the initial temperature is set such that the acceptance ratio of bad moves is equal to a prefixed value. In [5], the properties of the acceptance probability have been analyzed and an algorithm for computing a temperature that is compatible with a given acceptance ratio is presented. SA is used for solving the problem discussed in Section 4.1, and it has also been used in [3, 8] in the very first studies on the problem discussed in Section 4.3.

3.2 Genetic Algorithms

Genetic algorithms (GA) [17] mimic the evolution of a population of chromosomes. Two or more chromosomes can generate child chromosomes by applying a crossover operator and these child chromosomes undergo genetic mutations by a mutation operator. The chromosomes with better chances of survivability can procreate. Therefore, the population of such chromosomes improves generation after generation and converges to the optimal chromosomes. In GA, the way in which chromosomes are selected plays a crucial role for the convergence of the algorithm. If only the best chromosomes are chosen every time, the procedure will probably get stuck at a local optimum. On the other hand, if all chromosomes are allowed to procreate, the chance of finding the optimal chromosome becomes purely random. The mutation rate is also important in GA. A carefully determined small mutation rate may prevent GA from getting stuck at a local OPTIMUM, whereas a high mutation rate may produce arbitrarily poor results. GA has been used in [42] for solving the problem presented in Section 4.2, where the research space was reduced over a lattice.

3.3 Differential Evolution

Often, the global optimization problems in real life have a nondifferentiable objective function. In this case, one cannot take advantage of the gradient of the objective function, which may guide the search toward better solutions. The differential evolution (DE) [41] method extends the gradient search strategy to nondifferential objective functions. In DE, the direction that leads to one feasible point to another is defined as the differential variation. The DE method works with a population of points and has similarities with GA. The mutation operator is performed by moving a selected point along a differential variation. The crossover operator and the selection strategy are similar to those used in GA. One of the real-life applications of DE is, for instance, the resolution of Radio Network Design problems, where optimal locations for base station transmitters are searched for maximizing the coverage area with a minimum number of transmitters [28].

3.4 Harmony Search

The harmony search (HS), a novel meta-heuristic for global optimization, has recently been proposed, and it is inspired by jazz music improvisation [15, 24, 27]. As musicians seek harmonies, HS algorithm looks for the global optimum of a given objective function. If a musician improvises a good harmony, he does not forget this harmony, and he improvises on this harmony later in order to improve the quality of the music. In analogy, HS has a memory of harmonies, which is randomly generated when the algorithm starts. At each step, a new harmony is generated by taking notes from the harmonies in memory and is evaluated by the objective function. If the harmony is better than the worst one in memory, it replaces the worst one, otherwise it is discarded. Random notes are added to improvised harmonies to force variation, and pitch adjustments are performed on the notes for improving solutions and avoiding local optima. The optimization method has been applied, for example, to a traveling salesman problem, a specific academic optimization problem, and to a least-cost pipe network design problem [15].

3.5 Tabu Search

Tabu search (TS) [16] introduces a new idea in the meta-heuristic searches: keep and update memory of solutions that are already visited during the search process to prevent the search from repeating these solutions. The solutions in the list are considered as tabu and are not accepted during the search. The size of the tabu memory determines the success of the algorithm. Keeping a large tabu memory is computationally demanding and conservative, whereas a very small memory converges to a random walk and may fail to lead the search. In TS, the local optima are avoided by allowing nonimproving moves. TS has been applied, for instance, to the vehicle routing problem [4].

3.6 Methods Inspired by Animal Behavior

Over the years, meta-heuristic searches have been developed for solving global optimization problems that are inspired by animal behavior. Swarm intelligence can be defined as the collective intelligence that emerges from a group of simple entities, such as ant colonies, flocks of birds, termites, swarm of bees, and schools of fish [26]. The most popular swarm-inspired methods are ant colony optimization (ACO) and particle swarm optimization (PSO).

ACO [11] algorithms simulate the behavior of a colony of ants finding and conserving food supplies. At the beginning, all the ants move randomly in search of food. Over the course of the search, they start building the optimal path, which is the shortest way to fetch food. These paths are marked with the pheromone that ants leave on their way. Pheromone concentration is higher on better paths as these are the shorter paths. The central component of an ACO algorithm is a parameterized probabilistic model called the pheromone model. At each step of the procedure, all the pheromone values are updated in order to focus the search in regions of the search space containing high-quality points.

PSO [22] simulates the motion of a large number of insects or other organisms. PSO works on a population of random feasible points, called particles, and lets them *fly* through the problem space by following the current optimum particles. At each step, the speed of each particle toward the best solution found thus far is randomly changed. PSO has similarities with GA, but it does not have evolutionary operations such as crossovers and mutations.

ACO has been applied for predicting the conformations of proteins from their amino acid sequence by using a model based on hydrophobic forces in [40]. PSO has recently been applied for the same purpose in [43]. PSO has also been applied, for instance, to electromagnetic problems [37].

3.7 Monkey Search

Monkey search (MS) algorithm [31, 39] resembles a monkey looking for food by climbing up trees. The basic assumption for such behavior is that the monkey explores trees and learns which branches lead to better food resources. Food is represented by desirable solutions. The branches of the tree of solutions are represented as perturbations between two neighboring feasible solutions in the considered optimization problem. These perturbations can be totally random, or they can also be customized based on other strategies previously developed for global optimization.

The monkey climbs up and down the branches of a tree. Unknown branches are discovered using perturbations. Whenever a better solution is found, the monkey climbs down the unique path that connects this solution to the root, while marking the path with this better solution. After reaching the root, the monkey climbs back up the tree using the previously discovered branches. The path the monkey follows on the way up the tree is probabilistically determined at each branch based on the previous marks. This strategy is intended to lead the monkey to branches with better solutions. The monkey climbs until it reaches an undiscovered frontier and continues its search from there on with new unknown branches.

A predetermined number of best solutions updated by each successive tree is stored in memory. The first tree roots are randomly chosen in order to force the search in different spaces of the search domain, and then they are picked from the memory in order to force the convergence. The memory is updated whenever a new solution is better than the ones that are already in the set. However, a new good solution may be rejected if it is very close to the solutions in the memory. MS has been recently applied for solving the problems discussed in Sections 4.2 and 4.3.

4 Applications

In this section, we provide examples of real-life problems that are modeled as global optimization problems and solved by meta-heuristic methods. The three examples belong to different research fields, and they are able to show how different problem features and different requirements on the solutions can guide the choice toward one optimization method rather than other.

The first problem comes from the field of agriculture [1, 38]. It has been chosen among the other problems from agriculture because it shows how a simple optimization method can provide solutions accurate enough for our needs. It deals with forest inventories, and the aim is to keep inventory of forests with maximum precision and minimum cost. An optimization problem is formulated and solved by the SA algorithm for finding an approximation of the problem solution. A sophisticated algorithm for this problem is redundant, and therefore SA is a good choice for its simplicity and its flexibility.

The second problem emerges from chemistry. The stable conformations of clusters of atoms regulated by the Lennard-Jones (LJ) [25] energy are searched. The objective function is constructed to model the potential energy of the cluster. This function has many local minima in which an optimization method may get stuck. Methods like SA are too simple to optimize this kind of complex objective function, and hence more sophisticated algorithms must be used.

The last optimization problem we discuss is related to the protein folding problem. The objective in this problem is to find the stable and functional conformations of protein molecule inside the cells. The energy functions used in this study are more complex than the LJ energy function, which is sometimes one of the terms in the objective function [14]. Therefore, this problem needs sophisticated algorithms for finding good approximations of the optimal solution. The specific problem on which we want to focus is based on protein geometric features only. These features are used for analyzing geometric aspects of the protein molecules [3, 8]. The objective function is not complex like the ones modeling the energy in the molecule, yet it is not easy to find the optimum solution. Although the objective function is relatively simple, it still requires sophisticated heuristic methods such as the MS algorithm, which produces good results.

4.1 Forest Inventories

In extensive forest inventories [38], different sampling techniques are used for collecting necessary information with the aim of estimating tree population characteristics at a low cost. Large samples usually guarantee the precision of the estimates, but they can be very expensive, whereas small samples may not provide sufficient precision. The traditional approach is to consider the problem of selecting the appropriate sampling design, which is a global optimization problem. The objective function $C(n,N)$ represents the cost of the forest inventory, and it is subject to a number of constraints. In the double sampling approach [10], an estimate of the principle variable's mean Y, such as the forest volume, is obtained by using its relationship to a supplementary variable X, such as the basal area of trees. During the first phase, measurements of the supplementary variable X are taken on a sample of size N. After that, the principal variable Y is measured on a smaller subsample of size $n < N$. In the first example considered in [1], which is related to a forest area in Greece, the cost function is modeled as follows:

$$C(n, N) = c_w \sqrt{I_A N} + \bar{k}_1 \sqrt{N} + k_1 n.$$

In this model, c_w is the cost of walking a unit distance, I_A is the inventory area corresponding with c_w square units, $\bar{k}_1$ is the basal area measurement cost of a plot, and k_1 is the volume measurement cost of a plot. The objective function $C(n,N)$ represents the total time needed for performing the measurements. This problem is subject to constraints, which limit the research space, such as the bounds $[n_{min}, n_{max}]$ for the integer values that n and N can take.

In [1], this global optimization problem is solved by the SA algorithm. As the authors point out, this optimization problem can also be solved analytically, but the algorithmic approach is usually preferred. Indeed, finding an explicit solution can be quite difficult, whereas heuristic algorithms can easily and quickly provide optimal sampling designs, or at least a design very close to the optimal. Forest managers also need flexible tools for examining alternative scenarios by performing experiments using different cost functions and multiple restrictions.

The objective function $C(n,N)$ is nonlinear and is not quadratic, but it is twice differentiable, so that gradient and Hessian matrix can be computed. Methods exploiting this information may be used, but they can be quite complex. The SA algorithm is chosen because it can be applied to any problem with any objective function and any constraints. SA can also be adapted to solve different design problems with few modifications. In other words, SA is very flexible. It is one of the easiest optimization algorithms to implement, and hence it may have the disadvantage to get stuck at a local optimum, which could be far from the expected solution. However, some formulations of this design problem are particularly easy, as exhaustive searches are also applied for solving it. The exhaustive searches are usually impossible to perform, because they require all of the possible pairs (n,N) to be checked. Therefore, the heuristic algorithm, and in particular the SA algorithm, represent an alternative for optimizing the objective function in a short amount of time compared with exhaustive search. Cost and precision are conflicting terms in this problem. Therefore, the more important the precision is, the harder it is to solve this problem. In such cases, either a trade-off solution is required or the multiobjective optimization can be used [21].

In SA, perturbations on the current solution are applied for simulating the system mobility, but only some of them are allowed according to the temperature of the system. When the temperature is high, all the perturbations are allowed. As the temperature parameter t decreases, the perturbations corresponding with higher system mobility are rejected. In [1], if the perturbation leads to a better solution, it is accepted, whereas worse solutions are accepted with probability:

$$M((n, N), (n^*, N^*), t) = \exp\left(-\frac{C^g(n, N)\Delta C}{t}\right).$$

In this formulation, (n,N) represents the current solution of the design problem, (n^*,N^*) is a new generated solution, $\Delta C = C(n^*,N^*) - C(n,N)$, and g is a nonpositive parameter. If $g = 0$, the function M corresponds with the Metropolis function [29]. The use of exponent g aims to increase the algorithm convergence. All of the parameters involved in the SA algorithm are experimentally obtained. The computational results show that SA is able to identify optimum sampling designs. The SA algorithm is applied many times using different starting parameters. Even in the worse cases, SA finds a design very close to the optimal.

4.2 Lennard-Jones Clusters

Another problem we include here involves finding the stable conformations of cluster of atoms subject to the LJ energy [25, 33, 36, 42, 44]. We also discuss the issues related to the development of a software procedure for optimizing such clusters. The LJ energy function is a very simple energy function. This function is sufficiently accurate for describing the cluster structures of a noble gas, and it can provide approximations for clusters of nickel and gold. The LJ pairwise potential is given conventionally in reduced terms as follows:

$$v(r) = \frac{1}{r^{12}} - \frac{1}{r^6}.$$

In this equality, r is the interparticle Euclidean distance between two atoms of the cluster x_i and x_j. All of the atoms are considered to be identical. The objective is to find the global minimum of the potential energy defined by

$$E(X) = 4 \sum_{i=1}^{n-1} \sum_{j=i+1}^{n} v(r_{ij}),$$

where X is a vector containing all the atom spatial coordinates x_i. Despite its simplicity, the LJ energy is notoriously difficult to optimize, because the number of its local minima grows exponentially with the size n. For instance, a cluster containing $n = 13$ atoms has about 1000 local minima. Currently, the best clusters obtained to date by using different methods are collected in a public database [7].

Most of the methods proposed for optimizing this energy function are heuristic methods. Because the degrees of freedom of such clusters are huge, the optimization is usually performed over a lattice. It has been proved in [36] that, when the size of the cluster is less than 1600, the stable structures have an icosahedral structure. There are some exceptions, in which the clusters have lower energy when they have a decahedral or octahedral structure. In [42], a method for building a lattice structure on which the stable LJ clusters can be identified is presented. In this work, GA is applied to optimize the LJ clusters by using previously constructed lattices as research space. The developed software is able to find the optimal LJ cluster of small sizes in few seconds. This lattice structure helps investigate the structures of large clusters formed by 500 atoms and more. However, this kind of strategy is highly efficient only for the optimization problem for which it is designed. Moreover, this method for optimizing LJ clusters is not able to find new cluster conformations except for the icosahedral, decahedral, and octahedral motifs. Therefore, this method cannot locate the optimal LJ clusters if they do not belong to one of the used motifs.

The MS algorithm has been recently applied for optimizing LJ clusters. This is a meta-heuristic method more sophisticated than others and that exploits strategies applied in other meta-heuristic methods. It has been proved that it is able to manage this kind of problem better than do other meta-heuristics. In [31], small clusters on a continuum research space S_1 have been generated, whereas larger clusters have been studied in [39], where the research space S_2 is strongly reduced. In the first case, S_1 is defined as an n-dimensional sphere large enough to contain the whole cluster, where n is the cluster size. When the lattice is used, the cluster atoms are forced to have predetermined positions in the three-dimensional space with slight allowable movements around these positions. While a three-dimensional vector can be used for locating an atom in S_1, a more complex data structure is required when using S_2. Each atom position is located through the label of the corresponding lattice site and a three-dimensional vector showing its movement around the lattice site. These movement possibilities are very small and are needed to locally improve the solution quality. When the research space S_2 has been used, the MS algorithm has been divided in two stages. In the first stage, all the cluster atoms have no possibilities to move around the lattice site in order to locate the most favorable

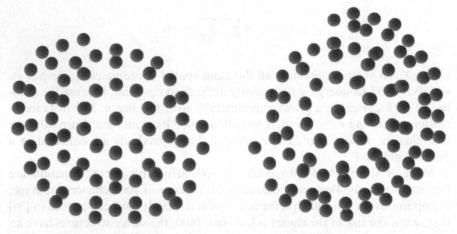

Fig. 1 LJ optimal clusters with $n = 100$ and 120 in the space S_2

sites faster. In the second stage, both changing lattice sites and moving around them are allowed.

The computational results presented in [31, 39] show that the MS algorithm is able to optimize the LJ clusters. When the research space is S_1, small clusters are generated with a size smaller than 30. These experiments are limited to simulations lasting no more than half an hour. When S_2 is used, larger clusters in a smaller time are generated. Figure 1 shows obtained clusters having size n equal to 100 and 120: the LJ energies related to these two clusters are −557.038375 and −687.019135, respectively.

4.3 Simulating Protein Conformations

The problem of finding the stable and functional conformation of a protein molecule can be formulated as an optimization problem. The energy function that is usually used in these studies depends on the sequence of amino acids forming the protein and on all of the interactions among the amino acid atoms. The number of forces involved is large, and therefore the function describing the energy in protein conformations can be very complex. The minimum of such an objective function should correspond with the conformation in which the energy is minimum, which is the stable and functional conformation of a protein [14].

In recent years, a new approach to protein folding has been introduced, in which the protein conformations are modeled by using just geometric features of such molecules [2]. The model considered in this section is proposed in [3, 19] and modified in [8]. The model is named the *tube model* because a protein conformation is modeled as a tube with a certain diameter such that the tube does not cross itself. This model cannot be used for predicting protein conformations, but it can just simulate conformations that may be similar to those proteins may actually have.

The tube model leads to the formulation of a global optimization problem, in which some geometric requirements on the protein conformations are imposed. The tube that models the protein conformation has particular diameter values when the protein secondary structure such as α-helices and β-sheets are considered. By using this geometric information and other general information, many optimization problems can be formulated in order to perform different simulations of protein conformations. More details on the tube model and on the different formulations of the optimization problem can be found in [32]. The main aim is to simulate a large number of protein conformations in a short amount of time. Therefore, simple requirements are imposed, and the generated conformations are checked to verify if some conformation that actually exists in nature is found.

One of the possible formulations is given as follows:

$$\min f(X)$$

where $X = \{x_1, x_2, ..., x_n\}$ is a protein conformation represented through the spatial coordinate of each C_α carbon atom, and f is the objective function that takes into account three geometric requirements. The function to be optimized is

$$f(X) = \gamma_1 \sum_{i=1}^{n-3} \sum_{j=i+3}^{n} d(x_i, x_j) +$$

$$\gamma_2 \sum_{i=1}^{n-2} \sum_{j=i+2}^{n} \exp_+(th - d(x_i, x_j)) + \gamma_3 \sum_{i=1}^{n-2} (d(x_i, x_{i+2}) - c)^2.$$

In this formulation, γ_1, γ_2, γ_3, th, and c are five positive and real constants. The function $\exp_+$ corresponds with the exponential function when its argument is non-negative, and it is zero otherwise; d represents the Euclidean distance.

The considered optimization problem simulates protein conformations that are compact and contain α-helices. The first term of $f(X)$ minimizes all the relative Euclidean distances. The second one avoids the case that the conformation falls on itself by keeping a threshold value of th. Finally, the last term forces each x_i and x_{i+2} to have a relative distance equal to a typical distance in α-helices. Even though only a C_α atom is considered for each amino acid, we represent the protein conformation by the backbone dihedral angles representation in order to avoid unnatural conformations. Hence, the objective function $f(X)$ is evaluated after a change of representation.

MS has been used for simulating protein conformations through the tube model. Because the simulation of a large amount of conformations is needed, high-definition conformations are not required; good approximations suffice. For this reason, some easier optimization method may also be used, which

Fig. 2 On the left, a
simulated protein
conformation; on the right,
the conformation of the 2cro
protein

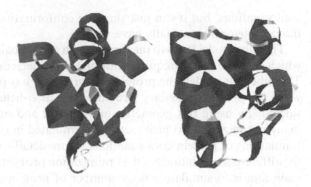

could be easier to implement and faster to converge. However, it has been
shown in [31] that MS performs better than other meta-heuristics and generates
protein conformations that meet the selected geometric requirements. The
simulated set contains conformations having the same geometric features of
protein molecules, and some of them are very close to proteins that actually
exist. In Fig. 2, the conformation of a real protein and one of the simulated
conformations are compared. The considered real protein is referred in the PDB
archive [35] by the code 2cro. The global shapes of the conformations are
similar, they have the same number of helices, and the helices have about the
same length. This is surprising if we think that the obtained conformation has
been simulated just using simple geometric requirements. For this reason, these
kinds of studies are very promising in the field of protein folding for protein
prediction. More details and comments on the protein conformations that can
be simulated by using the tube model can be found in [32].

5 Conclusions

Models describe real life through numbers, mathematical expressions, and
formulas. They represent the first step needed for solving real-life problems
using computers, which are able to understand a language based on mathematics. Implementation issues, such as the choice of the method for solving a
certain problem, are also very important. In this chapter, we focus on problems
modeled as global optimization problems and solved by meta-heuristic methods. The objective is to find the global optimal value of an objective function in
its domain, while subject to a number of constraints. Many methods have been
developed for solving these kinds of problems, and meta-heuristic methods
have proved to be good choices. Most of these meta-heuristic methods are
very flexible and easy to implement, whereas some of them can be more complex
but able to provide high-quality solutions.

A review of frequently used heuristic methods for global optimization is
provided in this chapter. These methods are usually inspired by natural

phenomena, such as the annealing physical process, the evolution process, the behavior of ant colonies, music improvisation, the behavior of monkeys, and so forth. These methods have been widely applied to many optimization problems arising from real-life problems. Different meta-heuristics produce better results for different kinds of problems. Choosing the right method to use among these meta-heuristic methods is one of the most important tasks in the development of a software procedure for solving global optimization problems. In this chapter, we discuss three particular optimization problems, which emerge from agriculture, chemistry, and biology. The first one is from agriculture, where the precision and cost of forest inventories are to be optimized. In some cases, this problem is simple enough to be efficiently solved by an SA algorithm. The second problem considered comes from chemistry, where a cluster of atoms subject to the Lennard-Jones energy is to be identified. The last problem is from biology, and protein conformations are to be simulated. The last two problems are solved by an MS algorithm, which produced better results compared with other meta-heuristics.

References

1. L. Angelis, G. Stamatellos, Multiple Objective Optimization of Sampling Designs for Forest Inventories using Random Search Algorithms, Computers and Electronics in Agriculture 42(3), 129–148, 2004.
2. D. Baker, A Surprising Simplicity to Protein Folding, Nature 405, 39–42, 2000.
3. J.R. Banavar, A. Maritan, C. Micheletti and A. Trovato, Geometry and Physics of Proteins, Proteins: Structure, Function, and Genetics 47(3), 315–322, 2002.
4. J. Brandao, A Tabu Search Algorithm for the Open Vehicle Routing Problem, European Journal of Operational Research 157(3), 552–564, 2004.
5. W. Ben-Ameur, Computing the Initial Temperature of Simulated Annealing, Computational Optimization and Applications 29(3), 369–385, 2004.
6. S. Cafieri, M. D'Apuzzo, M. Marino, A. Mucherino, and G. Toraldo, Interior Point Solver for Large-Scale Quadratic Programming Problems with Bound Constraints, Journal of Optimization Theory and Applications 129(1), 55–75, 2006.
7. Cambridge database: http://www-wales.ch.cam.ac.uk/CCD.html.
8. G. Ceci, A. Mucherino, M. D'Apuzzo, D. di Serafino, S. Costantini, A. Facchiano, and G. Colonna, Computational Methods for Protein Fold Prediction: an Ab-Initio Topological Approach, Data Mining in Biomedicine, Springer Optimization and Its Applications, Panos Pardalos et al. (Eds.), vol.7, Springer, Berlin, 2007.
9. A.R. Conn and N.I.M. Gould, Trust-Region Methods, SIAM Mathematical Optimization, 2000.
10. P.G. De Vries, Sampling for Forest Inventory, Springer, Berlin, 1986.
11. M. Dorigo and G. Di Caro, Ant Colony Optimization: A New Meta-Heuristic, in New Ideas in Optimization, D. Corne, M. Dorigo and F. Glover (Eds.), McGraw-Hill, London, UK, 11–32, 1999.
12. E. Feinerman and M.S. Falkovitz, Optimal Scheduling of Nitrogen Fertilization and Irrigation, Water Resources Management 11(2), 101–117, 1997.
13. R. Fletcher, Practical Methods of Optimization, Wiley, New York, Second Edition, 1987.
14. C.A. Floudas, J.L. Klepeis, and P.M. Pardalos, Global Optimization Approaches in Protein Folding and Peptide Docking, DIMACS Series in Discrete Mathematics and

Theoretical Computer Science, Vol. 47, 141–172, M. Farach-Colton, F. S. Roberts, M. Vingron, and M. Waterman, editors. American Mathematical Society, Providence, RI.

15. Z.W. Geem, J.H. Kim, and G.V. Loganathan, A New Heuristic Optimization Algorithm: Harmony Search, SIMULATIONS 76(2), 60–68, 2001.

16. F. Glover and F. Laguna, Tabu Search, Kluwer Academic Publishers, Dordrecht, 1997.

17. D.E. Goldberg, Genetic Algorithms in Search, Optimization & Machine Learning, Addison-Wesley, Reading, MA, 1989.

18. C.G. Han, P.M. Pardalos, and Y. Ye, Computational Aspects of an Interior Point Algorithm for Quadratic Programming Problems with Box Constraints, Large-Scale Numerical Optimization, T. Coleman and Y. Li (Eds.), SIAM, Philadelphia, 1990.

19. T.X. Hoang, A. Trovato, F. Seno, J.R. Banavar, and A. Maritan, Geometry and Symmetry Presculpt the Free-Energy Landscape of Proteins, Proceedings of the National Academy of Sciences USA 101: 7960–7964, 2004.

20. A.V.M. Ines, K. Honda, A.D. Gupta, P. Droogers, and R.S. Clemente, Combining Remote Sensing-Simulation Modeling and Genetic Algorithm Optimization to Explore Water Management Options in Irrigated Agriculture, Agricultural Water Management 83, 221–232, 2006.

21. D.F. Jones, S.K. Mirrazavi, and M. Tamiz, Multi-objective Meta-Heuristics: An Overview of the Current State-of-the-Art, European Journal of Operational Research 137, 1–9, 2002.

22. J. Kennedy and R. Eberhart, Particle Swarm Optimization, Proceedings IEEE International Conference on Neural Networks 4, Perth, WA, Australia, 1942–1948, 1995.

23. S. Kirkpatrick, C.D. Gelatt Jr., and M.P. Vecchi, Optimization by Simulated Annealing, Science 220(4598), 671–680, 1983.

24. K.S. Lee, Z. Geem, S.-H. Lee, and K.-W. Bae, The Harmony Search Heuristic Algorithm for Discrete Structural Optimization, Engineering Optimization 37(7), 663–684, 2005.

25. J.E. Lennard-Jones, Cohesion, Proceedings of the Physical Society 43, 461–482, 1931.

26. L. Lhotska, M. Macas, and M. Bursa, PSO and ACO in Optimization Problems, In E. Corchado, H. Yin, V.J. Botti, C. Fyfe (Eds.): Intelligent Data Engineering and Automated Learning - IDEAL 2006, 7th International Conference, Burgos, Spain, September 20–23, 2006, Proceedings. Lecture Notes in Computer Science 4224 Springer 2006, ISBN 3-540-45485-3.

27. M. Mahdavi, M. Fesanghary, and E. Damangir, An Improved Harmony Search Algorithm for Solving Optimization Problems, Applied Mathematics and Computation 188(22), 1567–1579, 2007.

28. S.P. Mendes, J.A.G. Pulido, M.A.V. Rodriguez, M.D.J. Simon, and J.M.S. Perez, A Differential Evolution Based Algorithm to Optimize the Radio Network Design Problem, E-SCIENCE '06: Proceedings of the Second IEEE International Conference on e-Science and Grid Computing, 2006.

29. N. Metropolis, A.W. Rosenbluth, M.N. Rosenbluth, A.H. Teller, and E. Teller, Equation of State Calculations by Fast Computing Machines, Journal of Chemical Physics 21(6): 1087–1092, 1953.

30. P.M. Morse, Diatomic Molecules According to the Wave Mechanics. II. Vibrational Levels, Physical Review 34, 57–64, 1929.

31. A. Mucherino and O. Seref, Monkey Search: A Novel Meta-Heuristic Search for Global Optimization, AIP Conference Proceedings 953, Data Mining, System Analysis and Optimization in Biomedicine, 162–173, 2007.

32. A. Mucherino, O. Seref, and P.M. Pardalos, Simulating Protein Conformations: the Tube Model, working paper.

33. J.A. Northby, Structure and Binding of Lennard-Jones clusters: $13 \leq N \leq 147$, Journal of Chemical Physics 87(10), 6166–6177, 1987.

34. P.M. Pardalos and H.E. Romeijn (eds.), Handbook of Global Optimization, Vol. 2, Kluwer Academic, Norwell, MA, 2002.

35. Protein Data Bank: http://www.rcsb.org/pdb/.
36. B. Raoult, J. Farges, M.F. De Feraudy, and G. Torchet, Comparison between Icosahedral, Decahedral and Crystalline Lennard-Jones Models Containing 500 to 6000 Atoms, Philosophical Magazine B60, 881–906, 1989.
37. J. Robinson and Y. Rahmat-Samii, Particle Swarm Optimization in Electromagnetics, IEEE Transations on Antennas and Propagation 52(2), 397–407, 2004.
38. C.T. Scott and M. Kohl, A Method of Comparing Sampling Designs Alternatives for Extensive Inventories, Mitteilungen der Eidgenossischen Forschungsanstalt fur Wald. Schnee and Landschaft 68(1), 3–62, 1993.
39. O. Seref, A. Mucherino, and P.M. Pardalos, Monkey Search: A Novel Meta-Heuristic Method, working paper.
40. A. Shmygelska and H.H. Hoos, An Ant Colony Optimisation Algorithm for the 2D and 3D Hydrophobic Polar Protein Folding Problem, BMC Bioinformatics 6, 30, 2005.
41. R. Storn and K. Price, Differential Evolution – A Simple and Efficient Heuristic for Global Optimization over Continuous Spaces, Journal of Global Optimization 11(4), 341–359, 1997.
42. Y. Xiang, H. Jiang, W. Cai, and X. Shao, An Efficient Method Based on Lattice Construction and the Genetic Algorithm for Optimization of Large Lennard-Jones Clusters, J. Physical Chemistry 108(16), 3586–3592, 2004.
43. X. Zhang, and T. Li, Improved Particle Swarm Optimization Algorithm for 2D Protein Folding Prediction, ICBBE 2007: The 1st International Conference on Bioinformatics and Biomedical Engineering, 53–56, 2007.
44. T. Zhou, W.-J. Bai, L. Cheng, and B.-H. Wang, Continuous Extremal Optimization for Lennard Jones Clusters, Physical Review E72, 016702, 1–5, 2005.

Modeling and Device Development for Chlorophyll Estimation in Vegetation

Vitaliy Yatsenko, Claudio Cifarelli, Nikita Boyko, and Panos M. Pardalos

Abstract Accurate estimation of leaf chlorophyll level by remote sensing is a challenging problem. Such estimation is especially needed in an ecologically dangerous environment. Our goal is to develop new methods that allow estimating chlorophyll concentration using remote sensing data for multiple kinds of soil and vegetation. The estimation is based on a training data set obtained from the leaf samples collected at various points on the earth's surface. A laboratory spectrophotometer was used to measure spectral reflectance curves in the visible and near-infrared ranges of the spectrum. The spectrometer was designed to comply with the strict measurement requirements essential for robust estimation. Optical indices related to leaf-level chlorophyll estimation were used as input data to test different modeling assumptions in open canopies where density of vegetation, soil, and chlorophyll content were separately targeted using a laboratory spectrometer. The goal of the research work is to estimate chlorophyll level based on spectrum characteristics of light reflected from the earth's surface. We have applied pattern recognition techniques as well as linear and nonlinear regression models. Unlike previously suggested approaches, our methods use the shape of the spectral curve obtained from measuring reflected light. The numerical experiments confirmed robustness of the model using input data retrieved from an ecologically dangerous environment.

1 Introduction

Vegetation is one of the most important targets for satellite monitoring as it is a necessary component of the biosphere. The status of the photosynthetic mechanism of plants is a sensitive and useful indicator of emerging abnormal or dangerous ecological processes. The level of chlorophyll contained in leaf structures is the most important parameter that is useful for detecting harmful

V. Yatsenko (✉)
Space Research Institute of National Academy of Sciences of Ukraine
and National Space Agency of Ukraine, Kiev 03187, Ukraine
e-mail: vyatsenko@gmail.com

P.J. Papajorgji, P.M. Pardalos (eds.), *Advances in Modeling Agricultural Systems*,   421
DOI 10.1007/978-0-387-75181-8_20, © Springer Science+Business Media, LLC 2009

ecological processes. Many scientific publications study plants' reaction to harmful environmental conditions [1–4, 16]. For example researchers discovered a decrease in chlorophyll level in the vicinity of the Chernobyl zone as well as in areas of gas leaks from factory chimneys. A decrease in chlorophyll level can be a result of viral infection, deficiency of nitrogen nutrition, abnormal humidity, and so on. Therefore, it is important to develop an inexpensive yet ubiquitous method for estimating chlorophyll concentration at arbitrary points on the earth's surface. The chlorophyll level is estimated using the spectral curve shape obtained from airplane or satellite recordings.

The University of Florida and scientists from the Space Research Institute of NASU (National Academy of Sciences of Ukraine) and NSAU (National Space Agency of Ukraine) have been jointly working on discovering informative parameters that can be used for estimating chlorophyll levels in leaves. This chapter is aimed at extracting the most informative region of the reflectance spectrum obtained from leaves. This range is known as the red edge of the spectrum (680–750 nm) where a rapid rise in the reflection value is observed. It was found that the intensity ratio of the two extrema in the first derivative of the spectrum curve correlates with chlorophyll concentration [5, 6] and therefore can serve an input for regression models.

The latest generation of the mobile spectral equipment is based on modern radiation receivers such as line and CCD matrixes. Such spectrometers with multi- and hyperspectral ranges are often installed in satellites and airplanes. Hyperspectral devices have about 300 spectral measuring channels in the range 400–1050 nm and a spectral resolution, of 2 nm [1, 2]. The equipment allows scanning both spatial and spectral information with high spatial resolution up to several meters. The development of such equipment makes it possible to practically apply our estimation approaches based on quantitative parameters of reflectance spectrum curve profile.

It should be noted that such new technical possibilities are frequently used in the old techniques of data processing in combination with spectral coefficients of brightness. Thus, significant efforts are being made to overcome internal drawbacks of such approaches by employing software means. Even the first derivative plots are only used for creation of new vegetation indices, for example, derivative indices. This approach has a serious drawback that almost annihilates its efficiency when measuring open crops: there is a need to minimize a distortion caused by an influence of the light reflected from open canopies. A comprehensible solution of this problem has not been found yet.

We have developed an integrated hardware–software complex for remote measuring of chlorophyll level based on the hyperspectral approach. The estimation technique is based on using quantitative parameters of the shape of a reflectance spectral curve. Such an integrated complex can be extended to analyze the information obtained from similar mobile embedded devices using thematic processing.

2 Methodological Approaches to Estimating Phytocenosis Parameters

We have developed innovative methodological procedures for quantitative estimation of the chlorophyll contents in plant leaves as well as other parameters important for testing state of plants. As opposed to other recognized approaches, our methods use quantitative parameters of the shape of a reflectance spectral curve. This method allows estimating a group of the most important phytocenosis characteristics in a relatively small spectral interval, 500–800 nm. Because less comprehensive spectral measures are needed, this method can be used with less expensive and more reliable spectral sensing equipment (Fig. 1).

2.1 Method Based on Use of the First Derivative

The shape of reflectance spectral curves of leaves in the area of the red edge is highly sensitive to the chlorophyll contents [2, 8]. There are two major sections of spectral reflectance curves that are sensitive to chlorophyll levels and are characterized with different curvatures. A decrease of the chlorophyll contents brings changes in the relative contribution of these sections. This can be observed as an increase in curvature at the red edge. The ratio of intensities in two basic maxima in the first derivative plot (I_1/I_2) at 722–725 nm and 700–705 nm can serve as a quantitative index of these changes. For example, the value I_1/I_2 varies in a range from 1.4 up to 0.5–0.6 for leaves of a winter wheat if chlorophyll concentration changes in the interval 8–15 mg/dm^2. This interrelation allows deriving linear and non-linear regression models for chlorophyll level calculations based on spectral characteristics [2, 3, 6].

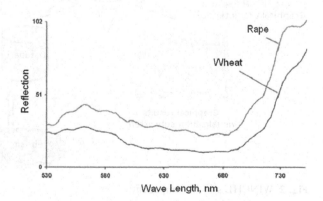

Fig. 1 Example of spectral curves for wheat and rape

The robustness of our method has been verified under conditions where open soil spots are present. This factor can essentially distort the shape of reflectance spectral curves of vegetation in the case where an open canopy reflects light that contributes to the measured spectrum. It is known that such a factor is extremely distorting if an approach is applied using coefficients of brightness. The presence of soil-reflected light can practically abolish the helpful information that is contained in the measured spectrum even in the case when proportion of uncovered soil area is relatively small [2]. We performed a simulation that has demonstrated that the divergence in the chlorophyll estimation performed at 100% and 25% projective soil covering with the highest reflection coefficient does not exceed 12%. In the case of 50% projective covering, the difference was no more than 5% and therefore fits into the accuracy limits of the regression equation [4, 7]. Thus, we are able to estimate chlorophyll level for both enclosed and open crops in a variety of configurations. This contrasts with the previously used traditional approaches that were based on factors of brightness.

2.2 Principal Components Analysis

The principal components analysis (PCA) method has been used for extraction of informative parameters from spectral curves. These parameters were used for the construction of regression equations describing chlorophyll concentration in vegetation [1, 14, 15]. PCA is performed on the symmetric

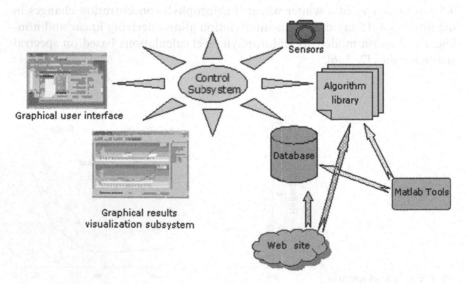

Fig. 2 WINCHL system components

covariance matrix. This matrix can be calculated from the matrix of a training data set obtained from spectral curves. PCA involves a mathematical procedure that transforms a number of possibly correlated variables into a smaller number of uncorrelated variables called *principal components*. These new components, which are arranged in decreasing order, have a useful property, in that they represent the spectral curve shape by the sum of partial contributions. Such a method of a spectral curve expansion on uncorrelated components allows reducing the number of informative parameters. Necessary information for construction of regression curves is obtained from such components. Eigenvalues of covariance matrix allow estimation the selected approximation quality. The developed method combined with the first derivative method allows fast estimation of chlorophyll content with high accuracy using ground, airborne, and satellite spectral measurements in the cases of incomplete projective covering (Fig. 2).

3 Support Vector Regression

In this section, a nonlinear method is used to estimate the content of chlorophyll. Support vectors machine algorithms [11, 12] are widely used both for regression and classification problems. SVM algorithms are based on the statistical learning theory [13] and their development is still one of the most important research directions in machine learning. The success of this approach is based on its exceptional generalization ability.

Let a data set be given as $(x_1, y_1), \ldots (x_l, y_l) \in R^d \times R$, where each element is represented by a set of variables stored in a vector $x \in R^d$ and a target or dependent variable y. In the case of linear function, the problem can be written as

$$f(x) = \langle w, x \rangle + b.$$

This idea can be formulated as a quadratic optimization problem (QP), but in order to have always a feasible solution for every input, some errors should be allowed. By using penalty term in the objective function, it is possible to obtain a trade-off between "sensitive parameters" and elements outside the regression boundaries (the flatness). The following formulation is called ε-SV:

$$\min \tfrac{1}{2}\|w\|^2 + C \sum_{i=1}^{l} (\xi_i, \xi_i^*)$$

Subject to

$$y_i - \langle w, x \rangle - b \leq \varepsilon + \xi_i$$
$$\langle w, x \rangle - y_i + b \leq \varepsilon + \xi_i$$
$$\xi_i, \xi_i^*.$$

Solving this QP usually involved the dual formulation due to independence from number of features and ability to apply nonlinear extension. The dual problem can be rewritten as [9, 10]

$$\min -\frac{1}{2} \sum_{i,j=1}^{l} (\alpha_i - \alpha_i^*)(\alpha_j - \alpha_j^*) \langle x_i, x_j \rangle - \varepsilon \sum_i^j (\alpha_i - \alpha_i^*) + \sum_i^j y_i(\alpha_i - \alpha_i^*)$$

Subject to

$$\sum_i^j (\alpha_i - \alpha_i^*) = 0$$

$$\alpha_i, \alpha_i^* \in [0, C].$$

There are few important qualities of this formulation. First, the complexity depends only on the number of elements and is independent from the number of features. Second, if the problem is formulated only in terms of inner products, it allows introducing implicit nonlinear mapping that extends the algorithm from the linear to nonlinear cases. Owing to these qualities, the method has a large variety of successful applications. Nonlinear extensions are introduced when linear separation of a data set is impossible. This is done via mechanism known as *kernel functions*. The most commonly used kernels are

Linear	$K(x_i, x_j) = (x_i, x_j)$
Polynomial	$K(x_i, x_j) = (x_i, x_j + 1)^d$
Gaussian	$K(x_i, x_j) = \exp\left(\frac{\|x_i - x_j\|}{\sigma}\right)$

The results of applying SVM regression model to the chlorophyll level estimation are represented by Table 1 and Fig. 3.

Table 1 Result for two SVM regression models

	SVR (530–750 nm)	SVR (530–680 nm)
STD error	0.308	0.334
R^2	0.967	0.964

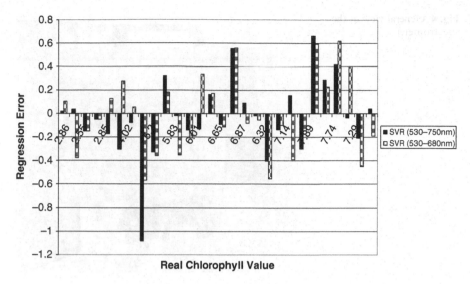

Fig. 3 SVM regression errors for the two models

4 Algorithms and Software

Recently developed methods for remote estimation of chlorophyll concentration were implemented in a software package. We have developed the algorithms based on first derivative, using experimental spectral curves smoothing techniques. We have also developed the evaluation of first derivative for maximum spectral resolution, as well as procedures for extrema searching. Various approaches have been applied to compute extrema. The best result was obtained by a procedure described in [3, 14, 15]. The developed software WINCHL (Fig. 2) is part of an integrated complex for vegetation parameters measurement in real-world conditions. In addition to chlorophyll concentration calculation, it includes additional software modules for providing various scientific computations.

5 Device for Remote Measurement of Vegetation Reflectance Spectra Under Field Conditions

A field-measuring device has been created to provide us with inputs for the vegetation condition testing [14, 15]. The general view of the device is shown in Fig. 4. The main difficulty related to designing this device was satisfying the two mutually contradicting requirements. On the one hand, a high spectral resolution (no less than 1–2 nm) was required. On the other hand, the error of output must have not exceeded 0.1% of output scale. Thus it was necessary to measure a large dynamic range of reflection coefficient in the area of the red edge within

Fig. 4 General view of the
spectrometer

the interval of 3% to 60% while illumination level could vary from 10,000 up to 120,000 lm. A specific device requirement to the field of vision (FOV) arises for ground measurements. The entrance channel FOV design that meets the requirements for airplane or space remote sensing is unacceptable for ground conditions due to measuring strip size several millimeters in width and more than 1 m in length at work from a height of 1.5 m.

An innovative design solution has been suggested for resolving this problem. The optical unit of the device is represented as a two-channel spectrophotometer, manufactured on the basis of polychromator with a flat diffraction grating, mirror spherical objectives/lenses, and CCD as a photo detector. The visual channel of the device FOV allows choosing a phytocenosis area, which needs to be measured. The recordings contain a complete information set, which consists of a recorded spectral data and auxiliary supporting information. The data can be written on a flash-memory card or uploaded directly to a computer that has installed the WINCHL package. The output information includes the data on temperature, illumination at the moment of measurement, as well as geographic coordinates of the position where the measurement took place. The optical unit is equipped with the appropriate sensors that digitize the data.

The field tests of the device were performed on wheat and rape crops (Fig. 1). The testing results of the areas with 100% projective covering were similar to the corresponding results obtained under laboratory conditions in a sense of a spectral curve shape. Differences in the area of the red edge at variations of the chlorophyll content have a similar nature. The higher reflection values of two spectral curves shown in Fig. 1 correspond with the spectrum measured on a field with a lower level of mineral feeding, which results in lower chlorophyll content. This method resulted in a lower level of noise.

Comparison of the single reflection spectrum record and averaged value of 10 repeated records of the same light flux entering the FOV of the device have shown that the value of differences in a range 530–750 nm does not exceed 4% of the minimum signal value at 685 nm and 1% of the maximum signal value at 750 nm. The curve lines of the first derivative essentially coincide with the average spectral curve and the single spectrum. Thus, the optical unit of our system functions as a hyperspectral remote spectrometer [14, 15]. This allows having a spectral resolution equal to 1.8 nm and potential spatial resolution equal to 2.5×2 mm. However, data are scanned as a single signal from the area equal to the device FOV (40×40 cm). This spatial resolution is sufficient for use in the field conditions. Our algorithms are embedded into the monitoring device that allows estimation of the chlorophyll content in biomass.

The obtained results can be the basis for creating similar onboard systems for aircraft and space craft, with respective changes to the entrance optical channel and, quite possibly, to the photo detector parameters.

6 Results

Remote testing of agrocenosis characteristics can become an effective tool for controlling and managing this kind of natural resources. To create an adequate agricultural monitoring system, it is beneficial to have the methods for estimating basic vegetation characteristics as well as the equipment that implements these methods. Hardware efficiency is increased if a single device is able to process multiple parameters simultaneously. It has been shown in the chapter that reflectance spectral curves of vegetation in the visible range can be used for testing such important parameters as level of nitrogen nutrition and humidity [3].

The comparative analysis of the two methods for chlorophyll level estimation was performed on the experimental data. It has been revealed that one of the methods is not sensitive to the additional reflection contributed by soil reflection in open canopies. On the contrary, the method based on principal components shows such sensitivity. The values of chlorophyll estimation obtained for the same object by these two methods allows estimating the value of a projective covering.

7 Conclusions

New modeling approaches to remote sensing of vegetation using component analysis and SVM regression are proposed. They allow estimating biochemical components based on the information about the type of plant, soil, and projective covering. It is shown that the quantitative system analysis of spectral reflectance curves is useful for developing new adaptive algorithms for extraction of information characteristics.

It has been shown that five principal components of the spectral curve contain most essential Shannon information. These components can be used as independent variables of nonlinear regression models. It is demonstrated that there is a nonlinear dependence between the vectors of principal components and chlorophyll concentration. The models of chlorophyll concentration estimations confirmed an ability to predict the concentration with a high degree of accuracy.

The similar level of classification accuracy with 100% covering was demonstrated by the method based on derivatives of spectral curves. At the same time, a systematic approach can be used for automatic detection of biochemical components when there is incomplete information about type of plant, soil, or projective covering.

In summary, the results indicate that the new technique is a practical and suitable for extracting the red edge position from hyperspectral data for explaining a wide range of chlorophyll concentrations. However, further research is needed to assess the accuracy of the proposed modeling technique for predicting leaf chlorophyll for a wide variety of plant species. In addition, the efficacy of the technique for predicting other plant parameters such as leaf area index and biomass also needs to be established.

References

1. P. A. Khandriga and V. A. Yatsenko. An application of the principal components method for chlorophyll content estimation in vegetation, System Technology, Dnepropetrovsk, Ukraine, 2(31):83–91, 2004.
2. S. M. Kochubey. Comparative analysis of information power of multispectral imaging and high-resolution spectrometry in the remote sensing of vegetation cover. Space Sciences and Technology, Kyiv, Ukraine, 5:41–48, 1999.
3. S. M. Kochubey. Estimating of the main characteristics of agricultural crops from reflectance spectrum of vegetation in the optical range. Space Sciences and Technology, Kyiv, Ukraine, 9:185–190, 2003.
4. S. M. Kochubey and P. Bidyuk. Novel approach to remote sensing of vegetation, Proc. of SPIE, Int. Conference "AeroSence. Technologies and Systems for Defence Security, Orlando (USA), 21–25 April 2003. vol. 5093, p. 181–188, 2003.
5. S. M. Kochubey, N. Kobets, and T. M. Shadchina. The quantitative analysis of shape of spectral reflectance curves of plant leaves as a way for testing their status. *Physiology and Biochemistry of Cultivar Plants*, 20:535–539, 1988.
6. S. M. Kochubey, N. Kobets, and T. M. Shadchina. Spectral properties of plants as a base of distant diagnostic methods. Naukova Dumka, Kyiv, 1990.
7. S. M. Kochubey, P. M. Pardalos, and V. A. Yatsenko. Method and the device for remote sensing of vegetation. Remote Sensing for Agriculture, Ecosystems, and Hydrology IV, Proc. of SPIE, Crete (Greece), 22-25 September 2002.-vol. 4879:243–251, 2002.
8. S. M. Kochubey, V. A. Yatsenko, and N. V. Gurinovich. Development of the Method and the Device for Remote Sensing of Vegetation, www.vegetation.kiev.ua.
9. Smola A. J. and Schölkopf B. A tutorial on support vector regression. NeuroCOLT technical report, Royal Holloway University of London, UK, 1998. http://www.kernel-machines.org.

10. B. Smola, A. Schölkopf. A tutorial on support vector regression. Statistics and Comput-
    ing, 14(3):199–222, 2004.
11. V. N. Vapnik. The Nature of Statistical Learning Theory. Springer-Verlag, New York,
    1995.
12. V. N. Vapnik. Statistical Learning Theory. Wiley, New York, 1998.
13. V. N. Vapnik, and A. Y. Chervonenkis. Theory of Pattern Recognition. Nauka, Moscow,
    1974.
14. V. Yatsenko, S. Kochubey, V. Donets, and T. Kazantsev. Hardware-software complex
    for chlorophyll estimation in phytocenoses under field conditions. Proc. of SPIE "Detec-
    tors and Associated Signal Processing II", Jena 13–14 September 2005, 5964:1–6, 2005.
15. V. Yatsenko, S. Kochubey, P. Khandriga, V. Donets, and P. Chichik. Optical spectro-
    meter and software for remote sensing of vegetations. Proc. CAOL 2005. 2nd Interna-
    tional Conference on Advanced Optoelecektonics and Laser, Yalta (Crimea, Ukraine),
    September 12–17, 2005, -IEEE, vol.2. P. 267–269.
16. Yatsenko, S. Kochubey, V. Donets, and O. Semeniv. New method of chlorophyll content
    estimation in phytocenoses and software realization. Space Sciences and Technology,
    Kyiv, Ukraine, 3:35–45, 2007.

Clustering and Classification Algorithms in Food and Agricultural Applications: A Survey

Radnaabazar Chinchuluun, Won Suk Lee, Jevin Bhorania,
and Panos M. Pardalos

Abstract Data mining has become an important tool for information analysis in many disciplines. Data clustering, also known as unsupervised classification, is a popular data-mining technique. Clustering is a very challenging task because of little or no prior knowledge. Literature review reveals researchers' interest in development of efficient clustering algorithms and their application to a variety of real-life situations. This chapter presents fundamental concepts of widely used classification algorithms including k-means, k-nearest neighbor, artificial neural networks, and fuzzy c-means. We also discuss applications of these algorithms in food and agriculture sciences including fruits classification, machine vision, wine classification, and analysis of remotely sensed forest images.

1 Introduction

Today, with the help of advancement in technologies, we are capable of accumulating a vast amount of data in different formats and databases. It becomes very difficult to identify useful information from this enormous amount of data generated. Knowledge discovery in databases (KDD) field is concerned with the development of methods and techniques for making sense of the data, and data mining is a step of KDD that consists of applying data analysis and discovery algorithms over the data [23]. In other words, data mining is the process of analyzing data from different perspectives and summarizing it into useful information. Data-mining techniques are widely applied in a variety of fields including finance, marketing, and sales planning (by learning about purchasing habits of customers), biotech/genetics, e-commerce, telecommunication, detecting fraud, Web, food science, precision agriculture, and so forth.

P.M. Pardalos (✉)
Department of Agricultural and Biological Engineering, University of Florida,
Gainesville, FL, USA
e-mail: pardalos@ufl.edu

P.J. Papajorgji, P.M. Pardalos (eds.), *Advances in Modeling Agricultural Systems*,
DOI 10.1007/978-0-387-75181-8_21, © Springer Science+Business Media, LLC 2009

The overall organization of the paper is as follows. In the next section, we present theoretical aspects of clustering algorithms. In particular, k-means, k-nearest neighbor, fuzzy c-means, and artificial neural networks are discussed. The section also includes the recent developments and some of the drawbacks of the respective algorithms. In Section 3, we discuss various applications of these algorithms including grading of fruits and vegetables, classification of wines, robotic harvesting and machine vision, and classification of remotely sensed images and other forest parameters. Section 4 concludes the paper.

2 Data Mining Algorithms

Data mining is the process of analyzing data using tools such as clustering, classification, feature selection, and outliers detection. Clustering techniques partition a given set of data into groups of similar samples according to some similarity criteria. Classification techniques determine classes of the test samples using known classification of training data set. Feature selection techniques select a subset of features responsible for creating the condition corresponding with any class. Clustering is generally an initial step of data mining, and it groups data into similar samples that can be used as a starting point of other techniques. Data clustering can be divided into two parts: hierarchical and partitional clustering [38]. Single link and complete link are examples of hierarchical clustering, whereas partitional clustering includes squared error algorithms (k-means), graph theoretic, mixture resolving (expectation maximization), mode seeking, and so forth. Bayesian classifier is a traditional statistical classification algorithm based on Bayes' theory. Bayesian classifiers use combination of conditional probability and posterior probabilities for classifying any information. Further details about Bayesian classifiers and applications can be found in [20,25,59,71,91]. Some other clustering techniques such as fuzzy clustering, artificial neural networks, nearest neighbor clustering, and evolutionary approach–based clustering are also becoming very popular tools for researchers. We discuss some of these clustering techniques in this section.

2.1 k-Means Algorithm

Clustering analysis plays an important role in scientific research and commercial applications. Therefore, the k-means [57] method is one of the most popular clustering methods that have been applied in a variety of fields including pattern recognition, information retrieval, document extraction, microbiology analysis, and so forth. The method is called k-means because it represents each of k number of clusters C_j ($j = 1, 2, \ldots, k$) by mean (or weighted average) of its points. The goal of this method is to classify a given data set through a certain number of clusters such that some metric relative to the centroids of the clusters is minimized. We can define our problem mathematically as follows.

Suppose that we are given a set X of a finite number of points in d-dimensional Euclidean space R^d, that is, $X = (x^1, \ldots, x^n)$ where $x^i \in R^d, i = 1, 2, \ldots, n$.

We aim at finding a partition $C_j, j = 1, 2, \ldots, k$:

$$X = \bigcup_{j=1}^{k} C_j, C_j \cap C_l = \phi$$

for all $j \neq l$, of X that minimizes the squared error function

$$f(C_1, C_2, \ldots, C_k) = \sum_{j=1}^{k} \sum_{x^i \in C_j} \| c^j - x^i \|^2,$$

where $\|\cdot\|$ denotes the Euclidean norm, and c^j is the center of the cluster C_j

$$c^j = \frac{1}{|C_j|} \sum_{x^i \in C_j} x^i, j = 1, 2, \ldots, k. \tag{1}$$

2.1.1 Algorithm k-Means

Step 1. Initialize the centroids $c^j_0, j = 1, 2, \ldots, k$. Set $q = 0$ (where q is iteration counter).

Step 2. Assign each point x^i ($i = 1, 2, \ldots, n$) to the cluster that has the closest centroid c^j ($j \in \{1, 2, \ldots, k\}$), that is,

$$j = \underset{1 \leq l \leq k}{\arg\min} \| x^i - c^l_q \|^2.$$

Step 3. When all points have been assigned, for $j = 1, 2, \ldots, k$, calculate the new position c^j_{q+1} of the centroid j using Equation (1).

Step 4. If $c^j_q = c^j_{q+1}$ for all $j = 1, 2, \ldots, k$, then stop, otherwise set $q = q + 1$ and go to Step 2.

The k-means algorithm is easy to implement, and its time complexity is of order n ($O(n)$), where n is number of patterns [38]. However, the solution obtained by using the k-means algorithm is one of the many local solutions and sensitive to the initial starting points. To find a better solution, we can run the algorithm several times and choose the best one as the optimal solution. Unfortunately, repetition with different random selections [20] appears to be the *de facto* method. Bradley and Fayyad [13] presented a procedure for computing a refined starting condition from a given initial one that is based on an efficient sampling technique for estimating the modes of a distribution, and their experiments presented that refined initial starting points indeed lead to improved solutions. Yager and Fillev [90] developed the mountain method, which is a simple and effective approach for approximate estimation of the cluster centers on the basis of the concept of a mountain function. It can be useful for obtaining the initial values of the clusters that are required by more complex cluster algorithms. Another drawback of the algorithm is that there are no efficient methods for defining initial number of partitions. Many alternative methods to improve k-means were published in literature. Krishna and Murty [48] proposed a new genetic k-means algorithm (GKA) for global search and faster convergence. Zhang et al. [95] proposed parallel k-means algorithm for higher efficiency.

Many algorithms similar to k-means have appeared in the literature [21,57]. In [28], Hansen and Mladenovis proposed J-means algorithm, which defines the neighborhood of the current solution by all possible centroid-to-entity relocations followed by corresponding changes of assignments. Moves are made in such neighborhoods until a local optimum is reached. k-Means algorithm sometimes gives result that are local optimal. Maulik and Bandyopadhyay [61] proposed genetic algorithm (GA)-clustering algorithm based on k-means and GAs to overcome the shortfall of local optima. More advances in the k-means algorithm and its application can be found in [17,43,52,55,85,96].

2.2 *Fuzzy c-Means Clustering*

The k-means algorithm can be classified as hard of crisp clustering technique, in which each object can be assigned to only one cluster. Fuzzy clustering relaxes this restriction and an object can belong to several clusters at the same time but with certain degrees of memberships. The most known fuzzy clustering method is the fuzzy c-means (FCM) method, introduced by Dunn [22] and later generalized by Bezdek [10]. FCM partitions a data set $X = (x_1, x_2, \ldots, x_n) \subset R^p$ of p features into c fuzzy subsets where $u_{i,k}$ is the membership of x_k in class i ($i = 1, 2, \ldots, c$). These classes are identified by their cluster centers v_i, ($i = 1, \ldots, c$). The objective of FCM is to find an optimal fuzzy c partition minimizing the objective function,

$$J_m(U, V; X) = \sum_{k=1}^{n} \sum_{i=1}^{c} (u_{ik})^m \|x_k - v_i\|^2, \tag{2}$$

where the value of fuzzy partition matrix U is constrained in the range [0,1] such that

$$\sum_{i=1}^{c} u_{ik} = 1, \forall k = 1, 2, \ldots, n, \tag{3}$$

and

$$\sum_{k=1}^{n} u_{ik} < n, \; \forall i = 1, 2, \ldots, c. \tag{4}$$

Here, $m \in [1, \infty)$ is an exponential weighting function that controls the fuzziness of the membership values. $\|\cdot\|$ is the Euclidean norm, and $V = (v_1, v_2, \ldots, v_c)$ is a matrix of unknown cluster centers $v_i \in R^p$ $(i = 1, \ldots, c)$. Fuzzy c-means algorithm to minimize (2) can be described as follows:

Step 1. Choose appropriate values for m, c, and a small positive number ε. Initialize randomly a fuzzy partition matrix U^0 and set iteration number $t = 0$.

Step 2. For given membership values $u_{ik}^{(t)}$, calculate the cluster centers $v_i^{(t)}$ $(i = 1, 2, \ldots, c)$ as

$$v_i^{(t)} = \frac{\sum_{k=1}^{n} (u_{ik}^{(t)})^m x_k}{\sum_{k=1}^{n} (u_{ik}^{(t)})^m}. \tag{5}$$

Step 3. Given a new cluster center from Step 2, update membership values $u_{ik}^{(t+1)}$ using

$$u_{ik}^{(t+1)} = \left[\sum_{j=1}^{c} \left(\frac{\left\| x_k - v_i^{(t)} \right\|^2}{\left\| x_k - v_j^{(t)} \right\|^2} \right)^{\frac{2}{m-1}} \right]^{-1} \tag{6}$$

Step 4. Repeat Steps 2 and 3 until $\left| U^{(t+1)} - U^{(t)} \right| \le \varepsilon$ or a predefined number of iterations is reached.

Methods discussed in previous sections are crisp/hard partitioning methods, which allow partitioning data into a specified number of mutually exclusive data sets only, whereas fuzzy methods are soft partitioning methods where an object can belong to one or more data sets/partitions. Similar to the crisp/hard partitioning methods, selection of initial matrix of centers plays an important role in convergence of FCM. Many times, FCM does not guarantee the global optimal solutions due to randomized initialization of cluster centers and matrix U. Moreover, FCM solutions are also sensitive to noise and outliers [89]. Hathway et al. [31] have proposed a modified FCM using 1 norm distance to increase robustness against outliers. Hung and Yang's *psFCM* algorithm [35] finds the actual clusters' centers and refines initial value of FCM. This technique reduces the computational time by a large amount. Kolen and Hutcheson modified FCM, which highly reduces the computation time by combining updates of the two matrices [46]. Many other improved FCMs and their applications can also be found in [8,27,32,44,65,66,94].

2.3 k-Nearest Neighbor Classification

The k-nearest neighborhood (KNN) method is adopted widely due to its efficiency. The objective of k-nearest neighborhood algorithm is to discover k nearest neighbors for the given instance according to the majority class of k-nearest neighbors. This instance is defined as the training data set, and it is used to classify each member of a "target" data set. A Euclidean distance measure is used to calculate how close each member of the training set is to the target data that is being examined. The k-nearest neighbor classification algorithm can be divided into two phases:

2.3.1 Training Phase

- Define a training set $S = \{(x_1, y_1), (x_2, y_2), ..., (x_n, y_n)\}$, where $x_i = (x_i^1, x_i^2, ..., x_i^d)$ is a d-dimensional feature vector of real numbers, for all $i = 1, ..., n$.
- Define class labels y_i corresponding with each x_i for all i, $y_i \in C$ where $C = (1, ..., N_c)$, N_c is the number of different classes.
- Task: determine y_{new} for x_{new}.

2.3.2 Testing Phase

- Find the closest point x_j to x_{new} w.r.t. Euclidean distance

$$\sqrt{(x_j^1 - x_{new}^1)^2 + ..., + (x_j^d - x_{new}^d)^2}.$$

- Classify by $y_{new} = y_j$

A serious drawback of this k-nearest neighbor technique is the computational complexity in searching the k-nearest neighbors among those available training samples [75]. Kuncheva [50] claims to achieve better computational efficiency and higher classification accuracy by using genetic algorithms as editing techniques. Bermejo and Cabestany [9] proposed a KNN algorithm with local Parzen window estimate to improve approximation quality. They also suggested an adaptive learning algorithm to allow fewer data points to be used in a training data set. Wu et al. [88] achieved a considerably accelerated

KNN classification procedure without sacrificing accuracy by incorporating two techniques that are template condensing and preprocessing. Many other techniques have been proposed to reduce the computational burden of *k*-nearest neighbor algorithms in [24,36,68].

2.4 Artificial Neural Networks

Artificial neural networks (ANNs) are self-adaptive statistical models based on analogy with the structure of the human brain. The term *artificial*, which is often dropped, differentiates ANNs from their biological counterpart. Neural networks are built from simple units, called neurons or cells by analogy with the real thing. Neurons are linked with each other by a set of weighted connections. The information to be analyzed is fed to the neurons of the input layer and then propagated to the neurons of the hidden layer (if any) for further processing. The result of this processing is then propagated to the next hidden layer and the process is continued until the output layer. Each unit receives some information from other units and processes this information, which will be converted into the output of the unit. There are no specific methods of choosing the network parameters such as number of hidden layers and type of activation function. Generally, one input node and one output node are chosen for each data class.

Training set is known *a priori* and is used to fine-tune the network for future similar records. While in the training phase, known data containing inputs and corresponding outputs are fed to the network, and the network learns to infer the relationship between these two.

Process of classification by ANN can be broadly defined as follows:

- Run a sample from training set by giving its attribute values as input.
- The summation of weights and activation functions are applied at each node of hidden and output layers, until an output is generated (feed-forward process).
- Compare output with the expected output from training set.
- If output does not match, go back layer to layer and modify arc weights and biases of nodes (back-propagation process).
- Run the next sample and process the same.
- Eventually, the weights will converge, and process stops.

Feed-forward topology is widely used in multilayer perceptions networks. Feed-forward network provides a general framework for representing non-linear functional mappings between a set of input variables and a set of output variables [11]. Below is a brief overview of feed-forward network for each training sample X in samples and for each hidden or output layer node j:

- Calculate input I_j to that node as

$$I_j = \sum_{i=1}^{d} w_{ji}x_i + w_{j0}. \tag{7}$$

- Calculate output O_j from that node as

$$O_j = \frac{1}{1 + e^{-I_j}}.$$

Back-propagation is a widely used algorithm for purpose of training the neural networks. Back-propagation algorithm can be divided in two stages. In the first stage, the derivatives of the error function with respect to the weights are evaluated. In the second stage, these derivatives are then used to compute the adjustments to be made to the weights by using gradient descent or any other optimization schemes. An overview of back-propagation can be given as below:

- For each node j in output layer, calculate the error as

$$\text{Err}_j = O_j(1 - O_j)(T - O_j)$$

- For each node j in hidden layer, calculate the error as

$$\text{Err}_j = O_j(1 - O_j)\left(\sum_k \text{Err}_k w_{jk}\right)$$

- For each weight w_{ij}, calculate weight increment as

$$\Delta w_{ij} = l \cdot \text{Err}_j \cdot O_i.$$

- Now, update the previous weight as

$$w_{ij} = w_{ij} + \Delta w_{ij}.$$

- For each bias θ, calculate bias increment as

$$\Delta \theta_j = l \cdot \text{Err}_j$$

- Finally, update bias with

$$\theta_j = \theta_j + \Delta \theta_j.$$

Neural networks are widely used in classification but there are still many unsolved issues in applying neural networks such as scalability, misclassification, convergence, higher mean square errors, and so forth. Many researchers have tried to overcome these issues and proposed a variety of neural networks with better performance. Jiang and Wah [42] proposed a new approach of constructing and training neural networks to overcome the problems including local minima and the slow convergence of the learning process. In their approach, feed-forward network was constructed based on the data clusters generated based on locally trained clustering (LTC) and then further trained using standard algorithms operating on the global training set that converges rapidly due to its inherited knowledge with good generalization ability from the global training. Ji and Ma [41] proposed a learning method based on combination of weak classifiers, which were found by a randomized algorithm, to achieve good generalization and fast training time on both the test problems and the real applications. They showed that if the weakness factor was chosen according to the critical value given by their theory, the combinations of weak classifiers could achieve a good generalization performance with polynomial space- and time-complexity. Yu et al. [92] proposed dynamic learning using derivative information instead of fixed learning rate to optimize back-propagation. The information gathered from the forward and backward propagation was used for dynamic learning; with this technique they achieved higher convergence rate and significant reduction in learning process. Probability of misclassification of any random sample can be termed as generalization error of a classifier. Many researchers have used ensemble methods to reduce misclassification or generalization errors [29,30]. Further information on classification errors, learning and generalization, and some of the recent developments in neural networks can be found in [49,84,93].

3 Applications

Data clustering algorithms mentioned in Section 2 have been widely applied to a variety of food and agricultural applications including pattern recognition of fruit shape [63], classification of fresh-cut star fruits [1], protection of citrus fruits from frost damage [79], harvest prediction models for cherry tomatoes [34], categorization of fried potato chips based on color classification [60], determining the geographic origin of potatoes [2], classification of cereal grains [67], and corn yield prediction and mapping system [87].

Recently, precision technologies have gained much attention in agricultural engineering. Among precision technologies, yield mapping and harvesting fruits from the trees widely use ANNs as a part of their classification algorithms. Grading of fruits, vegetables, and wines with machine vision allows consistency in quality. Most of the machine vision techniques use the above-mentioned classification algorithms. These techniques help in observation of forests and agricultural land, estimating biomass and carbon content in forests by analyzing remotely sensed images.

In this section, we discuss only some of the applications of data clustering algorithms in food and agricultural sciences as it is impossible to mention all of them. In particular, the grading of fruits and vegetables, classification of wines, robotic harvesting and machine vision, and analysis of remotely sensed forest images are considered.

3.1 Grading Methods of Fruits and Vegetables

Quality plays the most important role in pricing and marketing of fruits and vegetables. Fresh fruits such as apples and oranges are graded into quality categories according to their size, color, shape, and presence of defects. Usually this grading is decided manually by the empirical sense of grading by a worker's eye [64]. Recently, many methods using machine vision have been proposed to substitute the manual classification of fruits. The grading of apples using machine vision can be arbitrarily divided into four steps: the acquisition of images, their segmentation, their interpretation, and finally the fruit classification [54].

Various methods have been applied for the image acquisition purpose. Placing fruits on rollers and observing them with cameras while moving is a widely used and most common method of image acquisition. In Nakano's [64] work, an image data collecting system consisting of a turntable, a stage controller, and a monoaxle driver was developed. On the system, the whole image of an apple is rolled out as one scene on a computer monitor as perspective projections and the images were captured. Leemans and Destain [54] used an image acquisition system with a double row grading machine where the fruits were rolled on rollers. The system also consisted of lighting tubes and two cameras placed at the inner side and two cameras outside of the tunnel. Water core is an internal disorder, and apples affected by this cannot be stored for a long time. Shahin et al. [80] developed a classifier for sorting apples based on water core. Artificial neural networks, Bayesian classifier, and fuzzy classifiers were applied on line scan images of apple water core (captured with computed axial tomography [CAT] scanner), and their results showed that ANN, with an accuracy of 88%, performs better than the fuzzy classifiers (80%) and Bayesian classifier (79%).

3.1.1 Image Interpretation by k-Means Algorithm

Leemans and Destain [54] proposed a hierarchical grading method, with k-means algorithm, based on the presence of defects and applied to Jonagold apples. Several images covering the whole surface of the fruits were acquired by a prototype machine. These images were then segmented, and the features of the defects were extracted.

Blobs (defects, calyx and stem ends) were characterized by 15 parameters or features, which included five for the color, four for the shape, five for the texture, and one for its position (the angle between the nearest pole and the blob). This database was a dynamic table, having length equal to the number of blobs. The table was quite short for healthy fruits but included more than a hundred parameters for some kinds of defects.

They used k-means clustering (16 clusters) on the blob features of all the fruits of the training set. When the clusters were defined, the decision rules to put a blob in a cluster were computed by linear discriminant analysis. For each fruit and for each of these clusters, the sum of the *a posteriori* classification probabilities and the standard deviation of these probabilities were computed. A principal component analysis was carried out, and 16 first principal components representing 97% of the whole variation were used to compute a quadratic discriminant analysis to finally grade the fruits. They reported that the correct classification rate was 0.91 for accepted fruit and 0.55 for rejected fruit and 0.73 overall.

3.1.2 Image Interpretation by Neural Networks

Nakano [64] used neural networks (two layer perception, five hidden layers) to sort San-Fuji apples into five color and quality classes. In Nakano's work, an image data collecting system consisting of a turntable, a stage controller, and a monoaxle driver was developed. On the system, the whole image of an apple was rolled out as one scene on a computer monitor as perspective projections. The system calculated nine color characteristic data (the average color gradients $(\bar{R}, \bar{G}, \bar{B})$, the variances (V_R, V_G, V_B), and the chromatic coordinates (r,g,b) for the external appearance of the entire apple from the three primary colors in the following manner:

$$\bar{R} = R/n, V_R = \sum_{i=1}^{n} (R_i - \bar{R})^2/n, r = R/(R+G+B) \qquad (8)$$

where $R = \sum_{i=1}^{n} R_i$ and n is the number of total pixels in the image data.

Two neural network models were used in the study. Neural network A was used to classify a pixel at any part of an apple into "normal red," "injured color red," "poor color red," "vine," and "upper and lower background color." This neural network consisted of three layers (input layer, hidden layer, and output layer). The input layer had seven units: R_i, G_i, B_i from each pixel of the image data, the average data $\bar{R}$, $\bar{G}$, $\bar{B}$, and the position of the pixel Y in regard to the two-dimensional coordinates. The hidden layer had five units. The output layer had six units that indicated the results of the judged condition of the surface as mentioned above (normal red, injured color red, poor color red, vine, and upper and lower background color). Supervised learning for each pixel of an entire apple was performed. Test sample data was classified into the above six surface conditions using a neural network model that was structured by supervised learning.

Neural network B was used to classify the surface quality of the apples into five categories: superior (AA, more than 89% of the surface is deep red, and orange in the background), excellent (A, 70% to 89% area of the surface is red, and the background is yellowish orange), good (B, 50% to 69% of the surface is red, and the background is yellowish green), poor color (C, less than 50% of the surface is red, and the background is light green or uneven red-colored), and injured (D, injured part can be seen on the apple's surface). It also had three layers (input layer, hidden layer, and output layer). The input layer had 11 units including the nine characteristics in Equation (8), as well as two other factors: the color ratio (R_c) and the injured or noninjured surfaces:

$$R_c = R_n/n,$$

where R_n is the number of pixels that are judged to be normal red in neural network A. The factor, injured or noninjured, was defined as "1" or "0" from the results from neural network A. The function that gives the characteristics of each neuron is a logistic function as follows:

$$f(x) = 1/(1 + \exp(-x)),$$

where x is the input data of a neuron.

Supervised learning was done with output units that had five neurons (AA, A, B, C, D). When the quality of the tested apple was "A," the neuron in the output layer was set to [0,1,0,0,0] as the result of supervised learning. The neuron data in the output layer was a decimal figure that varied from 0.00 to 1.0. Nakano reported that the grade judgment ratio (the number of correctly judged apples / the number of tested apples) for the five categories was 92.5% for AA, 33.3% for A, 65.8% for B, 87.2% for C, and 75% for D.

Guyer and Yang [26] applied enhanced genetic artificial neural networks (EGANNs) to detect defects using spectral signatures of different tissues on cherry images. In their work, they used a genetic algorithm (GA) to evaluate weights in a multilayer feed-forward artificial neural network. They trained the network with cherry images representative of all the different types of surface defects, and pixels that belonged to different tissue types in these images were interactively selected as training pixel samples. In classification part, additional cherry samples were presented to the imaging system, individually, to collect spectral images at multiple wavelength bands. These samples were analyzed and classified by EGANN through a pixel-by-pixel image analysis. They achieved an average of 73% classification accuracy for correct identification and quantification of all types of cherry defects with no false positives or false negatives occurring.

Kondo et al. [47] used artificial neural networks for quality evaluation of *Iyokan* oranges. They extracted various features such as fruit color, shape, and roughness of fruit surface from the images and calculated $R:G$ color component ratio, Feret's diameter ratio, and textural features. While training the network, these features and weight of the fruit were entered to the input layers of neural networks, and sugar content or pH of the fruit was used as the values of the output layers. They applied this trained networks to various samples to predict the sugar content or pH from the fruit appearance.

Kiliç et al. [45] developed a computer vision system (CVS) for the quality inspection of beans based on size and color quantification of samples. In their work, hardware was developed to capture a standard image from the samples, and the software was coded in MATLAB for segmentation, morphologic operation, and color quantification of the samples. Moment analysis was performed to identify the beans based on their intensity distribution. They determined average, variance, skewness, and kurtosis values for each channel of RGB color format. Then artificial neural networks were used for color quantification of the samples. Their experiments achieved classification rate of 90.6%.

3.2 Machine Vision and Robotic Harvesting

A variety of classifiers and many different features have been used to identify the fruit from the background. Parrish and Goksel [70] used monochrome images enhanced by color filters to identify apples and to harvest them. A thinness ratio (R),

$$R = 4\pi A / P^2,$$

where P is the perimeter and A is the area of the blobs, was used to distinguish between noise, clusters of fruits, and single fruits. Color images were used for distinguishing the citrus fruit by setting a threshold in the hue value [82]. The threshold in hue was found using spectral reflectance curves of citrus fruit. Though many other features have been used to address the problem of citrus fruit identification, color still remains one of the most used features. The feature is simple to use and gives adequately good results. Apart from extracting different features to identify the fruit, different classifiers such as neural network-based classifier [62] and Bayesian [83] have also been used. Marchant and Onyango [59] described the performance of a Bayesian classifier compared with a neural network-based classifier in plant/weed/soil discrimination in color images. They used a probability distribution formed out of the training data and reported that given enough memory, the Bayesian classifier performed equally well as the neural network classifier. Plebe and Grasso [72] described the image processing system developed to date to guide automatic harvesting of oranges, which has been integrated in the first complete full-scale prototype orange picking robot. Majumdar and Jayas [58] used machine vision to classify bulk samples of cereal grains. Simonton and Pease [81] used machine vision to identify and classify major plant parts.

Annamalai et al. [3] and Pyadipati et al. [73,74] have used several algorithms including ANN and k-means for pattern classification of their machine vision system in precision agriculture. In [78], Regunathan and Lee have presented their machine vision system for citrus fruit identification and size determination of citrus fruits. Images of citrus fruits were captured and converted from RGB format to hue, saturation, and luminance (HLS) format. These HLS format images were classified using three different classifiers including ANN, Bayesian, and Fischer's linear discriminant algorithms. The distance between camera and citrus fruit was measured with ultrasonic sensors, and with the help of basic trigonometry, the actual size of fruits was calculated. However, their results showed that root mean square error for ANN classifier for both fruit size estimation and fruit count was lesser than for other two classifiers. Chinchuluun and Lee [16] used similar techniques for mapping yield of citrus fruits. After the processing images for illumination enhancement and noise removal, they applied a hue, saturation, and intensity (HSI) model. Images with HSI model were then classified in fruits, leaves, and other classes with the help of k-means algorithm. Pydipati et al. [73] used machine vision for citrus disease detection. They applied a texture analysis method termed the color co-occurrence method (CCM) to determine whether classification algorithms could distinguish diseased and normal citrus leaves. The samples were processed for edge detections with Canny edge detector, which is a multi-stage algorithm to detect edges in images, followed by noise removal and size reduction. Images were then converted to HSI format from RGB format before application of classification algorithms including back-propagation neural network, Mahalanobis minimum distance classifier, and neural network with radial basis functions. However, their experiments suggested that Mahalanobis statistical classifier and back-propagation neural network classifier performed equally well when using hue and saturation texture features.

3.3 Classification of Wines

Wine is one of the most widely consumed beverages in the world and has very obvious commercial values as well as social importance. Therefore, the evaluation of the quality of wine plays a very important role for both manufacture and sale. Historically, the quality or geographic origin of wine was determined only through tasting by wine experts. However, recently more advanced instruments have become available. Therefore, it seems reasonable to determine the geographic origins of wine by reliable chemical analysis techniques in combination with modern chemometric methods instead of by traditional wine experts.

Concerning the analysis methods of constituents in wine, there are many reports such as gas chromatography (GC) for the analysis of aroma compounds in wine, high-performance liquid chromatograph (HPLC) for the determination of phenolic acids, spectrophotometry for Fe, Mn, and preservatives containing benzoic acid, sorbic acid dehydroacetic, and ethyl parahydrobenzoate.

The inductive coupled plasma optical emission spectrometer (ICP-OES) provides a powerful means for fast analysis of a number of elements of the Periodic Table. Linskens and Jackson [56] have presented a monograph that describes various methods of wine analysis.

In Sun et al. [86], three-layer ANN model with back-propagation of error was used to classify wine samples in six different regions based on the measurements of trace amounts of B, V, Mn, Zn, Fe, Al, Cu, Sr, Ba, Rb, Na, P, Ca, Mg, and K using an ICP-OES. In their work, 15 element concentration values (mg/L) were measured at 17 different wavelengths for each wine sample (p), and the 17-dimensional concentration vector $a_p = (a_{p1}, ..., a_{p17})$ served as the input pattern; so the input layer contained 17 input nodes plus 1 bias node. The output layer represented the six different geographic origins (six nodes). During the learning process of the network, a series of input patterns (i.e., 17-dimensional concentration vectors) with their corresponding expected output patterns (i.e., the true class of wine samples) were presented to the network to learn, with the connection weights between nodes of different layers adjusted in an iterative fashion by the error back-propagation algorithm. In order to obtain an optimum artificial neural network model, various network architectures were tested. The number of hidden nodes was an adjustable parameter that was optimized by reducing the number used until the network prediction performance deteriorated or the best prediction accuracy was found. They reported that 10 to 30 hidden nodes plus one bias node gave the best prediction performance for the wine samples. They also applied cluster analysis, principal component analysis, a Bayes discrimination method, and Fisher discrimination method and compared the results with that obtained by artificial neural network model. The report said that a satisfactory prediction result (100%) by an artificial neural network using the jackknife leave-one-out procedure was obtained for the classification of wine samples containing six categories.

Recently, Beltrán et al. [7] presented the results of Chilean red wine classification, considering the varieties Cabernet Sauvignon, Merlot and Carmenérè from different valleys, years, and vineyards. The classification was based on the information contained in phenolic compound chromatograms obtained from an HPLC-DAD. Different feature extraction techniques including the discrete Fourier transform, the wavelet transform, the class profiles, and the Fisher transformation were analyzed together with several classification methods such as quadratic discriminant analysis, linear discriminant analysis, k nearest neighbors, and probabilistic neural networks. The information contained in the chromatograms corresponds with phenolic compounds of small molecular weight obtained through a high performance liquid chromatograph attached with an aligned photo diode detector (DAD). Leave-one-out (LOO) validation procedure was used. In this sense, one sample was left out and the classification system was trained using 171 remaining samples. Then the sample left out was presented to the classifier to determine to which class it belongs. They assumed this classification procedure was the best method to be used in cases where the amount of information was low as in their case. For each combination of feature extraction and classifier, 172 tests were performed, computing the average correct classification rate and standard deviation; the best results were obtained when using as feature extraction method the combination of wavelet transform of the resampled chromatogram together with the computation of the correlation coefficients in the time domain and the probabilistic neural network classifier, reaching correct classification rates of 94.77% on an average.

The classification of aged wine distillates is a nonlinear, multicriteria decision-making problem characterized by overwhelming complexity, nonlinearity, and lack of objective information regarding the desired qualitative characteristics of the final product. Many times an appropriate mathematical model cannot be found for the evaluation of aged wine distillates estimations with emphasis on the properties of the aroma and the taste. The most efficient solution for such problem is to develop adequate and reliable expert systems based on fuzzy logic and neural networks.

Raptis et al. [76] proposed a fuzzy classifier and a neural network for the classification of wine distillates for each of two distinct features of the products, namely the *aroma* and the *taste*. The fuzzy classifier is based on the fuzzy k-nearest neighbor algorithm whereas the neural system is a feed-forward sigmoidal (logistic function) multilayer network that is trained using a back-propagation method.

Castineira et al. [15] used k-nearest neighbor algorithm as one of the methods to determine authenticity of white wine from German wine-growing regions based on wine content. They considered 13 elements (ultra trace, trace, and major components) such as Li, B, Mg, Ca, V, Mn, Fe, Co, Zn, Rb, Sr, Cs, and Pb to distinguish and determine the origin of the wine. Diaz et al. [19] used ANN to differentiate red wines from Canary Islands according to the island of origin. ANN classified wines according their geographic origin based on a data matrix formed by 11 elements (K, Na, Li, Rb, Ca, Mg, Sr, Fe, Cu, Zn, and Mn) of Canary red wines.

3.4 Classification of Forest Data with Remotely Sensed Images

With the advancement of satellite remote sensing technologies, forest and agricultural land observation has become very convenient. These sensors produce potentially useful data in enormous quantity on daily basis. This quantity of data also presents a challenge of data interpretation and classification to the researchers. Researchers have been widely using techniques such as k-means, k-nearest neighbor (k-NN), and artificial neural network for classification of remotely sensed forest images.

Atkinson and Tatanall [4] described applications of neural networks in the field of remote sensing. Reese et al. [77] used k-nearest neighbor algorithm for estimation of forest parameters such as wood volume, age, and biomass. They used satellite data, digital map data, and forest inventory data with descriptions from Landsat thematic mapper (TM), Satellite probatoire d'Observation de la

Terre (SPOT) data, and Swedish National Forest Inventory (NFI) field data. After removing outliers from NFI data, parameters such as total wood volume, wood volume for different tree species, biomass, and age were calculated at pixel level within the forest mask. From Equation (9), they calculated estimated forest parameter value ($\hat{v}$) at each pixel (p) as a weighted mean value of the reference plots' (j) forest parameter (v_j) of the k-nearest sample in spectral space. Weights (w) assigned to each of k samples, in this method, were proportional to the inverse squared Euclidean distance (d) between pixel to be estimated and reference plot (Equation 10). In this work, Reese et al. used a number of nearest neighbors (k) of five to prevent overgeneralizing of data

$$\hat{v}_p = \sum_{j=1}^{k} w_{j,p} \cdot v_{j,p}, \tag{9}$$

where

$$w_{j,p} = \frac{1}{d_{j,p}^2} \Bigg/ \sum_{i=1}^{k} \frac{1}{d_{i,p}^2} \tag{10}$$

and

$$d_{1,p} \leq d_{2,p} \leq \ldots \leq d_{k,p}.$$

Their work showed that pixel-level accuracy was quite poor and hence they suggested aggregation of estimation over larger areas to reduce the errors. They showed the application of such estimations in moose and bird habitat studies, county-level planning activities, and computation of statistics on timber volume.

Classification of regenerating forest stages is a part of estimating process of biomass and carbon content in forests. Kulpich [51] used artificial neural networks for classifying regenerating forest stages with the use of remotely sensed images. Kulpich used optical thematic mapper and synthetic aperture radar (SAR) bands to classify different regenerating classes, using the forest age map as the reference. Their results showed that higher classification accuracy could be achieved with SAR and TM bands in combination compared with SAR bands alone, and ANN achieved an overall accuracy of 87% for mature forest, pasture, young (0–5 years) and intermediate (6–18 years) regenerating forest class.

Diamantopoulou [18] applied three-layer feed-forward cascade correlation ANN models for bark volute estimation of standing pine trees. The author used four variables, outside-bark diameter at breast height, inside-bark diameter at breast height, outside-bark volume of the tree (v), and bark correction factor for the prediction of pine bark volume. Feed-forward and supervised ANN was selected, and Kalman filter was applied for inferring missing information.

Kalman filter gave an estimate value of missing information such that the error was statically minimized. ANN method was compared with nonlinear regression analysis models, and results proved that ANN was superior to the regression models. ANN achieved more accurate estimation of bark volume with estimation errors reduced to 7.28% of mean bark volume. Diamantopoulou suggested ANN as a promising alternative to the traditional regression models as ANN had the ability to overcome problems in forest data such as nonlinear relationships, non-Gaussian distributions, outliers, and noises.

4 Conclusions

In this chapter, we have presented a number of selected classification algorithms. These algorithms include k-means, k nearest neighbor, fuzzy c-means, and artificial neural networks. We have also discussed some of the limitations of each of the algorithms and recent developments to make these algorithms more efficient and accurate. There exist several real-world applications of these algorithms. We presented a few applications related to food and agricultural sciences.

References

1. Abdullah, M.Z., Mohamad-Saleh, J., Fathinul-Syahir, A.S., Mohd-Azemi, B.M.N., 2006. Discrimination and classification of fresh-cut starfruits (Averrhoa carambola L.) using automated machine vision system. J. Food Eng. 76, 506–523.
2. Anderson, K.A., Magnuson, B.A., Tschirgi, M.L., Smith, B., 1999. Determining the geographic origin of potatoes with trace metal analysis using statistical and neural network classifiers. J. Agric. Food Chem. 47(4), 1568–1575.
3. Annamalai, P., Lee, W.S., Burks, T.F., 2004. Color vision system for estimating citrus yield in real-time. ASAE Paper No. 043054. St. Joseph, MI: ASAE.
4. Atkinson, P.M., Tatnall, A.R.L., 1997. Neural networks in remote sensing. Int. J. Remote Sens. 18(4), 699–709.
5. Baraldi, A., Blonda, P., 1999. A survey of fuzzy clustering algorithms for pattern recognition—Part I. IEEE Trans. Sys. Man Cyber. – Part B, Cybernetics 29(6), 778–785.
6. Baraldi, A., Blonda, P., 1999. A survey of fuzzy clustering algorithms for pattern recognition—Part II. IEEE Trans. Sys. Man Cyber. – Part B, Cybernetics 29(6), 786–801.
7. Beltrán, N.H., Duarte-Mermoud, M.A., Bustos, M.A., Salah, S.A., Loyola, E.A., Peña-Neira, A.I., Jalocha, J.W., 2006. Feature extraction and classification of Chilean wines. J. Food Eng. 75, 1–10.
8. Benati, S., 2006. Categorical data fuzzy clustering: an analysis of local search heuristics. Comput. Oper. Res. 35(3), 766–775.
9. Bermejo, S., Cabestany, J., 2000. Adaptive soft k-nearest-neighbour classifiers. Pattern Recognit. 33, 1999–2005.
10. Bezdek, J., 1981. Pattern Recognition with Fuzzy Objective Function Algorithms. New York: Plenum.
11. Bishop, C.M., 1995. Neural Networks for Pattern Recognition. Oxford: Oxford University Press.

12. Boginski, V., Butenko, S., Pardalos, P.M., 2006. Mining market data: A network approach. Comput. Oper. Res. 33(11), 3171–3184.
13. Bradley, S., Fayyad, M., 1998. Refining initial points for k-means clustering. In: J. Shavlik (Ed.), Proceedings of the 15th International Conference on Machine Learning (ICML98). San Francisco: Morgan Kaufmann, pp. 91–99.
14. Busygin, S., Prokopyev, O.A., Pardalos, P.M., 2005. Feature selection for consistent biclustering via fractional 0.1 programming. Comb. Optim. 10, 7–21.
15. Castiñeira Gómez, M.D.M., Fledmann, I., Jakubowski, N., Andersson, J.T., 2004. Classification of German white wines with certified brand of origin by multielement quantitation and pattern recognition techniques. J. Agric. Food Chem. 52, 2962–2974.
16. Chinchuluun, R., Lee, W.S., 2006. Citrus yield mapping system in natural outdoor scenes using the Watershed transform. ASAE Paper No. 063010. St. Joseph, MI: ASAE.
17. Chung, K.L., Lin, J.S., 2007. Faster and more robust point symmetry-based k-means algorithm. Pattern. Recognit. 40(2), 410–422.
18. Diamantopoulou, M.J., 2005. Artificial neural networks as an alternative tool in pine bark volume estimation. Comput. Electron. Agric. 48, 235–244.
19. Díaz, C., Conde, J.E., EstéVez, D., Olivero, S.J.P., Trujillo, J.P.P., 2003. Application of multivariate analysis and artificial neural networks for the differentiation of red wines from the canary islands according to the island of origin. J. Agric. Food Chem. 51, 4303–4307.
20. Duda, R.O., Hart, P.E., 1973. Pattern Classification and Scene Analysis. New York: John Wiley & Sons.
21. Duda, R.O., Hart, P.E., Stork, D.G., 2001. Pattern Classification (2nd edition). New York: Wiley.
22. Dunn, J., 1974. A fuzzy relative of the ISODATA process and its use in detecting compact well separated clusters. J. Cyber. 3(3), 32–57.
23. Fayyad, U., Piatetsky-Shapiro, G., Smyth, P., 1996. From data mining to knowledge discovery in databases. AI Magazine – AAAI 17(3), 37–54.
24. Fukunaga, K., Narendra, P.M., 1975. A branch and bound algorithm for computing k-nearest neighbors. IEEE Trans. Comput. 24(7), 750–753.
25. Granitto, P.M., Verdes, P.F., Ceccatto, H.A., 2005. Large-scale investigation of weed seed identification by machine vision. Comput. Electron. Agric. 47, 15–24.
26. Guyer, D., Yang, X., 2000. Use of genetic artificial neural networks and spectral imaging for defect detection on cherries. Comput. Electron. Agric. 29, 179–194.
27. Hammah, R.E., Curran, J.H., 2000. Validity measures for the fuzzy cluster analysis of orientations. IEEE Trans. Patt. Anal. Mach. Intel. 22(12), 1467–1472.
28. Hansen, P., Mladenovis, N., 2001. J-means: a new local search heuristic for minimum sum of squares clustering. Pattern Recognit. 34, 405–413.
29. Hansen, L.K., Salamon, P., 1990. Neural network ensembles. IEEE Trans. Pattern Anal. Mach. Intell. 12(10), 993–1001.
30. Hashem, S., Schmeiser, B., 1995. Improving model accuracy using optimal linear combinations of trained neural networks. IEEE Trans. Neural Netw. 6(3), 792–794.
31. Hathaway, R., Bezdek, J., Hu, Y., 2000. Generalized fuzzy c-means clustering strategies using L norm distances. IEEE Trans. Fuzzy Syst. 8(5) 576–582.
32. Hathaway, R., Bezdek, J., 2001. Fuzzy c-means clustering of incomplete data. IEEE Trans. Syst. Man Cyb. – Part B, Cybernetics 31(5) 735–744.
33. Holmstrom, H., Nilsson, M., Stahl, G., 2002. Forecasted reference sample plot data in estimations of stem volume using satellite spectral data and the k NN method. Int. J. Remote. Sens. 23(9), 1757–1774.
34. Hoshi, T., Sasaki, T., Tsutsui, H., Watanabe, T., Tagawa, F., 2000. A daily harvest predict ion model of cherry tomatoes by mining from past averaging data and using topological case-based modeling. Comput. Electron. Agric. 29, 149–160.

35. Hung, M., Yang, D., 2001. An efficient fuzzy c-means clustering algorithm. In: Proceedings IEEE International Conference on Data Mining, IEEE Computer Society pp. 225–232.
36. Hwang, W.J., Wen, K.W., 1998. Fast k classification algorithm based on partial distance search. Electron. Lett. 34(21), 2062–2063.
37. Jain, A.K., Dubes, R.C., 1988. Algorithms for Clustering Data. Englewood Cliffs, NJ: Prentice Hall.
38. Jain, A.K., Murty, M.N., Flynn, P.J., 1999. Data clustering: a review. ACM Comput. Surv. 31(3), 264–323.
39. James, M., 1985. Classification Algorithms. London: Collins Professional and Technical Books.
40. Jayas, D.S., Paliwal, J., Visen, N.S., 2000. Multi-layer neural networks for image analysis of agricultural products. J. Agric. Eng. Res. 77(2), 119–128.
41. Ji, C., Ma, S., 1997. Combinations of weak classifiers. IEEE Tran. Neural Netw. 8(1), 32–42.
42. Jiang, X., Harvey, A., Wah, K.S., 2003. Constructing andtraining feed-forward neural networks for pattern classification. Pattern Recognit. 36, 853–867.
43. Kanungo, T., Mount, D., Netanyahu, N., Piatko, C., Silverman, R., Wu, A., 2000. An efficient k-means clustering algorithm: analysis and implementation. IEEE Trans. Pattern Anal. Mach. Intell. 24(7), 881–892.
44. Karayiannis, N.B., 1997. A methodology for constructing fuzzy algorithms for learning vector quantization. IEEE Trans. Neural Netw. 8(3), 505–518.
45. Kiliç, K., Boyaci, İ.H., Köksel, H., Küsmenoglu, İ., 2007. A classification system for beans using computer vision system and artificial neural networks. J. Food Eng. 78(3), 897–904.
46. Kolen, J., Hutcheson, T., 2002. Reducing the time complexity of the fuzzy c-means algorithm, IEEE Trans. Fuzzy Syst. 10(2), 263–267.
47. Kondo, N., Ahmad, U., Monta, M., Murase, H., 2000. Machine vision based quality evaluation of Iyokan orange fruit using neural networks, Comput. Electron. Agric. 29, 135–147.
48. Krishna, K., Murty, M., 1999. Genetic k-means algorithm. IEEE Trans. Syst. Man Cyber. – Part B, Cybernetics 29(3), 433–439.
49. Kulkarni, S.R., Lugosi, G., Venkatesh, S.S., 1998. Learning pattern classification – a survey. IEEE Trans. Inf. Theory 44(6), 2178–2206.
50. Kuncheva, L.I., 1997. Fitness functions in editing k-NN reference set by genetic algorithms. Pattern Recognit. 30(6), 1041–1049.
51. Kuplich, T.M., 2006. Classifying regenerating forest stages in Amazônia using remotely sensed images and a neural network. Forest Ecol. Manage. 234(1–3), 1–9.
52. Laszlo, M., Mukherjee, S., 2006. A genetic algorithm using hyper-quadtrees for low-dimensional k-means clustering. IEEE Trans. Pattern. Anal. Mach. Intell. 28(4) 533–543.
53. Law, M.H.C., Topchy, A.P., Jain, A.K., 2004. Multiobjective data clustering. In: Proceedings of the 2004 IEEE Computer Society Conference on Computer Vision and Pattern Recognition, CVPR 2004, 2, IEEE Computer Society, pp. 424–430.
54. Leemans, V., Destain, M.F., 2004. A real time grading method of apples based on features extracted from defects. J. Food Eng. 61, 83–89.
55. Likasa, A., Vlassis, N., Verbeek, J.J., 2003. The global k-means clustering algorithm. Pattern Recognit. 36(2), 451–461.
56. Linskens, H.F., Jackson, J.F., 1988. Wine analysis. In: Modern Methods of Plant Analysis. New Series, Volume 6, Springer.
57. MacQueen, J.B., 1967. Some methods for classification and analysis of multivariate observations. In: Proceedings of 5th Berkeley Symposium on Mathematical Statistics and Probability. Berkeley: University of California Press, Vol 1, pp. 281–297.
58. Majumdar, S., Jayas, D.S., 1999. Classification of bulk samples of cereal grains using machine vision. J. Agric. Eng. Res. 73(1), 35–47.

59. Marchant, J.A., Onyango, C.M., 2003. Comparison of a Bayesian classifier with a multi-layer feed-forward neural network using the example of plant/weed/soil discrimination. Comput. Electron. Agric. 39, 3–22.
60. Marique, T., Kharoubi, A., Bauffe, P., Ducattillion, C., 2003. Modeling of fried potato chips color classification using image analysis and artificial neural network. J. Food Eng. Phys. Prop. 68(7), 2263–2266.
61. Maulik, U., Bandyopadhyay, S., 2000. Genetic algorithm-based clustering technique. Pattern Recognit. 33, 1455–1465.
62. Molto, E., Pla, F., Juste, F., 1992. Vision systems for the location of citrus fruit in a tree canopy. J. Agric. Eng. Res. 52, 101–110.
63. Morimoto, T., Takeuchi, T., Miyata, H., Hashimoto, Y., 2000. Pattern recognition of fruit shape based on the concept of chaos and neural networks. Comput. Electron. Agric. 26, 171–186.
64. Nakano, K., 1997. Application of neural networks to the color grading of apples. Comput. Electron. Agric. 18, 105–116.
65. Nascimento, S., Mirkin, B., Moura-Pires, F., 2000. A fuzzy clustering model of data and fuzzy c-Means. The Ninth IEEE International Conference on Fuzzy Systems, IEEE Transaction on Fuzzy Systems, Vol. 1, pp. 302–307.
66. Pal, N.R., Bezdek, J.C., 1995. On cluster validity for the fuzzy c-means model. IEEE Trans. Fuzzy. Syst. 3(3), 370–379.
67. Paliwal, J., Visen, N.S., Jayas, D.S., 2001. Evaluation of neural network architectures for cereal grain classification using morphological features. J. Agric. Eng. Res. 79(4), 361–370.
68. Pan, J.S., Qiao, Y.L., Sun, S.H., 2004. A fast k nearest neighbors classification algorithm. IEICE Trans. Fund. Electron. Commun. Comput. E87-A(4), 961–963.
69. Papajorgji, P.J., Pardalos, P.M., 2005. Software Engineering Techniques Applied to Agricultural Systems. New York: Springer.
70. Parrish Jr., A.E., Goksel, A.K., 1977. Pictorial pattern recognition applied to fruit harvesting. Trans. ASAE 20(5), 822–827.
71. Pernkopf, F., 2005. Bayesian network classifiers versus selective k-NN classifier. Pattern Recognit. 38(1), 1–10.
72. Plebe, A., Grasso, G., 2001. Localization of spherical fruits for robotic harvesting. Mach. Vision. Appl. 13, 70–79.
73. Pydipati, R., Burks, T.F., Lee, W.S., 2005. Statistical and neural network classfiers for citrus disease detection using machine vision. Trans. ASAE 48(5), 2007–2014.
74. Pydipati, R., Burks, T.F., Lee, W.S., 2006. Identification of citrus disease using color texture features and discriminant analysis. Comput. Electron. Agric. 52, 49–59.
75. Qiao, Y.L., Pan, J.S., Sun, S.H., 2004. Improved k nearest neighbor classification algorithm. The 2004 IEEE Asia-Pacific Conference on Circuits and Systems, 6–9 Dec. IEEE Transaction on Circuits and Systems, Vol. 2, pp. 1101–1104.
76. Raptis, C.G., Siettos, C.I., Kiranoudis, C.T., Bafas, G.V., 2000. Classification of aged wine distillates using fuzzy and neural network systems. J. Food Eng. 46, 267–275.
77. Reese, H., Nilsson, M., Sandström, P., Olsson, H., 2002. Applications using estimates of forest parameters derived from satellite and forest inventory data. Comput. Electron. Agric. 37, 37–55.
78. Regunathan, M., Lee, W.S., 2005. Citrus yield mapping and size determination using machine vision and ultrasonic sensors. ASAE Paper No. 053017. St. Joseph, MI: ASAE.
79. Robinson, C., Mort, N., 1997. A neural network system for the protection of citrus crops from frost damage. Comput. Electron. Agric. 16, 177–187.
80. Shahin, M.A., Tollner, E.W., McClendon, R.W., 2001. Artificial intelligence classifiers for sorting apples based on water core. J. Agric. Eng. Res. 79(3), 265–274.
81. Simonton, W., Pease, J., 1993. Orientation independent machine vision classification of plant parts. J. Agric. Eng. Res. 54(3), 231–243.

454   R. Chinchuluun et al.

82. Slaughter, D.C., Harrell, R.C., 1987. Color vision in robotic fruit harvesting. Trans. ASAE 30(4), 1144–1148.
83. Slaughter, D.C., Harrell, R.C., 1989. Discriminating fruit for robotic harvest using color in natural outdoor scenes. Trans. ASAE 32(2), 757–763.
84. Solazzia, M., Uncinib, A., 2004. Regularising neural networks using flexible multivariate activation function. Neural. Netw. 17(2), 247–260.
85. Su, M., Chou, C., 2001. A modified version of the K-means algorithm with a distance based on cluster symmetry. IEEE Trans. Pattern Anal. Mach. Intell. 23(6), 674–680.
86. Sun, L.X., Danzer, K., Thiel, G., 1997. Classification of wine samples by means of artificial neural networks and discrimination analytical methods. Fres. J. Anal. Chem. 359, 143–149.
87. Uno, Y., Prasher, S.O., Lacroix, R., Goel, P.K., Karimi, Y., Viau, A., Patel, R.M., 2005. Artificial neural networks to predict corn yield from compact airborne spectrographic imager data. Comput. Electron. Agric. 47, 149–161.
88. Wu, Y., Ianakiev, K., Govindaraju, V., 2002. Improved k-nearest neighbor classification. Pattern Recognit. 35, 2311–2318.
89. Xu, R., 2005. Survey of clustering algorithms. IEEE Trans. Neural Netw. 16(3), 645–678.
90. Yager, R.R., Filev, D.P., 1994. Approximate clustering via the mountain method. IEEE Trans. Syst. Man Cyber. 24(8), 1279–1284.
91. Yager, R.R., 2006. An extension of the naive Bayesian classifier. Inf. Sci. 176(5), 577–588.
92. Yu, X., Chen, G., Cheng, S., 1995. Dynamic learning rate optimization of the back propagation algorithm. IEEE Trans. Neural Netw. 6(3), 669–677.
93. Zhang, G.P., 2000. Neural networks for classification: a survey. IEEE Trans. Syst. Man Cyber. 30(4), 451–462.
94. Zhang, J.S., Leung, Y.W., 2004. Improved possibilistic c-means clustering algorithms. IEEE Trans. Fuzzy. Syst. 12(2), 209–217.
95. Zhang, Y., Xiong, Z., Mao, J., Ou, L., 2006. The study of parallel k-means algorithm. In: Proceedings of the 6th World Congress on Intelligent Control and Automation, IEEE Transaction on Intelligent Control and Automation, Vol. 2, pp. 5868–5871.
96. Zhong, S., 2005. Efficient online spherical K-means clustering. In: Proceedings of International Joint Conference on Neural Networks, IEEE Transaction on Neural Networks, Vol. 5, pp. 3180–3185.

Mathematical Modelling of Modified Atmosphere Package: An Engineering Approach to Design Packaging Systems for Fresh-Cut Produce

Elena Torrieri, Pramod V. Mahajan, Silvana Cavella, Maria De Sousa Gallagher, Fernanda A.R. Oliveira, and Paolo Masi

Abstract Consumer demand for freshness and for convenience food has led to the evolution and increased production of fresh-cut fruits and vegetables. Moreover, this may represent a way to increase the consumption of fresh fruits and vegetables and therefore be a benefit for the crops-sector economy. Because the increase in convenience for the consumer has a detrimental effect on product quality, attention must be focused on extending shelf-life while maintaining quality. Modified atmosphere packaging (MAP) is a packaging technology that, by making qualitative or quantitative changes to the atmosphere composition around the product, can improve product preservation. However, MAP must be carefully designed, as a poorly designed system may be ineffective or even shorten product shelf-life. Thus, whereas in the past a trial-and-error approach to packaging of food was predominant, nowadays the need has emerged for an engineering approach to properly design a package to improve product shelf-life. Therefore, to ensure an appropriate gas composition during the product's shelf-life, a model should take into account all the variables that play a critical role, such as product respiration and its mass; packaging material and its geometry; and environmental conditions such as temperature, relative humidity, and gas composition.

In this chapter, a procedure to design a package for fresh-cut produce is described. An engineering approach is used to develop a mathematical model with the experimental data obtained for product respiration and package permeability and to solve the mass balance equations using computer software. A case study on fresh-cut apples is reported to validate the model.

1 Fresh-Cut Produce

In recent years, consumer demand for freshness and convenience has led to the development and increased production of fresh-cut fruits and vegetables. The International Fresh-cut Produce Association (IFPA) estimates sales in the U.S.

E. Torrieri (✉)
University of Naples, Federico II, Italy
e-mail: Elena.torrieri@unina.it

P.J. Papajorgji, P.M. Pardalos (eds.), *Advances in Modeling Agricultural Systems,*
DOI 10.1007/978-0-387-75181-8_22, © Springer Science+Buisness Media, LLC 2009

fresh-cut produce market at approximately $12.5 billion. Packed salads are still the workhorses of the fresh-cut industry with 2003 retail sales measured at $3 billion. However, advancing fast is cut fruit, currently a $0.3 billion category at retail but forecast to exceed $1 billion in the next 3 to 4 years [1]. In Europe, this market grew explosively in the early 1990s [2]. Giovannetti [3] reports that fresh-cut products represent 8% of the French and English markets and 7% of the Italian market.

Fresh-cut produce as defined by the IFPA is "any fruit or vegetable or combination thereof that has been physically altered from its original form, but remains in a fresh state" [4]. Regardless of commodity, it has been trimmed, peeled, washed, and cut into 100% usable product that is subsequently bagged or pre-packaged to offer consumers high nutrition, convenience, and value while still maintaining freshness. *Fresh state* and *minimal processing* are the main drivers for these products. Additionally, modern consumers tend to be health conscious, and their increasing interest in the role of food for maintaining and improving human well-being has led to constant growth in the fresh-cut fruits and vegetables industry [5].

The large diffusion among the consumers of this kind of product may also result in a strong benefit for the crop-sector economy as the availability of ready-to-eat and ready-to-use fruits and vegetables will certainly increase the domestic and outdoor consumption occasions. Moreover, their extended shelf-life may favour also enhancement of typical production, which can be sold in extraregional or even extranational markets.

Minimal processing includes separation operations, such as peeling, coring, selecting, sorting, and grading, as well as operations to reduce size, such as chopping, slicing, dicing, granulating, and shredding. The result is a convenient fresh-like product that can be prepared and consumed in a short time. However, the increase in convenience for the consumer has a detrimental effect on the product quality [6, 7, 8, 9]. The consequences of minimal processing on fresh produce are an increase in the respiration rate, production of ethylene, oxidative browning, water loss, and degradation of the membrane lipids [10]. Thus, attention must focus on extending shelf-life by maintaining quality and ensuring food safety throughout the postharvest chain, namely handling, packaging, storage, and distribution.

During storage and distribution, the packaging of a fresh-cut produce plays a critical role in quality preservation and shelf-life extension. Packaging is essential in protecting the product against the outside environment and preventing mechanical damage and chemical or biological contamination. It allows the control of water loss, which is an important symptom of loss of quality, and controls the flux of oxygen and carbon dioxide, building up and/or maintaining an optimal atmosphere within the packaging headspace [11]. This last function brings us to the modified atmosphere packaging (MAP) concept.

2 Modified Atmosphere Packaging

Wang [12] reported an ancient Chinese poem that describes the fondness of an empress for the litchi fruits in the Tang Dynasty in the 8th century. She demanded that freshly picked litchi fruits be transported to her from Chang-An (now Xi-An) in southern China, approximately 1000 km away, by continuous horseback riding, the fastest possible way through rugged terrain at that time. It was said that the carriers discovered that litchi would keep better if the fruits were sealed inside the hollow centres of bamboo stems along with some fresh leaves. However, the poem did not explain why. We now know that the atmospheres within the bamboo centres, modified by the respiration of litchi fruits and the fresh leaves, were probably the key factor responsible for keeping the freshness of these litchi fruits. Despite the huge techonological improvements in the transport, manufacturing, and communication sectors, there has been hardly any application of technology and engineering in the packaging of fresh-cut produce.

Vacuum technology, the first type of MAP to be used, was introduced in the 1960s in the United States to package chicken and was also used for raw meat and luncheon meat packaging. Gas-flush was started later that decade for coffee, and the technology was extended to a wide array of package forms, including flexible packages, cups, and trays. The 1980s and 1990s saw considerable expansion in the number of MAP machines, and it was during this period that it began to be applied to flexible packages of fresh products.

MAP of fresh produce relies on modifying the atmosphere inside the package, achieved by the natural interplay between two processes, the respiration of the product and the transfer of gases through the packaging, which leads to an atmosphere richer in CO_2 and poorer in O_2. This atmosphere can potentially reduce the respiration rate, ethylene sensitivity, and physiologic changes [13, 14]. Active and passive MAP are the two systems generally recognised for packaging of fresh-cut produce. Their main differences are shown in Table 1. One of the main disadvantages of passive MAP is represented by the long time necessary to reach a condition of dynamic steady-state, close to the optimal gas composition. To reach the optimal storage condition in such a long time can be detrimental to the stability of the package. In contrast, by packing a product by means of active MAP, the equilibrium time is very short. Nevertheless, the improvement in the equilibrium time is paid with an increment in the cost.

Nowadays, minimising the time required to achieve equilibrium coupled with the creation of the atmosphere best suited for the extended storage of a given product are the main objectives of MAP design. Since the 1980s, MAP has evolved from a *pack-and-pray* procedure, which may have economic and safety hazard consequences, to a trial-and-error approach, which is an extremely time-consuming procedure. It later became clear that an efficient use of MAP technology should be based on knowledge of the effect of gas on product quality in order to find the optimal conditions for each product.

Table 1 Type of MAP for fresh-cut produce

	Passive	Active
Definition	Modification of the gas composition inside the package due to interplay between the product respiration rate and the gas exchange rate through the package	Modification of the gas composition inside the package by replacing, at the moment of packaging, the air with a specific gas mixture either by drawing a vacuum or filling a gas mix
Equilibrium time	1–2 days to 10–12 days	1–2 h
Products suitable for	Mushrooms, carrots, strawberry, spinach	Cut apples, dry fruits
Cost	No extra cost involved if the package is properly designed	Extra investment is required for special machinery; i.e., gas, gas mixer, packaging machine for MAP
Requirements for labeling	No	Yes

However, there are several factors that affect the optimal gas composition for a given product. Recommended concentrations of O_2 and CO_2 for some fresh-cut products are reported in Fig. 1 [15]. Different gas compositions correspond with different products, and only in-depth understanding of the effects of the gases on the deterioration process ensures the effectiveness of MAP technology in

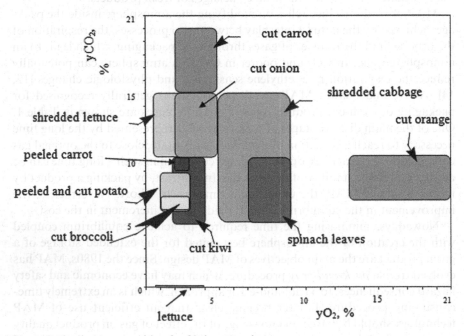

Fig. 1 Optimal gas composition for fresh-cut produce

extending product shelf-life. Moreover, as different products vary in their behaviour and as modified atmosphere packages will be exposed to a dynamic environment, each package has to be optimized for a specific requirement [16, 17]. Hence, the necessity to follow a design procedure to ensure the preservation of the product during storage. An inappropriately designed MAP system may be ineffective or even shorten the storage life of a product: if the desired atmosphere is not established rapidly, the package draws no benefit; if O_2 and O_2/CO_2 levels are not within the recommended range, the product may experience serious alterations and its storage life may be shortened. The design of a package depends on a number of variables: the characteristic of the product; its mass; the recommended atmospheric composition; the permeability of the packaging material to gases and its dependence on temperature; and the respiration rate of the product as affected by different gas compositions and temperature. Therefore, to ensure an optimal gas composition during product shelf-life, an engineering approach has to be followed to model all the variables that play a critical role. Simulation of a MAP system is the most appropriate method to allow proper MAP design and thereby obtain a successful commercial product.

3 An Engineering Approach to the Package Design

Sound use of MAP entails consolidated knowledge of the food–package–environmental interaction: when a product is packed, it is surrounded by a mixture of gases, the composition of which depends on the interactions between the food product, the package material, and the environment [18, 19, 20, 21]. If the packed food is a fresh-cut product, such interactions concern the respiration metabolism of the product (Fig. 2): the product exchanges gas with the surrounding atmosphere, consuming O_2 and producing CO_2. The ratio between the CO_2 production rate and O_2 consumption rate is the respiration quotient (RQ). Because of the respiration process, an O_2 and CO_2 concentration gradient between the headspace and the environment is generated. Thus, a gas flow is activated through the packaging material due to film permeability to O_2 and CO_2. The ratio between the permeability to CO_2 and the permeability to O_2 is known as selectivity (β). From the above considerations, to design a package for a fresh-cut product, one clearly needs to deal with an open system in dynamic conditions. Therefore, to ensure a suitable gas composition during the product's shelf-life, a model should take into account the steps shown in Fig. 3. The first and most important factor is the respiration rate. The mathematical model for respiration rate is usually a function of O_2, CO_2, and temperature. The second important factor is the type of packaging material and its permeability in terms of O_2 and CO_2. Permeability changes with temperature; hence, a mathematical model is required to predict the permeability change in the packaging material at the given temperature, which is a function of temperature.

Another factor is the best atmosphere required to extend product shelf-life, which varies from product to product and is widely covered in the literature. A

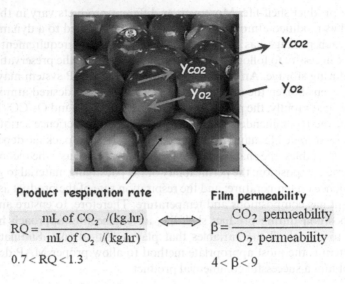

Product respiration rate   Film permeability

$$RQ = \frac{\text{mL of } CO_2 \ /(\text{kg.hr})}{\text{mL of } O_2 \ /(\text{kg.hr})} \iff \beta = \frac{CO_2 \text{ permeability}}{O_2 \text{ permeability}}$$

$$0.7 < RQ < 1.3 \qquad\qquad 4 < \beta < 9$$

Fig. 2 Basic principles of MAP

real challenge is how to achieve this atmosphere for a given type of product and packaging conditions. The answer is to pursue either a layman's approach of trial-and-error experiments or an engineering approach where mathematical equations depicting product respiration rate and package permeability could be used to solve the mass balance equations of a packaging system using available computer software.

3.1 Modelling the Gas Transport Through a Polymeric Film

Permeability is one of the most important properties of packaging material for MAP design. It involves mass transfer through polymeric film, or *permeation*, associated with a partial pressure differential of a gas or vapour between the two sides of the package (Fig. 2). Permeation is a composite phenomenon that involves diffusion and sorption processes at the same time. Diffusion obeys Fick's first law (Eq. 1) under steady-state conditions and Fick's second law (Eq. 2) under unsteady-state conditions. Assuming that diffusion takes place only in one direction:

$$J = -D\frac{dc}{dx} \tag{1}$$

$$\frac{dc}{dt} = -D\frac{d^2c}{dx^2} \tag{2}$$

where J, D, c, and x are the flow per unit of cross section, the diffusivity, the concentration of the permeant, and the distance across which the permeant has to

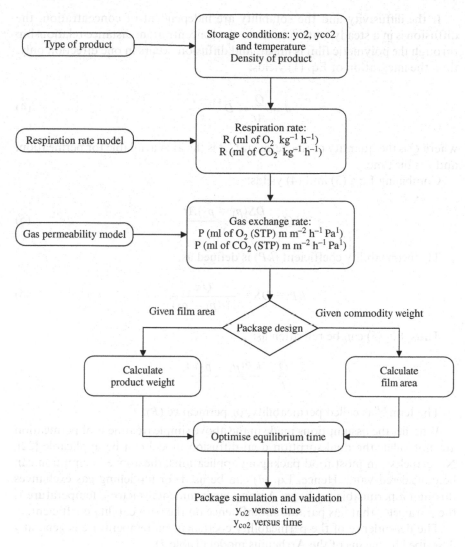

Fig. 3 Procedure for designing MAP for fresh-cut produce

travel, respectively. The sorption of a gas component into a packaging material under conditions where the gas concentration is lower than its maximum solubility (saturation) obeys Henry's law:

$$Sp = c \qquad (3)$$

where p is the partial pressure of the gas, c is the gas concentration at the surface of the packaging material, and S is the solubility of the gas in the packaging material.

If the diffusivity and the solubility are independent of concentration, the diffusion is in a steady-state condition, the concentration–distance relationship through the polymeric film is linear, and diffusion occurs in one direction only, then the integration of Eq. (1) yields:

$$J = \frac{Q}{At} = D\frac{c_1 - c_2}{x} \tag{4}$$

where Q is the quantity of the permeant, A is the area available for permeation, and t is the time.

Combining Eqs. (3) and (4) yields:

$$Q = \frac{DS(p_1 - p_2)At}{x}. \tag{5}$$

The permeability coefficient (kP) is defined as:

$$kP = DS = \frac{Qx}{At(p_1 - p_2)}. \tag{6}$$

Thus, Eq. (5) can be rewritten as:

$$\frac{Q}{t} = \frac{kP(p_1 - p_2)A}{x}. \tag{7}$$

The term $\frac{kP}{x}$ is called permeability, or permeance (P).

Whether the assumptions made in the above simple treatment of permeation are not valid, the dual-sorption dual-diffusion model can be applicable [22]. Nevertheless, in most food packaging applications, the above assumption can be considered valid. Hence, Eq. (7) can be used for modelling gas exchanges through a permeable package. Among environmental factors, temperature is the parameter that has paramount influence on the permeability coefficient.

The dependence of the permeability coefficient on temperature is generally described by means of the Arrhenius model (Table 2).

Although flexible polymeric films are widely used in MAP, in some cases, such as for high-respiring products or for products that require high CO_2/O_2 concentrations, or in the case of films that are good barriers to water vapour, it is difficult to develop a polymeric film with an appropriate selectivity (β). The use of perforated films represents a valid alternative to these limitations. For perforated films, the gas transmission rate is controlled by the number and size of the perforations. In Table 2 are reported some empirical models used to predict permeability coefficients for a given number of holes or tube sizes. Moreover, combination of a high flux membrane patch with perforations can offer a powerful means to control the composition of the O_2 and CO_2 in the package [26].

Table 2 Gas transport through polymer film model

Model Equation	Independent Variables	Ref.
$kP = kP_{ref}\exp\left[-\frac{Ea}{R}\left(\frac{1}{T} - \frac{1}{T_{ref}}\right)\right]$	Temperature (T)	[11]
$\frac{Q}{t} = (c - c_a)[AkP/x + A_h D/(x + R_h)]$	Surface and radius of pores (A_h, R_h)	[23]
$P^* = a_1 + a_2(D_p)^2 + a_3(L_p) + a_4(T) + a_5(D_p)^4$ $+ a_6(L_p)^2 + a_7(T)^2 + a_8\left(D_p^2\right)(L_p)$ $+ a_9\left(D_p^2\right)(T) + a_{10}(L_p)(T)$	Tube diameter (D_p) and length (L_p) and temperature (T)	[24]
$K_{O_2} = a \times D^b \times L^c$ $K_{CO_2} = \beta \times a \times D^b \times L^c$ $\beta = \frac{K_{CO_2}}{K_{O_2}}$	Tube diameter (D) and length (L)	[25]
$P'_{O_2} = \left[P_{O_2} + \frac{\pi\, R_H^2 \times 16.4 \times 10^{-6}}{(e + R_H)} \times N_H\right]$ $P'_{CO_2} = \left[P_{CO_2} + \frac{\pi\, R_H^2 \times 20.6 \times 10^{-6}}{(e + R_H)} \times N_H\right]$	Radius (R_H) and number of holes (N_H), thickness of the film (e),	[26]

3.2 Modelling the Respiration Process

Respiration in the cells of fruits and vegetables is the metabolic process involving the breakdown (catabolism) of complex organic compounds such as sugars, organic acids, amino acids, and fatty acids into lower-molecular-weight molecules with the accompanying production of energy (ATP and heat). Aerobic respiration includes reactions in which organic compounds are oxidized to CO_2 and the absorbed O_2 is reduced to H_2O. The summation of all reactions for the aerobic respiratory breakdown of the most common substrate, glucose, is

$$C_6H_{12}O_6 + 6O_2 \rightarrow 6CO_2 + 6H_2O + \text{Energy}. \tag{8}$$

When 1 mole of glucose is oxidized in the respiration process, CO_2, H_2O, and 36 moles of ATP are formed [11]. If carbohydrates such as glucose and fructose are completely oxidized in the aerobic respiration process of fresh commodities, the volume of O_2 uptake per unit time is about equal to the volume of CO_2 release per unit time. However, if organic acids are oxidized in fruits and vegetables, more CO_2 is generated than the volume of O_2 that is consumed [13]. The respiratory quotient (RQ), the ratio of the CO_2 released per unit time to the O_2 consumed per unit time, is useful for assessing the types of substrates used in the respiration process of a fruit or vegetable. Normal RQ values in the literature are reported as ranging from 0.7 to 1.3 (Table 3). The RQ is much greater than 1.0 when anaerobic respiration takes place [27].

Several mathematical models have been proposed for describing the effect of oxygen and carbon dioxide on the respiration process [27]. Some of them are empirical models (linear, quadratic, exponential), whereas the most widely used

Table 3 Various types of substrates involved in the respiration reaction

Substrate	Respiration Reaction	RQ	Example
Palmitic acid	$C_{16}H_{32}O_2 + 11O_2 \rightarrow C_{12}H_{22}O_{11} + 4CO_2 + 5H_2O$	0.36	Oil seeds
Malic acid	$C_4H_6O_5 + 3O_2 \rightarrow 4CO_2 + 3H_2O$	1.33	Apple
Glucose	$C_6H_{12}O_6 + 6O_2 \rightarrow 6CO_2 + 6H_2O$	1.00	Mango

one is a semifundamental model based on the Michaelis–Menten enzyme kinetic (Table 4). The parameters of the respiration rate model are estimated by nonlinear regression analysis of the experimental data obtained at different combinations of temperature, O_2, and CO_2. The usual methods of respiration rate determination are closed or static system, flow-through system, and permeable system. Fonseca et al. [27] performed an extensive review of the methods used for measuring and modelling the respiration process of fresh fruits and vegetables.

Fonseca *et al.* [27] performed an extensive review of the methods used for measuring and modelling the respiration process of fresh fruits and vegetables.

Temperature is one of the most critical factors for the respiration rate and in turn for product shelf life [44, 45]. Although almost all foods are sensible to temperature change, fresh-cut produce suffer particularly from temperature oscillation during storage. Recently, the most widely used model used to quantify the effect of temperature on respiration rate is the Arrhenius one:

$$R = R_{ref} \cdot \exp\left[-\frac{Ea}{R_c}\left(\frac{1}{T} - \frac{1}{T_{ref}}\right)\right] \tag{9}$$

where R_{ref} is the respiration rate at the reference temperature (T_{ref}), E_a is the activation energy, and R_c is the gas constant.

The Arrhenius model has also been used to describe the effect of temperature on the Michaelis–Menten parameters, when the effect of O_2, CO_2, and temperature have been modelled simultaneously (Table 4). Activation energy values range from 29.0 to 92.9 kJ mol^{-1} for common fruits and vegetables [46] and from 67 kJ/mol to 220 kJ mol^{-1} for common ready-to-eat fruits and vegetables [17, 37, 40].

To suitably design a MAP for fresh-cut product, it is important to compare the activation energy of the respiration process to the activation energy of the permeation process. Given that the activation energy of the most common permeable film used for food packaging ranges from 20 to 40 kJ/mol, it is not difficult to appreciate how critical temperature fluctuation can be for the atmosphere composition and in turn for product shelf-life. Jacxsens et al. [17] modelled the effect of temperature fluctuations on the respiration rate and gas permeability of packaging film containing fresh-cut produce to design an equilibrium modified atmosphere (EMA) package for fresh-cut produce subjected to temperature change. The authors reported that with the proposed packaging system, in a temperature range of between 2°C and 10°C, the gas

Table 4 Respiration rate models

Model	Model Equation	Independent Variables	Products Studied	Ref.
Linear	$R = \mu O_2$	O_2	Mango	[23]
Exponential	$R = A_1 \exp(-A_2 \cdot [CO_2]) \cdot [O_2]$	O_2, CO_2	Fresh-cut apples, kiwi, banana, prickly pear	[28, 29]
Michaelis–Menten	$R = \dfrac{V_{max} \cdot y_{O_2}}{K_m + y_{O_2}}$	y_{O_2}, y_{CO_2}	Broccoli, burlat cherry, blueberry, cauliflower, sweet cherry	[17, 30, 31, 32, 33, 34, 35, 36]
Competitive	$R = \dfrac{V_{max} \cdot y_{O_2}}{k_m \left(1 + \frac{y_{CO_2}}{k_{mc}}\right) + y_{O_2}}$	y_{O_2}, y_{CO_2}	Broccoli	[37]
Uncompetitive	$R = \dfrac{V_{max} \cdot y_{O_2}}{k_m + y_{O_2}\left(1 + \frac{y_{CO_2}}{k_{mu}}\right)}$	y_{O_2}, y_{CO_2}	Red bell pepper, strawberry, fresh apples	[23, 38, 39]
Noncompetitive	$R = \dfrac{V_{max} \cdot y_{O_2}}{\left(k_m + y_{O_2}\right) \cdot \left(1 + \frac{y_{CO_2}}{k_{mn}}\right)}$	y_{O_2}, y_{CO_2}	Endive, pear	[40, 41]
Michaelis–Menten + Arrhenius	$R = \dfrac{R_{max_0} \exp\left(\frac{E_{a_1}}{R}\left(\frac{1}{T} + \frac{1}{T_0}\right)\right) \cdot y_{O_2}}{k_{m_0} \exp\left(\frac{E_{a_2}}{R}\left(\frac{1}{T} + \frac{1}{T_0}\right)\right) + y_{O_2}}$	y_{O_2}, y_{CO_2}, temperature	Apple	[42]
Uncompetitive + Arrhenius	$R = \dfrac{\alpha_1 \exp(\alpha_2 T) y_{O_2}}{\phi_1 \exp(\phi_2 T) + y_{O_2}\left(1 + (y_{CO_2})/\gamma_1 \exp(\gamma_2 T)\right)}$	y_{CO_2}, temperature	Gelaga kale, apple	[27, 39]
Weibull + Arrhenius	$R = \left[R_{eq,ref} + (R_{o,ref} - R_{eq,ref}) \cdot e^{-\left(\frac{t}{\tau_{ref} \times e^{\frac{E}{R}\left(\frac{1}{T} - \frac{1}{T_{ref}}\right)}}\right)^{\beta}}\right] \cdot e^{\frac{E}{R}\left(\frac{1}{T} - \frac{1}{T_{ref}}\right)}$	Temperature, time	Shredded carrots	[43, 44]

composition inside the package was sufficiently low and, at the same time, above the minimum required O_2 level. By contrast, in the case of temperature abuse (12°C and 15°C), the atmosphere tended to become hypoxic. As reported by Talasila et al. [35], in order for a packaging film to be suitable under varying surrounding temperature, the change in its permeability to O_2 and CO_2 should compensate the change in the product respiration rate with variation in surrounding temperature. Indeed, temperature-stable gas conditions inside the package occur when film permeability matches the respiration rate of the packed product, as reported for capsicums by Chen et al. [47]. When the respiration rate is more affected by temperature fluctuation than by the permeability, the gas concentration at equilibrium, designed to keep a specific optimal gas concentration at a certain constant temperature, will differ from the desired optimal concentration whether the packed product is stored under different temperature conditions. As a result, when using such film to package the food product, with fluctuation temperature conditions in the distribution chain, the product will consume more O_2 than the amount of O_2 diffused to the package, an imbalance that will eventually cause an anaerobic atmosphere inside the package. However, this effect will not be so explicit when the activation energy of the respiration rate tends to decrease with aging of the food material and takes on a value of the same order of magnitude as the activation energy value for permeability of the packaging films such as reported for cut endive by Van de Velde and Hendrickx [48].

3.3 Material Balance in MAP

Figure 2 shows respiring product stored in a package consisting of a polymeric film. The simplest concept is to let the polymeric film serve as the regulator of O_2 flow into the package and of the CO_2 flow out. Assuming that there is no gas stratification inside the package and that the total pressure is constant, the unsteady material balance equations are

$$V_f \frac{dy_{O_2}}{dt} = \frac{kP_{O_2}}{x} A \cdot \left(y_{O_2}^{out} - y_{O_2}^{in} \right) - R_{O_2} M \qquad (10)$$

$$V_f \frac{dy_{CO_2}^{in}}{dt} = \frac{kP_{CO_2}}{x} A \cdot \left(y_{CO_2}^{in} - y_{CO_2}^{out} \right) + R_{CO_2} M \qquad (11)$$

where V_f is the free volume, y is the gas concentration (in volumetric concentration), x is the thickness of polymeric film, kP is the permeability of the package, A is the package surface, and R is the respiration rate expressed in volume of gas generated/consumed per unit time and weight of product (M); the subscripts O_2 and CO_2 refer to O_2 and CO_2, respectively.

The gas exchanges takes place until a dynamic steady-state is reached. The above equations describe the dynamic behaviour of the MAP system when the CO_2 evolution rate equals the efflux rate of CO_2 through the package and the O_2

consumption rate equals the influx rate of O_2 through the package. At steady state, the accumulation term of Eqs. (10) and (11) is zero, and these equations are reduced to:

$$\frac{kP_{O_2} \cdot A}{x} \cdot \left(y_{O_2}^{out} - y_{O_2}^{eq}\right) = M \cdot R_{O_2}^{eq} \tag{12}$$

$$\frac{kP_{CO_2} \cdot A}{x} \cdot \left(y_{CO_2}^{eq} - y_{CO_2}^{out}\right) = M \cdot R_{co_2}^{eq}. \tag{13}$$

In most commercial packaging situations, steady-state or dynamic behaviour is approached within 2 days.

Thus the equilibrium O_2 and CO_2 concentration are given by Eqs. (14) and (15), respectively:

$$y_{O_2}^{eq} = y_{O_2}^{out} + \frac{R_{O_2}^{eq} x M}{kP_{O_2} A} \tag{14}$$

$$y_{CO_2}^{eq} = y_{CO_2}^{out} + \frac{R_{CO_2}^{eq} x M}{kP_{O_2} A} \tag{15}$$

where the superscripts *eq* and *out* refer, respectively, to the gas concentration inside the package at the equilibrium (steady state) and the external gas composition. Thus the success of the MAP technique depends on the ability to predict the O_2 and CO_2 permeability of the film package and the respiration rate of the product [20, 27, 32, 37, 49]. As said before, both these phenomena are dependent on environmental factors, such as temperature, O_2, CO_2, and relative humidity (RH). Thus to properly design MAP, Eqs. (10) to (15) must be coupled with a mathematical model that considers the effect of environmental factors on permeability and respiration processes (Table 2 and Table 4).

3.4 Packaging Design Procedure

The steps involved in MAP design are summarized in Fig. 4. The first step, termed *package design*, involves determination of intrinsic properties (i.e., respiration rate, product mass and density, optimal storage conditions) and package characteristics (i.e., film permeability, package geometry) and their optimization to ensure optimal conditions inside the package during storage. Then the *simulation* step is carried out to predict O_2 and CO_2 concentrations inside the package as a function of time. Simulation allows prompt visualization of different scenarios related to variability in factors, such as respiration rate or temperature variation during storage. The last step is the *validation* of the model under real-life distribution chain.

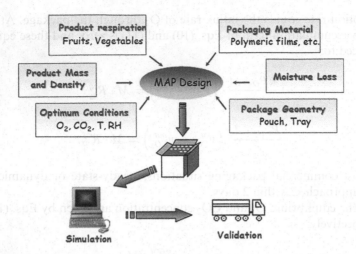

Fig. 4 Steps involved in MAP design

The important design variables may be divided into packaging factors, environmental factors, and commodity factors (Table 5).

Once the product has been selected and the environmental conditions of the distribution chains are established, some of these variables are fixed: the surrounding gas composition and the temperature, the product density and production rate of CO_2 and consumption rate of O_2, and the recommended atmosphere composition to be obtained in the package to extend product shelf-life. These variables must satisfy design Eqs. (10) to (15) and, hence, the system has four design variables, that is, only four of the remaining variables, M, V, A, x, kP_{O_2}, and kP_{CO_2}, can be specified arbitrarily.

Table 5 Variables involved in MAP design

	Variables
Product	Product mass
	Desired gas composition
	Respiration rate
Package	Area available for gas exchange
	Thickness
	Volume
	Number of holes or perforation
	Diameter of hole/perforation
	Length of perforation
	Permeability to O_2 and CO_2
Environmental	Temperature
	External gas composition
	Relative humidity

3.4.1 Selection of Packaging Material

Selection of suitable polymeric material requires analysis of the mass balance at steady state (Eqs. 12 or 13) to calculate O_2 or CO_2 permeability for a fixed product weight and film. Let us assume that a factory wishes to sell fresh-cut fruits in single portions of 100–150 g in a semirigid pack whose size must suit the product.

Because almost all fruits and vegetables have an O_2 consumption rate similar to the CO_2 production rate (i.e., RQ = 1), permeability to O_2 and CO_2 estimated by solving Eqs. (12) and (13) will be almost the same. However, because almost all packaging materials have a higher permeability to CO_2 than to O_2, it is not possible to select a material on the basis of both these values. A possible solution may be to select the polymeric material on the basis of permeability to the gas that has more effect on product shelf-life. Usually, for fresh-cut fruits it is important that the O_2 at equilibrium is above the critical value of fermentation. Whatever, this simple procedure could lead to an uncorrected prediction. Thus, a different solution allows selection of the packaging material on the basis of the selectivity value (β). Combining Eqs. (12) and (14), the following equation is obtained:

$$\beta = RQ \frac{\left(y_{O_2}^{out} - y_{O_2}^{eq}\right)}{\left(y_{CO_2}^{eq} - y_{CO_2}^{out}\right)} \tag{16}$$

where β is the selectivity of the film and RQ is the respiration quotient. However, given that for fresh-cut produce the RQ ranges between 0.7 and 1.3, whereas the selectivity value of the most common package used in food packaging ranges from 3 to 6, it is not easy to find a suitable package material, and in turn it will not be possible to get the exact atmosphere composition at equilibrium. As a result, the selectivity must be calculated by selecting a range of equilibrium gas composition (minimum and maximum) other than the optimal one but still protective for the product, seeking to avoid O_2 levels below the minimum required and CO_2 levels too high for the product [21]. At times, the permeability ratio required is unavailable, such as for β equal to 1. Hence in this circumstance, the use of perforation-mediated MAP has been proposed [50]. Ratti et al. [36] showed that a diffusion channel was able to maintain the desired concentration of O_2 during the storage of fresh cauliflower with enough flexibility to account for fluctuations in the storage temperature. For perforation or microperforated-mediated MAP, the packaging variables that must be optimized are the number, length, height, and diameter of tubes or holes [21].

Once the right permeability value or selectivity has been selected, the package material should be chosen among the available market materials by identifying the one that has the most similar permeability properties and selectivity to the estimated one. However, in most cases, it is not possible to find a material that has exactly the right permeability or selectivity. Thus a packaging material with permeability close to the desired one has to be chosen, then the arbitrarily fixed variables have to be optimized: product weight, packaging volume, and surface area.

3.4.2 Selection of Product Weight or Film Area

Product amount and film area represent design variables that must be optimized. However, in specific cases, such as for semipermeable packaging, the film area is fixed by the package size, and product weight remains the only variable. As both variables together cannot be optimized, it is necessary to set one variable and optimize the other. Thus, product weight can be calculated by solving Eqs. (12) or (13) by using the known product respiration rate, the desired equilibrium gas composition, and a defined area. The same procedure can be followed to estimate the area for a defined product weight. The estimated weight must be compared with the defined surface area to be sure that design is possible.

3.4.3 Optimization of the Volume

The last aspect that must be considered is the package volume optimization. Naturally, a range of volumes has to be discarded as they are unsuitable for the weight and density of the product to be packed. For example, to pack 0.5 kg of a fresh-cut product (i.e., such as endive), the package volume would have to be greater than 500 mL. By contrast, if the volume of the package has to be smaller than 500 mL, the solution obtained for product weight is incorrect and has to be changed. The volume variable is important because a change in volume affects the equilibrium time, that is, the time to reach a steady-state gas concentration inside the package. Rather than volume, the variable that must be optimized is the weight/volume ratio. For instance, packing a small quantity of product results in an increased headspace volume. This increases the time needed for the product to bring the package gas levels to the target modified atmosphere conditions. This is far from favourable, as it takes a long time to achieve the optimum modified atmosphere conditions. However, during a short period at ambient temperature, a package with a small void volume could rapidly generate anaerobic conditions, whereas a package with a large headspace volume will take more time before establishing anaerobic conditions [20]. The volume parameter can be optimized using a trial-and-error approach by setting a volume and predicting oxygen and carbon dioxide by solving numerically Eqs. (10) and (11). An example of the effect of volume variation for a set product weight on the equilibrium time is reported in Fig. 5. The time to reach equilibrium decreases from 15 days to 3.3 days as the volume is reduced from 600 mL to 300 mL.

Fig. 5 Effect of volume
variability on equilibrium
time

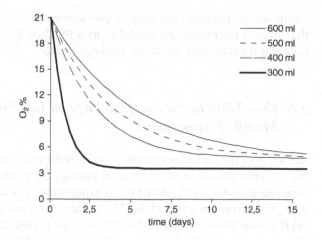

3.5 *Package Simulation*

The simulation of the gas composition is an important tool in package design because it reduces the experimental time needed to consider the multiple solutions available for the packaging optimization. We introduce the simulation problem to ascertain whether equilibrium time is appropriate for product shelf-life. Simulation is the last but most important part of package design. The aim is to predict O_2 and CO_2 changes inside the package to verify that the variables were correctly optimized.

To predict gas changes over time, we need to solve Eqs. (10) and (11). Although these equations are ordinary differentially equations (ODEs) that cannot be solved analytically, several numerical methods can be used by programming them into one of the many programs commercially available, such as SAS, MATLAB, FORTRAN, or SIMILE. In several works, it is reported that the ODE equations were solved analytically but without supplying any information on the method used [40, 41, 51, 52]. The Adams–Moulton method was used by Lee et al. [37, 53] to numerically integrate the ODE and simulate the gas environment in permeable bags containing apples and prepared vegetable salad fish. The same numerical integration method was used also by Song et al. [35] to predict simultaneously the change in gas composition, RH, and temperature in model packages with fresh produce. ODEs were solved numerically using Simpson's integration rule programmed in SAS by Van De Velde and Hendrickx [48] to predict gas composition of cut Belgian endive packed with low density polyethylene (LDPE). One of the most widely used numerical integration methods reported in the literature is the fourth order Runge–Kutta method [28, 29, 34, 38, 54, 55]. To allow for the effect of environmental variables on produce respiration rate and permeability, Eqs. (10) and (11) should be solved combined with models that predict the respiration rate as a function of gas composition, temperature, and, in particular cases, time, and with the models that predict packaging permeability as a function of temperature and tube or hole size if perforated-mediated package is used.

3.6 Variability in Product/Package on Equilibrium Modified Atmosphere

Accuracy of the packaging design procedure depends on the accuracy with which the variables related to the product or package are estimated. For example, let us suppose that the respiration rate at equilibrium gas composition is estimated with 20% error ($3.6 \pm 0.7 \, \text{mL} \, \text{kg}^{-1} \, \text{h}^{-1}$). Thus, three possibilities can be identified: (1) the respiration rate is equal to the average; (2) the respiration rate is equal to the maximum; (3) the respiration rate is equal to the minimum.

Product weight is calculated on the basis of the average value. As a result, different scenarios may occur, as reported in Fig. 6. If the respiration rate is equal to the minimum value, the equilibrium gas composition is equal to 6.7% O_2 and 14.7% CO_2, whereas if the respiration rate is equal to the maximum, the equilibrium gas composition is equal to 2.2% O_2 and 18.8% CO_2. Thus, the error is quite high and the consequence for product shelf-life can be different: if the critical O_2 level for the aerobic respiration process is higher than 2, the respiration process will shift to fermentation with deleterious consequences for product quality. Moreover, if the k_m value of the respiration process is low, for example lower than 1, an O_2 concentration of 6.7% can no longer be protective for the product, resulting in a reduction in shelf life. Thus, it is important to predict the product respiration rate with high sensitivity. Let us consider that the respiration rate is accurately predicted but the error is in the packaging operation at the filling step. For example, product weight can change with a variability of 20%.

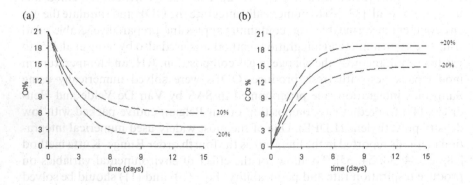

Fig. 6 Effect of variability in respiration rate on EMA: (a) O_2, (b) CO_2

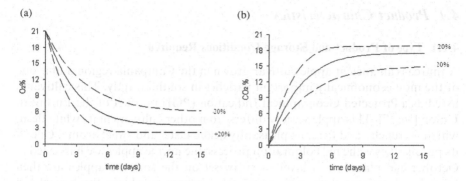

Fig. 7 Effect of the variability in the product weight on EMA: **(a)** O_2, **(b)** CO_2

In Fig. 7, we report the different options predicted when considering the average weight (0.180 kg) $\pm 20\%$. Also in this situation, a variation in product weight may produce a lower level of O_2 than the critical value. Thus the optimization process must take into consideration the error associated with the prediction or with process operation.

4 Case Study

A case study is presented to illustrate all the steps of the design procedure for fresh-cut Annurca apples shown in Fig. 8. The details of respiration rate model are reported by Torrieri et al. [42].

Fig. 8 Fresh-cut Annurca apples packed in modified atmosphere package

4.1 Product Characteristics

4.1.1 Type of Product and Storage Conditions Required

Annurca is an ancient apple cultivar grown in the Campania region and is one of the most economically important varieties in southern Italy. This cultivar is listed as a Protected Geographical Indication (PGI) product of the European Union [56, 57]. This apple variety differs from other cultivars in its white flesh, which is crunchy and firm, its pleasantly acidic taste, and sweet aroma. One of its peculiarities is the red colouration process: the unripe apples are harvested in October and placed on a layer of straw set on the soil; the apples are then turned and sprayed daily with water for 1 month to obtain the typical red colour. Annurca apple fruits used for testing were collected from S. Agata dei Goti (Benevento, Italy). during October 2004. It had the following character-istics at the harvest: fruits diameter of 60 mm, °Brix of 13.7 ± 0.5, and firmness of 176 ± 15 N. The product characteristics after the red coloration process were °Brix of 14.7 ± 0.5 and firmness of 140 ± 20 N. For the selected commodity, optimal gas composition is 3% to 5% O_2 and 15% to 20% CO_2 at 5°C.

4.1.2 Mathematical Model for Respiration Rate

The first step for designing the MAP is to determine the respiration rate at different gas compositions. Thus, Annurca apple fruits were cleaned by using tap water, peeled, cored, and sliced, and the respiration rate was measured by using a modified closed system [42].

The respiration rate was determined at steady state at different O_2 concen-trations (2%, 5%, 10%, 15%, 21%) to estimate Michaelis–Menten parameters (V_{max} and k_m). The test was replicated at different temperatures (5°C, 10°C, 15°C, 20°C) to estimate the Arrhenius model parameters (R_{ref}, E_a). The sample of fresh-cut Annurca apple slices (1 kg) was inserted in the jar (4.0×10^3 mL) and hermetically closed. Subsequently, a gas mixture was flushed through the jar at a constant flow rate (6000 mL h^{-1}) for a time (1–4 hours as function of temperature), which was necessary stabilization. After this time, the jar was closed, and O_2 and CO_2 changes monitored. The experiments lasted 8 hours for all the temperature levels. The temperature and relative humidity ($\sim$100%) inside the jar were monitored by means of a data logger (Escort Data Login Systems Ltd, Naples, Italy). The rate of O_2 consumption and CO_2 production in the course of the experiment was estimated by the following equations:

$$R_{O_2}(t) = -\frac{V_f}{M \cdot 100} \cdot \frac{dy_{O_2}}{dt} \tag{17}$$

$$R_{CO_2}(t) = \frac{V_f}{M \cdot 100} \cdot \frac{dy_{CO_2}}{dt} \tag{18}$$

where M is the sample mass (kg); V_f(mL) is the free volume inside the jar, which is $V_f = V - M/M\rho\rho$; V is the volume of the jar, and ρ is the apparent density of the Annurca apples (810 kg mL^{-1}). The respiration quotient (RQ) was calculated by dividing the CO_2 production rate, R_{CO_2}, by the O_2 consumption rate, R_{O_2}.

In Fig. 9 is reported the effect of O_2 concentration and temperature on respiration rate expressed as O_2 consumption rate. The RQ was equal to an average value of 1.2 for an O_2 concentration exceeding 1%. The influence of O_2 concentration and temperature was modelled as:

$$R_{O_2} = \frac{R_{max_0} \cdot \exp\left(-\frac{E_a}{R} \cdot \left(\frac{1}{T} - \frac{1}{T_{ref}}\right)\right) \cdot y_{O_2}}{k_{m_0} \cdot \exp\left(-\frac{E_a}{R} \cdot \left(\frac{1}{T} - \frac{1}{T_{ref}}\right)\right) + y_{O_2}} \tag{19}$$

$$R_{CO_2} = RQ \cdot R_{O_2}. \tag{20}$$

The CO_2 production rate was estimated by considering an RQ equal to 1.2. The parameters were estimated by nonlinear regression by least squares methods.

These parameters were used to estimate the respiration rate at the equilibrium gas composition (i.e., 5% O_2 and 15% CO_2). In Table 6, the estimated parameters and the relevant statistical data are reported.

4.2 Package Characteristics

A semiflexible package was chosen to pack the fresh-cut Annurca apples. A tray rather than a flexible package was assumed to be more suitable for consuming

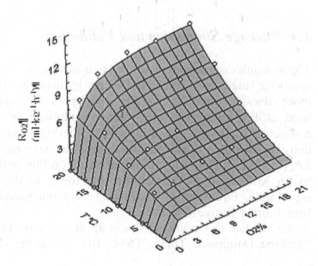

Fig. 9 Surface plot of the changes of respiration rate (R_{O_2}) as a function of temperature and oxygen (O_2) concentration at constant level of CO_2 (0%). The dots represent the experimental data with the standard deviation; the continuous surface represents the predicted data (model, Eq. 19)

Table 6 Parameter estimates of the global model describing the influence of oxygen and temperature on the respiration rate (Eq. 19) and relevant statistical data

Model Parameters	Estimate ± SE
R_{max0} (mL kg^{-1} h^{-1})	6.7 ± 0.1
k_{m0} (%)	0.68 ± 0.07
E_a (kJ/mol)	51 ± 1
R^2_{adj}	0.99
Fit SE	0.27

SE, standard error.

a ready-to-eat product, such as the fresh-cut Annurca apple. The size of the package was width 0.145 m, length 0.230 m, and height 0.015 m. Once the tray had been chosen, the surface area was equal to the tray-top surface (0.145 × 0.230 = 0.033 m^2), whereas the volume was set at the minimum available for the chosen tray (500 mL). The packaging material to seal the tray was chosen on the basis of the permeability to O_2 calculated by solving Eq. (12) once an arbitrary product weight had been set (0.150 kg). Of the commercial material, OPP/LDPE was chosen with a thickness of 100 μm. Its permeance was 1.0×10^{-4} mL O_2 m^{-2} h^{-1} Pa^{-1}, which is closer to the desired one for a given product.

4.3 Variable Optimization: Product Weight

The respiration rate at the equilibrium gas composition and film permeability were used to optimize the product weight to be packed to reach the optimal gas composition. Thus product weight was calculated by solving Eq. (12). The product mass to be packed to reach the equilibrium gas composition was found to be 0.15 kg, 0.13 kg, 0.11 kg, and 0.1 kg, respectively, at 5°C, 8°C, 14°C, and 20°C.

4.4 Package Simulation and Validation

The dynamic exchange of O_2 and CO_2 inside the package was modelled with the symbolic language of dynamic systems by using SIMILE modelling environment (www.simulistics.com). The model parameters reported in Table 6 were used as data input to estimate the consumption rate of O_2 of cut apples as a function of O_2 and temperature. Commodity variables, such as weight and density, were also inserted as data input, as were packaging variables (kP_{O_2}, kP_{CO_2}, V, A, x). The effect of temperature on film permeability was estimated by an Arrhenius-type relationship, and the E_a of the process was inserted as input data. Equations (10) and (11) were integrated numerically using the fourth order Runge–Kutta formula.

To validate the model, fresh-cut apples were packed by using a packaging machine (Minipack Torre, TSM, 105, Cava dei Tirreni, Salerno, Italy).

Polystyrene tray laminated with a multilayer barrier film ($V = 5.0 \times 10^3$ mL; CoopBox, Bologna, Italy) and sealed with a film of oriented polyethylene (OPP)/LDPE was used. The packed product was stored at 5°C, 8°C, 14°C, and 20°C for 15 days, and O_2 and CO_2 inside the package were periodically monitored by using a gas analyser (PBI Dansensor, Milan, Italy).

Figure 10 reports the gas exchange within the headspace of fresh-cut apples using the designed package as predicted by Eqs. (10) and (11). With the exception of samples stored at 20°C, the equilibrium concentration was reasonably well predicted by the model in question. At 20°C, the development of mold could have affected the results of the test. For the other samples, there is good agreement between experimental and predicted values, proving that the Michaelis–Menten noncompetitive model with parameters varying against temperature with an Arrhenius equation accurately predicts the effect of O_2 and temperature on the respiration rate of fresh-cut Annurca apple slices.

Nevertheless, gas composition within the recommended values was reached in approximately 14 days at 5°C and 8°C and after 11 days at 14°C (Fig. 10).

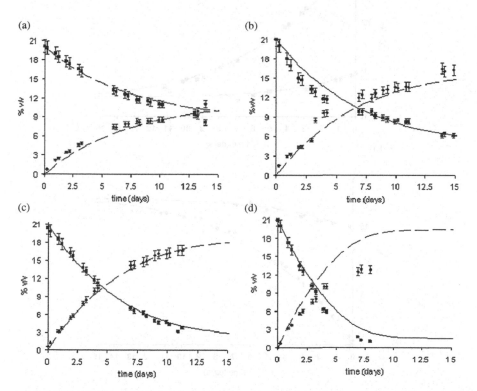

Fig. 10 Effect of temperature on EMA: (a) 5°C, (b) 8°C, (c)14°C, and (d) 20°C. The symbols show the experimental data (■ O_2; ◆ CO_2), and the lines show the values predicted by Eqs. (10) and (11) combined with Michaelis–Menten model and the Arrhenius model (solid lines, O_2; dashed lines, CO_2)

In all the situations, the time to reach equilibrium is too high in relation to product shelf-life. Indeed, a shelf life of 7 days is suggested at 5°C to 6°C. Thus, the selected variables (i.e., package volume and product weight) do not guarantee an appropriate EMA package, as the time to reach equilibrium was too long. A possible solution is to change the ratio between product weight and package volume or use protective atmosphere technology. For example, at 5°C, to reduce the time to reach the equilibrium within the package, the weight-to-volume ratio must be 0.5. As it was not easy to change the packaging volume due to technological limitations, we chose to flow the equilibrium gas composition inside the headspace at packaging, that is, to use protective packaging technology. In Fig. 11, we report the gas composition changes of fresh-cut

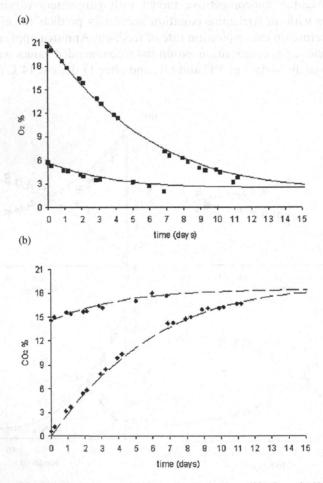

Fig. 11 Effect of flushing the package with the optimal mixed gas (5% O_2 and 15% CO_2) on the equilibrium time of the package at 14°C: (a) O_2, (b) CO_2. The symbols show the experimental data (■ O_2; ◆ CO_2), and the lines show the values predicted by Eqs. (10) and (11) combined with Michaelis–Menten model and Arrhenius model (solid lines, O_2; dashed lines, CO_2)

apples packed in a protective package of 5% O_2 and 15% CO_2, experimentally measured and predicted by the model. In Fig. 11, the gas composition changes of fresh-cut apples packed in air is also reported. The equilibrium gas composition is not exactly that predicted, but it lies within an acceptable range, and in addition the product was stored under a protective atmosphere throughout the storage. Moreover, the O_2 level always stayed above 1% thereby avoiding anoxia inside the package.

Nomenclature

A	surface area (m^2)
c	permeant concentration (mL (STP) mL^{-1})
CO_2	carbon dioxide
D	diffusivity ($m^2 h^{-1}$)
D_p	perforation diameter (m)
E_a	activation energy (kJ mol^{-1})
J	flow rate (mL h^{-1})
k_m	Michaelis–Menten equation constant (%)
kP	permeability coefficient (mL (STP) m $m^{-2} h^{-1} Pa^{-1}$)
P	permeance coefficient (mL (STP) $m^{-2} h^{-1} Pa^{-1}$)
L	tube/perforation length (m)
M	product mass (kg)
N_H	perforation num-ber
p	partial pressure (Pa)
O_2	oxygen
Q	quantity of permeant (mL)
R	respiration (consumption/production) rate (mL $kg^{-1}h^{-1}$)
R_c	universal gas constant (J $mol^{-1} k^{-1}$)
RH	relative humidity (%)
RQ	respiration quotient
S	solubility (mL (STP) $mL^{-1} Pa^{-1}$)
T	temperature (°C or K)
t	time (h)
V	volume (mL)
V_{max}	Michaelis–Menten equation constant (mL $kg^{-1} h^{-1}$)
x	film thick-ness(m)
y	volumetric concentration (% v/v)
Greek sym-bols	
β	selectivity
δ	product density (kg mL^{-3})
Superscripts	
eq	gas composition inside the package at equilibrium
in	internal gas composi-tion
out	external gas composi-tion
Subscripts	
Ref	reference

f   free
0   at reference tempera-ture
adj   adjustment

References

1. International Fresh-cut Produce Association-IFPA (2004). Fresh-cut produce fuels an American on-the-go. Available at http://www.fresh-cut.org/.
2. Ahvenainen, R. (1996). New approaches in improving the shelf life of minimally processed fruits and vegetables. Trends in Food Science & Technology, 7, 179–187.
3. Giovannetti, F. (2003). Solo quarta e quinta gamma tengono. Terra e Vita, 18, 67–70.
4. International Fresh-cut Produce Association (IFPA) and the Produce Marketing Association (PMA). (1999). Handling Guidelines for the Fresh-cut Produce Industry, 3rd edition. IFPA, Alexandria, VA, pp. 5, 7.
5. Ragaert, P., Verbeke, W., Devlieghere, F. & Debevere, J. (2004). Consumer perception and choice of minimally processed vegetables and packaged fruits. Food Quality and Preferences, 15, 259–270.
6. Bolin, H.R. & Huxsoll, C.C. (1991). Effect of preparation procedures and storage parameters on quality retention of salad-cut lettuce. Journal of Food Science, 5, 1319–1321.
7. Zagory, D. (1999). Effects of post-processing handling and packaging on microbial populations. Postharvest Biology and Technology, 15, 313–321.
8. Zhu, M., Chu, C.L., Wang, S.L. & Lencki, R.W. (2001). Influence of oxygen, carbon dioxide, and degree of cutting on the respiration rate of rutabaga. Journal of Food Science, 66(1), 30–37.
9. Rajkowski, K.T., & Baldwin, E.A. (2003). Concerns with minimal processing in apple, citrus, and vegetables. In: J.S. Novak, G.M. Sapers & V.K. Juneya, eds. Microbial Safety of Minimally Processed Foods. CRC Press, Boca Raton, FL, pp. 33–51.
10. Brecht, J.K. (1995). Physiology of lightly processed fruits and vegetables. Hortscience, 30, 18–21.
11. Robertson, G.L. (1993). Packaging of horticultural products. In: Food Packaging—Principle and Practice. Dekker, New York.
12. Wang, C.Y. (1990). Physical and biological effect of controlled atmosphere on fruits and vegetables. In: M. Calderon & R. Barki-Golan, eds. Food Preservation by Modified Atmospheres. CRC Press, Boca Raton, FL, pp. 197–224.
13. Kader, A.A., Zagory, D. & Kerbel, E.L. (1989). Modified atmosphere packaging of fruits and vegetables. CRC Critical Reviews in Food Science and Nutrition, 28 (1), 1–30.
14. Saltviet, M.E. (1993). A summary of CA and MA requirements and recommendations for the storage of harvested vegetables. In: G.D. Blanpied, J.A. Barstch & J.R. Hicks, eds. Proceedings of the Sixth International Controlled Atmosphere Research Conference, New York, USA: Ithaca Vol. II. pp. 800–818.
15. Fonseca, S.C. (2001). Development of perforation-mediated modified atmosphere packaging for extending the shelf life of shredded galega kale. PhD Dissertation, Universidade Catolica Portuguesa, Porto, Portugal.
16. Chau, K. & Talasila, P. (1994). Design of modified atmosphere packages for fresh fruits and vegetables. In: R. Singh & F. Oliveria, eds. Minimal Processing of Foods and Process Optimization. CRC Press, Boca Raton, FL, pp. 407–416.
17. Jacxsens, L., Devlieghere, F., De Rudder, T. & Debevere, J. (2000). Designing an equilibrium modified atmosphere package for fresh-cut vegetables subjected to changes in temperature. Lebensm.-Wiss. U. Technology, 33, 178–187.

18. Fonseca, S.C., Oliveira, F.A.R., Frias, J.M., Brecht, J. K. & Chau, K.V. (2002). Modelling respiration rate of shredded galega kale for development of modified atmosphere packaging. Journal of Food Engineering, 54, 299–307.
19. Del Nobile, M.A., Ambrosino, M.L., Sacchi, R. & Masi, P. (2003). Design of plastic bottles for packaging of virgin olive oil. Journal of Food Science, 68(1), 170–175.
20. Hertog, M.L.A.T.M. & Banks, N.H. (2003). Improving MAP through conceptual models. In: Ahvenainen R., ed. Novel Food Packaging Techniques. Woodhead Publishing limited, Cambridge, England, CRC Press, Boca Raton, FL, pp. 351–376.
21. Mahajan, P.V., Oliveira, F.A.R., Montanez, J.C. & Frias J. (2007). Development of user-friendly software for design of modified atmosphere packaging for fresh and fresh-cut produce. Innovative Food Science and Emerging Technologies 8, 84–92.
22. Masi, P. & Paul, D.R. (1982). Modelling gas transport in packaging applications. Journal of Membrane Science, 12, 137–151.
23. Fishman, S., Rodov, V., & Ben-Yehoshua, S. (1996). Mathematical model for perforation effect on oxygen and water vapour dynamics in modified atmosphere packages. Journal of Food Science, 61(5), 956–961.
24. Emond, J.P., Casteigne, F., Toupin, C.J. & Desilets, D. (1991). Mathematical modelling of gas exchange in modified atmosphere packaging. American Society of Agricultural Engineering, 34, 239–245.
25. Fonseca, S.C., Oliveira, F.A.R., Lino, I.B.M., Brecht, J.K. & Chau, K.V. (2000). Modelling O2 and CO2 exchange for development of perforation-mediated modified atmosphere packaging. Journal of Food Engineering, 43, 9–15.
26. Paul, D.R. & Clarke, R. (2002). Modelling of modified atmosphere packaging based on designs with a membrane and perforations. Journal of Membrane Science, 208, 269–283.
27. Fonseca, S.C., Oliveira, F.A.R. & Brecht, J.K. (2002). Modelling respiration rate of fresh fruits and vegetables for modified atmosphere packaging: a review. Journal of Food Engineering, 52, 99–119.
28. Rocculi, P., Del Nobile, M.A., Romani, S., Baiano, A. & Dalla Rosa, M. (2006). Use of a simple mathematical model to evaluate dipping and MAP effects on aerobic respiration of minimally processed apples. Journal of Food Engineering, 76, 334–340.
29. Del Nobile, M.A., Licciardello, F., Scrocco, C., Muratore, G. & Zappa, M. (2007). Design of plastic packages for minimally processed fruits. Journal of Food Engineering, 79, 217–224.
30. Andrich, G., Zinnai, A., Balzini, S., Silvestri, S. & Fiorentini, R., (1998). Aerobic respiration rate of Golden Delicious apples as a function of temperature and PO_2. Postharvest Biology and Technology, 14, 1–9.
31. Peppelembos, H.W. & Van't Leven J. (1996). Evaluation of four types of inhibition for modelling the influence of carbon dioxide on oxygen consumption of fruits and vegetables. Postharvest Biology and Technology, 7, 27–40.
32. Lakakul, R., Beaudry, R.M. & Hernandez, R.J. (1999). Modelling respiration of apple slices in modified-atmosphere packages. Journal of Food Science, 64, 105–110.
33. Torrieri, E., Cavella, P. & Masi, P. (2006). Progettazione di un imballaggio per un prodotto di IV gamma: influenza della temperatura sulla velocità di respirazione. 7° Congresso Italiano di Scienza e Tecnologia degli Alimenti, Ciseta 7. In: Ricerche ed innovazioni nell'industria alimentare, Pinerolo, Italy Vol VII. Chirotti Editori, pp. 891–895.
34. Salvador, M.L., Jaime, P. & Oria, R. (2002). Modelling of O_2 and CO_2 exchange dynamics in modified atmosphere packaging of burlat cherries. Journal of Food Science, 67(1), 231–235.
35. Song, Y., Vorsa, N. & Yam, K.L. (2002). Modelling respiration-transpiration in a modified atmosphere packaging system containing blueberry. Journal of Food Engineering, 53, 103–109.

36. Ratti, C., Raghavan, G.S.V. & Gariepy, Y. (1996). Respiration rate model and modified atmosphere packaging of fresh cauliflower. Journal of Food Engineering, 28, 297–306.
37. Lee, D.S., Haggar, P.E., Lee, J. & Yam, K.L. (1991). Model for fresh produce respiration in modified atmospheres based on principles of enzyme kinetics. Journal of Food Science, 56(6), 1580–1585.
38. Talasila, P.C., Chau, K.V., & Brecht, J.K. (1995). Design of rigid modified atmosphere packages for fresh fruits and vegetables. Journal of Food Science, 60(4), 758–769.
39. Mahajan, P.V. & Goswami, T.K. (2001). Enzyme kinetics based modelling of respiration rate for apple. Journal of Agricultural Engineering Research, 79(4), 399–406.
40. Charles, F., Sanchez, J. and Gontard, N. (2005). Modelling of active modified atmosphere packaging of endive exposed to several post-harvest temperatures. Journal of Food Science, 70(8), 443–449.
41. Nahor, H.B., Schotsmans, W., Sheerlinck, N. & Nicolai, B.M. (2005). Applicability of existing gas exchange models for bulk storage of pome fruit: assessment and testing. Postharvest Biology and Technology, 35, 15–24.
42. Torrieri, E. Cavella, S. & Masi P. (2007). Modelling respiration rate of Annurca Apple for development of modified atmosphere packaging. International Journal of Food Science and Technology (in press).
43. Iqbal, T., Oliveira, F.A.R., Mahajan, P.V., Kerry, J.P., Gil, L., Manso, M.C., & Cunha, L.M. (2005). Modeling the influence of storage time on the respiration rate of shredded carrots at different temperatures under ambient atmosphere. Acta Horticulturae (ISHS), 674, 105–111.
44. Iqbal, T., Oliveira F.A.R., Torrieri, E. & Sousa, M.J. (2003). Mathematical modelling of the influence of temperature on the respiration rate of shredded carrots stored in ambient air. 12th World Congress of Food Science & Technology, Chicago, IL, July 16–19 [book of abstracts].
45. Torrieri, E., Sousa, M.J., Masi, P., Kerry, J.P. & Oliveira F.A.R. (2004). Influence of the temperature on the quality of shredded carrots. Special Issue of Italian Journal of Food Science, Chirotti Editori, 336–348.
46. Exama, A., Arul, J., Lencki, R.W., Lee L.Z. & Toupin, C. (1993). Suitability of plastic films for modified atmosphere packaging of fruits and vegetables. Journal of Food Science, 58, 1365–1370.
47. Hertog M.L.A.T.M., & Banks N.H. (2000). The effect of temperature on gas relations in MA packages for capsicums (Capsicum annuum L., cv. Tasty): an integrated approach. Postharvest Biology and Technology, 20(1), 71–80.
48. Van de Velde, M.D. & Hendrickx M. (2001). Influence of storage atmosphere and temperature on quality evolution of cut Belgian endives. Journal of Food Science, 66, 1212–1218.
49. Mannapperuma, J.D. & Singh, R.P. (1994). Modelling of gas exchange in polymeric packages of fresh fruits and vegetables. In: R.P. Singh & F.A.R. Oliveira, eds. Process Optimisation and Minimal Processing of Foods. NY: CRC Press, pp. 437–445.
50. Emond, J.P. & Chau, K.V. (1990). Use of perforations in modified atmosphere packaging. American Society of Agricultural Engineers, Paper No. 90, 6512.
51. Fishman, S., Rodov, V., Peretz, J. & Ben-Yehoshua, S. (1996). Mathematical model for perforation effect on oxygen and water vapour dynamics in modified-atmosphere packages. Journal of Food Science, 61(5), 956–961.
52. Charles, F., Sanchez, J., & Gontard, N. (2003). Active modified atmosphere packaging of fresh fruits and vegetables: modelling with tomatoes and oxygen absorber. Journal of Food Science, 68(5), 1736–1742.
53. Lee, K.S., Park, I.S., & Lee, D.S. (1996). Modified atmosphere packaging of a mixed prepared vegetable salad dish. International Journal of Food Science and Technology, 31, 7–13.

54. Talasila, P.C., Chau, K.V. & Brecht, J.K. (1995). Modified atmosphere packaging under varying surrounding temperature. Transactions of ASAE, 38(3), 869–876.
55. Makino, Y., Iwasaki, K. & Hirata, T., (1996). A theoretical model for oxygen consumption in fresh produce under an atmosphere with carbon dioxide. Journal of Agricultural Engineering Research, 65, 193–203.
56. Council Regulation (EC) No. 2081/92 of 14 July 1992 on the protection of geographical indications (PGI) and designations of origin (PDO) for agricultural products. Official Journal of the European Communities, L208 24/07/1992.
57. Council Regulation (EC) No. 2081/92 of 14/07/1992. Publication of the request for the "melannurca campana" PGI (C138/2005) 'Melannurca Campana' No. EC: IT/00193/27. 4.2001. Official Journal of the European Communities, C138 7/06/2005.